Springer Theses

Recognizing Outstanding Ph.D. Research

Aims and Scope

The series "Springer Theses" brings together a selection of the very best Ph.D. theses from around the world and across the physical sciences. Nominated and endorsed by two recognized specialists, each published volume has been selected for its scientific excellence and the high impact of its contents for the pertinent field of research. For greater accessibility to non-specialists, the published versions include an extended introduction, as well as a foreword by the student's supervisor explaining the special relevance of the work for the field. As a whole, the series will provide a valuable resource both for newcomers to the research fields described, and for other scientists seeking detailed background information on special questions. Finally, it provides an accredited documentation of the valuable contributions made by today's younger generation of scientists.

Theses may be nominated for publication in this series by heads of department at internationally leading universities or institutes and should fulfill all of the following criteria

- They must be written in good English.
- The topic should fall within the confines of Chemistry, Physics, Earth Sciences, Engineering and related interdisciplinary fields such as Materials, Nanoscience, Chemical Engineering, Complex Systems and Biophysics.
- The work reported in the thesis must represent a significant scientific advance.
- If the thesis includes previously published material, permission to reproduce this must be gained from the respective copyright holder (a maximum 30% of the thesis should be a verbatim reproduction from the author's previous publications).
- They must have been examined and passed during the 12 months prior to nomination.
- Each thesis should include a foreword by the supervisor outlining the significance of its content.
- The theses should have a clearly defined structure including an introduction accessible to new PhD students and scientists not expert in the relevant field.

Indexed by zbMATH.

More information about this series at https://link.springer.com/bookseries/8790

Cristina Martin Perez

Search for the Higgs Boson Produced in Association with Top Quarks with the CMS Detector at the LHC

Doctoral Thesis accepted by
Institut Polytechnique de Paris, France

Springer

Author
Dr. Cristina Martin Perez
Institute for Particle Physics
and Astrophysics
ETH Zurich
Zürich, Switzerland

Supervisor
Dr. Alexandre Zabi
Laboratoire Leprince Ringuet
CNRS-IN2P3 École Polytechnique
Institut Polytechnique de Paris
Paris, France

ISSN 2190-5053 ISSN 2190-5061 (electronic)
Springer Theses
ISBN 978-3-030-90205-6 ISBN 978-3-030-90206-3 (eBook)
https://doi.org/10.1007/978-3-030-90206-3

This Springer imprint is published by the registered company Springer Nature Switzerland AG
The registered company address is: Gewerbestrasse 11, 6330 Cham, Switzerland

Supervisor's Foreword

Our understanding of the Universe, and its evolution from the earliest stage of development, is dependent on the precision with which we can describe the fundamental interactions. Identifying and modelling the fundamental laws of Nature represents the core of modern particle physics research. The masses of elementary particles themselves are the result of their interactions with the Brout-Englert-Higgs field. The discipline of particle physics has been in constant evolution for the past 50 years with many discoveries that led to the establishment of one of humanity's greatest scientific achievements, the Standard Model. The Standard Model represents the most predictive and precise model ever built, and is the perfect starting point towards a better understanding of Nature's deepest secrets. The fascinating world of high-energy physics combines gracefully the theoretical developments inspired from all physics disciplines and experimental developments systematically pushing the limits of modern technology. The discovery of the Higgs boson by the ATLAS and CMS experiments at CERN in 2012 was a major step forward leading to the evidence of the very first scalar field in the Universe. The characterisation of this field is among today's most important scientific endeavours and constitutes the heart of Cristina Martin Perez's Ph.D. research, described in this manuscript.

The years following the discovery of the Higgs boson have allowed the scientific community to measure its couplings with elementary particles, and in particular fermions. The top quark is the heaviest fermion and the measurement of its couplings to the Higgs boson is of particular interest. A direct measurement of this coupling can be achieved through the observation of processes involving Higgs boson production along with one (tH) or a pair of top quarks ($t\bar{t}H$) in the proton-proton collisions generated by the Large Hadron Collider at CERN. Given the very small cross-section associated with these processes, these observations were considered almost impossible in an intense hadronic environment with an overwhelming background. The final states are equally challenging to identify and the use of leptons present in the decay products, the so-called multilepton final state, is essential to conduct this measurement. The work reported in this manuscript elucidates how the use of modern signal extraction methods, combined with the large quantity of collision events recorded by the CMS experiment, could result in the observation of $t\bar{t}H$ in the multilepton channel. Another aspect of this work describes a primordial step towards any data analysis performed at colliders: the selection of interesting collisions amongst the millions produced per second. The first stages of selection are performed by a dedicated hardware system referred to as the trigger system. The architecture used to implement such systems has grown in complexity over the past decades. Modern technology provides powerful computation and high-speed interconnections to efficiently handle the large data volumes coming from highly segmented particle physics detectors. Sophisticated algorithms, often based on machine learning approaches, can be implemented in the heart of the trigger system. The manuscript describes how these algorithms are designed and how their performance can be optimised. The high-luminosity LHC, where the instantaneous luminosity of the collider will be increased sevenfold, represents a much greater challenge for the trigger system. The design of dedicated trigger algorithms to target specific physics signatures can be achieved through the precise reconstruction of physics objects.

Cristina Martin Perez has conducted her Ph.D. thesis work within the CMS group of the Ecole Polytechnique de Paris, Institut Polytechnique de Paris, under my supervision. Cristina's strong motivation for particle physics research led her to achieve excellence in the field with an everlasting determination to obtain brilliant scientific results. She has shown the rare ability to conduct both complex data analysis and technical work at the highest level. Cristina has made outstanding individual contributions to important topics in modern high-energy physics. These contributions are summarised in this clear and pedagogical manuscript, matching her skills in journalism, a subject in which she majored in before starting her career in physics. It is rare that technical subjects such as trigger algorithm development and modern signal extraction techniques are described at this level of detail in a Ph.D. manuscript, and this will certainly benefit students in the field of fundamental science.

Paris, France
July 2021

Dr. Alexandre Zabi
Director of research

Supervisor's Foreword

Our understanding of the Universe, and its evolution from the earliest stage of development, is dependent on the precision with which we can describe the fundamental interactions. Identifying and modelling the fundamental laws of Nature represents the core of modern particle physics research. The masses of elementary particles themselves are the result of their interactions with the Brout-Englert-Higgs field. The discipline of particle physics has been in constant evolution for the past 50 years with many discoveries that led to the establishment of one of humanity's greatest scientific achievements, the Standard Model. The Standard Model represents the most predictive and precise model ever built, and is the perfect starting point towards a better understanding of Nature's deepest secrets. The fascinating world of high-energy physics combines gracefully the theoretical developments inspired from all physics disciplines and experimental developments systematically pushing the limits of modern technology. The discovery of the Higgs boson by the ATLAS and CMS experiments at CERN in 2012 was a major step forward leading to the evidence of the very first scalar field in the Universe. The characterisation of this field is among today's most important scientific endeavours and constitutes the heart of Cristina Martin Perez's Ph.D. research, described in this manuscript.

The years following the discovery of the Higgs boson have allowed the scientific community to measure its couplings with elementary particles, and in particular fermions. The top quark is the heaviest fermion and the measurement of its couplings to the Higgs boson is of particular interest. A direct measurement of this coupling can be achieved through the observation of processes involving Higgs boson production along with one (tH) or a pair of top quarks ($t\bar{t}H$) in the proton-proton collisions generated by the Large Hadron Collider at CERN. Given the very small cross-section associated with these processes, these observations were considered almost impossible in an intense hadronic environment with an overwhelming background. The final states are equally challenging to identify and the use of leptons present in the decay products, the so-called multilepton final state, is essential to conduct this measurement. The work reported in this manuscript elucidates how the use of modern signal extraction methods, combined with the large quantity of collision events recorded by the CMS experiment, could result in the observation of $t\bar{t}H$ in the multilepton channel. Another aspect of this work describes a primordial step towards any data analysis performed at colliders: the selection of interesting collisions amongst the millions produced per second. The first stages of selection are performed by a dedicated hardware system referred to as the trigger system. The architecture used to implement such systems has grown in complexity over the past decades. Modern technology provides powerful computation and high-speed interconnections to efficiently handle the large data volumes coming from highly segmented particle physics detectors. Sophisticated algorithms, often based on machine learning approaches, can be implemented in the heart of the trigger system. The manuscript describes how these algorithms are designed and how their performance can be optimised. The high-luminosity LHC, where the instantaneous luminosity of the collider will be increased sevenfold, represents a much greater challenge for the trigger system. The design of dedicated trigger algorithms to target specific physics signatures can be achieved through the precise reconstruction of physics objects.

Cristina Martin Perez has conducted her Ph.D. thesis work within the CMS group of the Ecole Polytechnique de Paris, Institut Polytechnique de Paris, under my supervision. Cristina's strong motivation for particle physics research led her to achieve excellence in the field with an everlasting determination to obtain brilliant scientific results. She has shown the rare ability to conduct both complex data analysis and technical work at the highest level. Cristina has made outstanding individual contributions to important topics in modern high-energy physics. These contributions are summarised in this clear and pedagogical manuscript, matching her skills in journalism, a subject in which she majored in before starting her career in physics. It is rare that technical subjects such as trigger algorithm development and modern signal extraction techniques are described at this level of detail in a Ph.D. manuscript, and this will certainly benefit students in the field of fundamental science.

Paris, France
July 2021

Dr. Alexandre Zabi
Director of research

Abstract

This thesis presents a study of the Higgs boson production in association with one top quark (tH) or two top quarks ($t\bar{t}H$) with the proton-proton collisions at 13 TeV provided by the LHC at the CMS detector at CERN (Geneva). These production modes are important to characterize the coupling of the Higgs boson to the top quark (y_t), directly accessible upon measurement of the cross sections of these processes. Additionally, the processes are sensitive to the existence of physics beyond the Standard Model, as variations of the predicted couplings result in a measurable change of the value of the cross sections.

The production of these processes is very rare; hence, maximizing the selection efficiency is essential. The first part of the thesis work has been devoted to the optimization of the algorithm to select τ leptons that decay into hadrons (τ_h) in the Level-1 (L1) trigger of CMS. The work is focused on the improvement of the trigger during Run 2 (2016–2018) and Run 3 (2022–2024) of the LHC and to the development of a new trigger concept for the High-Luminosity LHC (HL-LHC), foreseen to start in 2027. The data collected by these triggers is pivotal to enlarge the sensitivity to tH and $t\bar{t}H$ processes.

The second part of the thesis work is dedicated to the search of tH and $t\bar{t}H$ processes with the subsequent decays $H \rightarrow \tau\tau$, $H \rightarrow W^*W$ et $H \rightarrow Z^*Z$, making use of the data produced in Run 2, corresponding to an integrated luminosity of 137 fb^{-1}. The presence of multiple background processes makes this analysis relatively complex. For this reason, multivariate techniques are employed to extract the signal making use of machine learning algorithms and the Matrix Element Method (MEM). The measured cross sections normalized to the Standard Model prediction amount to $0.92^{+0.26}_{-0.23}$ for $t\bar{t}H$ and to $5.67^{+4.05}_{-3.98}$ for tH, in agreement with the Standard Model. The value of y_t normalized to the Standard Model prediction is constrained to the intervals $-0.9 < y_t < -0.7$ or $0.7 < y_t < 1.1$ at 95% confidence level.

Acknowledgements

This doctoral thesis is the outcome of three years of challenges and unequalled personal growth. I am very grateful to all the people who have given me their support throughout this journey; you have all contributed to this milestone one way or another, and you certainly deserve a space in this manuscript.

First, I would like to thank the president of the jury Luca Malgeri, the referees Lorenzo Feligioni and Dirk Zerwas, and the examiners María Luisa Cepeda, Sabine Crépé-Renaudin, Paris Sphicas and Alex Tapper for the review of the manuscript. The interest you have shown towards my work and the feedback you have given me have been extremely motivating and rewarding.

I am very grateful to my supervisor, Alex, for believing in me and putting me in the spotlight right since the beginning with a remarkable ability to identify opportunities. You have found me a place in the collaboration and you have made my work meaningful with your passion and enthusiasm.

Despite being based at CERN during the thesis, I felt very closely the unconditional support of my colleagues at LLR up until the preparation of my defence. I am thankful to Yves, Florian and Roberto for the frequent discussions and their shared expertise. I cannot thank Olivier enough for his pedagogic insight, his patience and his encouragement; it has been a real pleasure to work with you. I am very thankful to Jean-Baptiste for his guidance, practicality and availability. Big thanks to Thomas for his precious advice during my first steps and his commitment throughout all phases. To Gilles, Andrea and Michael: thank you for providing me with the technical tools and support to conduct this work. I would like to thank all the members of the administration at LLR for their instrumental work behind the scenes.

I am very lucky to have shared the teaching experience with Alex and the TREX *moniteurs*. In person or virtually, we always gave the maximum during those tiring Fridays that we managed to get through with good mood and, above all, a lot of sushi.

To all my fellow Ph.D.s and postdocs at the lab: it was a blast to go along this experience with you. I am grateful for the memories we built and I wish you all the success you deserve. Special thanks to Chiara for the emotional support, to Matteo for his positive energy and the uncountable laughs, to Artur for his pragmatism and advice, to Marina for her human quality, and to Inna for her strength.

I would like to thank all the brilliant people at CERN and, especially, those with whom I worked more closely; their hard work and passion have been determinant to deliver the results of this thesis. It was specially enriching to work beside Marco, Sergio, Pietro, Josh, Binghuan, Xanda and Karl in $t\bar{t}H$ multilepton analysis. Likewise, thanks to Sandeep and Pantelis for making such a great trigger team.

Big thanks to all the people outside particle physics that gave me extra strength and showed curiosity towards my work. In particular, thanks to Niki, Fouz, Soraya, Ali and my beloved *sosa-la-cosa*. Your friendship has been an essential pillar during these years and far before.

My deepest gratitude goes to Kevin, who has lived this adventure in first person. You have carried the weight of my fears, you have celebrated my success as own and you have filled my days with the very much needed patience and optimism. I will never find the words to thank you enough.

Gracias a mi padre, mi madre y mi hermana por haber creído en mí a cada paso de mi sinuosa trayectoria. Sin vuestro apoyo y los valores que me habéis enseñado jamás habría llegado donde estoy hoy. Esta tesis es para vosotros.

Contents

Chapter 2
Physics of the Higgs Boson and the Top Quark

The Standard Model (SM) of particle physics was developed throughout the second half of the XXth century and provides a description of the elementary particles and their fundamental interactions [1, 2]. Its theoretical framework is built upon the mathematical foundations of quantum field theory (QFT) and gauge symmetries, refined by the constant back and forth between theory and experiment. It is well corroborated by the experimental observations, and its predictive power was further consolidated with the discovery of the Higgs boson by the ATLAS and CMS experiments at the Large Hadron Collider (LHC) at CERN, Geneva (Switzerland), on the 4th July 2012 [3, 4], 50 years after it was first postulated.

Being the two most massive particles known to date, the top quark (t) and the Higgs boson (H) are an instrumental probe of the SM. They interact via the so-called *Yukawa interaction*, which can be directly tested by studying the rate at which the Higgs boson is produced in association with either one (tH) or two top quarks ($t\bar{t}H$) in the proton-proton (pp) collisions at the LHC. This is expressed in terms of the *cross section* (σ) of these processes, namely the probability that they will take place in a collision. The value of these cross sections has been measured with a limited precision by the ATLAS and CMS experiments, and further sensitivity is required to unambiguously determine the presence or absence of new physics beyond the SM (BSM) in this sector. The measurement of the cross sections of these processes with the CMS detector, providing a direct access to the value of the Yukawa coupling of the Higgs with the top quark, is the main topic of this manuscript.

This chapter discusses the importance of the study of the $t\bar{t}H$ and tH processes in the context of the SM. Potential interpretations in the context of physics beyond the SM are also included. Section 2.1 provides a theoretical overview of the SM particle content and interactions. In Sect. 2.2, the associated production of the Higgs boson with top quarks is presented and its phenomenology at collider experiments is described, including a summary of the latest measurements performed at the

C. Martin Perez, *Search for the Higgs Boson Produced in Association with Top Quarks with the CMS Detector at the LHC*, Springer Theses,
https://doi.org/10.1007/978-3-030-90206-3_2

LHC. The LHC accelerator and the CMS detector used to conduct the measurement described in this manuscript are presented in Chap. 3.

2.1 The Standard Model of Particle Physics

Two types of particles are included in the SM: the building blocks of matter, also known as *matter* particles, and the intermediate interaction particles, or *force carriers*. The first group is composed by *fermions* of spin 1/2, whereas the second group is composed by *bosons* of spin 1, which are the particles exchanged by the fermions during interactions. At the core of the SM stands the idea that fermions and bosons are excited states of their corresponding fields, defined in space-time. Any system is described by a *Lagrangian density*, or simply Lagrangian, which encodes the propagation of these fields and the interactions between them, based on a basic underlying symmetry, the *gauge invariance*. This invariance is expressed mathematically as $SU(3)_C \times SU(2)_L \times U(1)_Y$, where the first component $SU(3)_C$ corresponds to the *strong* interaction and the components $SU(2)_L \times U(1)_Y$ express the *electromagnetic* and *weak* interactions, unified as the electroweak interaction. The SM does not include the description of the *gravitational* interaction, but this force can be neglected at the considered energies: its intensity is 25 orders of magnitude lower than the weak force, the weakest within the SM. An overview of the SM particle content can be found in Fig. 2.1; the properties of the fermions and bosons, along with the mathematical foundations of their interactions, are explained in what follows. The generation of the mass of the bosons and fermions is explained by the phenomenon of spontaneous symmetry breaking of the electroweak theory, which results from the postulation of the existence of the Higgs boson; it is presented in Sect. 2.1.2, followed by an overview of the phenomenology of the Higgs boson.

2.1.1 Particle Content and Interactions

Fermions

Matter is described by fermions, particles of spin $s = 1/2$ satisfying the Fermi-Dirac statistics and the Pauli exclusion principle. They are subcategorized in *quarks* and *leptons*. Experimental observations show that 12 fermions exist, namely 6 quarks and 6 leptons. They are divided in three *families*, or *generations*, each containing two quarks with electric charges of $+2/3$ and $-1/3$, and two leptons with electric charges of -1 and 0. For each fermion there is a corresponding *antiparticle*, which is identical in every aspect except that it has opposite internal quantum numbers, such as the electric charge.

The first family of quarks is composed by the up (u) and down (d) quarks, with masses of 2.2 MeV [5] and 4.7 MeV [5], respectively; they constitute ordinary matter. Their analogues in the subsequent families are the charm (c) and strange (s) quarks

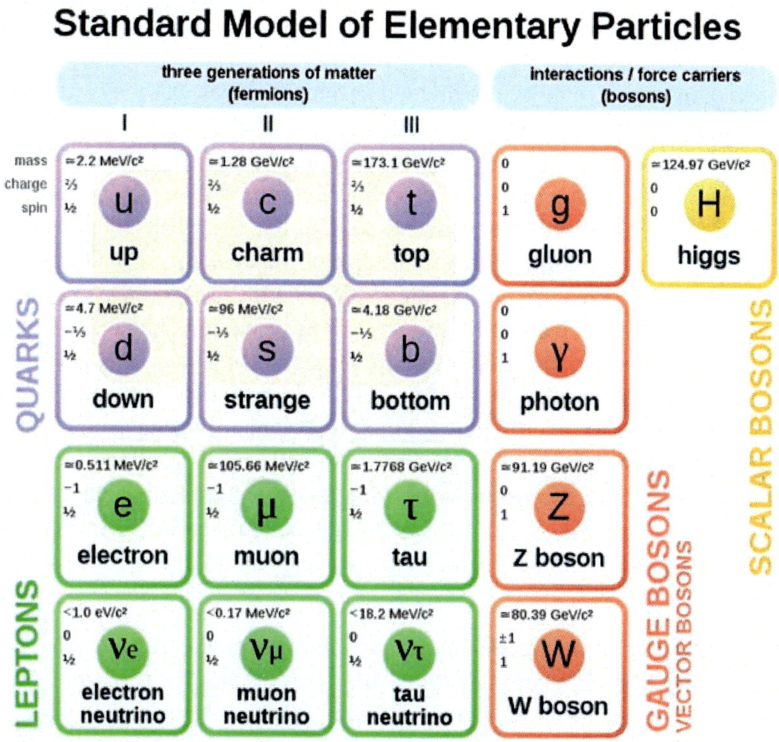

Fig. 2.1 Diagram representing the elementary particles of the Standard Model. Matter is constituted by three generations of quarks (in purple) and leptons (in green), while the interactions amongst them are governed by the gauge bosons (in red). The Higgs boson (in yellow) is responsible for the masses of the particles

in the second family, with masses of 1.3 GeV [5] and 93 MeV [5], respectively, and the top (t) and bottom (b) quarks in the third family, with masses of 172.9 GeV [5] and 4.18 GeV [5], respectively. The quarks belonging to these last two families are only produced in high energy collisions and, being the most massive particles, they undergo cascade decays into fermions of the first family.

Quarks are the only SM particles that are subject to the three forces: the electromagnetic, the weak and the strong. Each quark carries a quantum number called *flavour*, which is subject to the electroweak interaction, and a quantum number called *colour*, which is subject to the strong interaction. The latter is described by the Quantum Chromodynamics (QCD) theory. A property of this theory is the *colour confinement*, through which quarks do not exist as free states and can only be experimentally observed as bound states. Hence, they form *mesons*, which are quark-antiquark states, and *baryons*, which are composed by three quarks. Both bound states are denoted as *hadrons*, and they constitute the only bound states found in nature. Although quarks are confined in hadrons, they are *asymptotically free* particles, meaning the strong coupling becomes weaker when the momentum transfer is large. This property allows the fundamental interactions between them to be studied in proton colliders such as the LHC.

The leptons, as the quarks, are divided in three families, but they are only subject to the electromagnetic and the weak interactions. The charged leptons of the three families are the *electron* (e), *muon* (μ) and *tau* lepton (τ), respectively. The electron is the lightest one, with a mass of 511 keV [5], and is stable. The muon has a mass of 105.7 MeV [5] and a lifetime of 2.2 μs [5], but is still detectable at the LHC experiments given its momentum and the size of the detectors. Finally, the τ lepton, with a mass of 1.78 GeV [5] and a lifetime of 2.9×10^{-13} s [5], can only be detected through its decay products, which can be leptons or hadrons given its high mass.

Each lepton is paired to a neutrino of the same flavour (ν_e, ν_μ, ν_τ), which is electrically neutral and is massless in the classical SM formulation. However, the observation of neutrino flavour oscillations [6] implies that neutrinos have non-zero masses, with upper limits of $m(\nu_e) < 2$ eV [5], $m(\nu_\mu) < 0.19$ MeV [5] and $m(\nu_\tau) < 18.2$ MeV [5]. Being electrically neutral, neutrinos interact with the matter only via the weak force, and consequently they are not directly detectable at collider experiments. Their presence can nonetheless be inferred via the energy imbalance of the event.

Bosons

The SM describes the fundamental interactions via force carriers, which have spin $s = 1$ and obey Bose-Einstein statistics. These are commonly called *gauge bosons*, as their existence arises from a local gauge invariance mechanism. The strong nuclear force is mediated by eight *gluons* (g); they are massless, electrically neutral and carry colour quantum number. A consequence of the gluons having colour is that they interact not just with the quarks but also with themselves. The *photon* (γ) is the force-carrier associated to the electromagnetic interaction. It is massless and has no electric charge, and it does not interact with itself. The $W^\pm$ and Z *bosons* are the mediators of the weak nuclear force. They have masses of 80.4 GeV [5] and 91.2 GeV [5], respectively, and can self-interact. The range of the various interactions is directly related to the mass of the corresponding gauge bosons. The infinite range of the electromagnetic interactions corresponds to an interaction mediated by a massless gauge boson. The short range of the weak interactions ($\sim 10^{-16}$ cm) corresponds to the exchange of a gauge particle with mass of the order of $\sim$100 GeV. The strong interaction is not infinite, as should correspond to the exchange of a massless gluon, due to the property of confinement.

The recently discovered *Higgs boson* has a unique role in the SM, as it is the only elementary scalar particle ($s = 0$) discovered in nature. It is electrically neutral and, contrary to the gauge bosons, it does not originate from a local gauge invariance mechanism. It is responsible for the so-called *electroweak symmetry breaking* mechanism, which explains how gauge bosons acquire masses, since an explicit mass term would make the theory non-renomalizable. The same Higgs boson couples to the fermions to confer them mass via the Yukawa interaction; the interaction strength of the Higgs boson to the top quark is the topic of this thesis. The discovery of the Higgs boson constitutes a historical breakthrough in particle physics, not only because it was the last tile of the SM, but because it opens up the study of a whole new scalar sector.

Strong interaction

The strong interaction is governed by QCD, the theory that describes the interactions between quarks and gluons. QCD is a non-abelian gauge theory based on a local gauge symmetry group called $SU(3)_C$, where the subscript C stands for colour. This is the quantum number associated to this group, and it can take three values: *red*, *green* and *blue*. Quarks being spin-1/2 fermions, they satisfy the Dirac equation and hence the free-field Lagrangian is given by the Dirac Lagrangian

$$\mathcal{L} = \bar{q}(i\gamma^{\mu}\partial_{\mu} - m)q \ , \tag{2.1}$$

where q corresponds to the quark field, m to its mass and γ^{μ} to the Dirac matrices. The symbol ∂_{μ} denotes the partial derivative with respect to the spacetime coordinates. Under an $SU(3)_C$ transformation, the quark field transforms as

$$q \rightarrow q' = e^{-ig_s \frac{\lambda^a}{2}\alpha_a} q \ , \tag{2.2}$$

where g_s is a real constant and α_a are the parameters of the transformation. The term $\lambda^a/2$ corresponds to 3×3 traceless herminitian matrices, the so-called *Gell-Mann* matrices, which generate the group.

In order for the Lagrangian in Eq. 2.1 to be invariant under the transformation in Eq. 2.2, the derivative ∂_{μ} has to be re-defined to the to the so-called *covariant derivative* D_{μ} as

$$D_{\mu} = \partial_{\mu} + ig_s \frac{\lambda_a}{2} G^a_{\mu} \ , \tag{2.3}$$

where the gauge vector fields G^a_{μ} correspond to the eight gluons that mediate the strong force. They have to compensate the additional term in Eq. 2.3 to maintain the local invariance of the Lagrangian, hence under $SU(3)_C$ they transform as

$$G^a_{\mu} \rightarrow G'^a_{\mu} = G^a_{\mu} + \partial_{\mu}\alpha^a + g_s f_{abc}\alpha^b G^c_{\mu} \ . \tag{2.4}$$

The f_{abc} are the structure constants of the $SU(3)_C$ group, which encode the commutation rules $[\lambda_a/2, \lambda_b/2] = i f_{abc}\lambda_c/2$.

After the introduction of the gluon vector fields, the Lagrangian has to be completed with a term describing their propagation. This is done with a kinetic energy term of the type $-\frac{1}{4} G^{\mu\nu}_a G^a_{\mu\nu}$, where $G^a_{\mu\nu}$ is the gluon field strength, defined as $G^a_{\mu\nu} = \partial_{\mu}G^a_{\nu} - \partial_{\nu}G^a_{\mu}$. However, this term of the Lagrangian would not be gauge invariant, so $G^a_{\mu\nu}$ should instead be expressed as

$$G^a_{\mu\nu} = \partial_{\mu}G^a_{\nu} - \partial_{\nu}G^a_{\mu} - g_s f_{abc} G^b_{\mu} G^c_{\nu} \ . \tag{2.5}$$

The last term in Eq. 2.5 is at the origin of the cubic and quartic self-interactions of the gluon fields. Such interactions between the force mediators are a general property of non-abelian gauge theories. The overall QCD Lagrangian is

$$\mathcal{L}_{\text{QCD}} = i\bar{q}\gamma^{\mu}\partial_{\mu}q - m\bar{q}q - g_s\bar{q}\gamma^{\mu}\frac{\lambda_a}{2}qG_{\mu}^{a} - \frac{1}{4}G_a^{\mu\nu}G_{\mu\nu}^{a} \tag{2.6}$$

with the summation over all quark fields involved. The first term in Eq. 2.6 represents the free-field propagation of the quark, whereas the second term is the mass term. The third term arises from the introduction of the covariant derivative and describes interaction of the gluon with a quark and an antiquark. The strength of the interaction is parametrized by the constant g_s, usually redefined as the *strong coupling constant* $\alpha_s = g_s^2/4\pi$. This constant has the property of asymptotic freedom: it becomes very small when the energy transfer is large enough, leading to a quasi-free behaviour of the quarks and gluons. Finally, the fourth term in Eq. 2.6 represents the propagation of the gluons; upon expansion, it leads to 3-gluons and 4-gluons self-interactions. Gluons must be massless, since adding a mass term $m^2G_a^{\mu}G_{\mu}^{a}$ would lead to a gauge-invariant Lagrangian.

Electroweak interaction

The electroweak interaction is a unified description of the electromagnetic and weak interactions. It is built upon the same local gauge invariant as for QCD, but this time imposing a symmetry under $\text{SU}(2)_\text{L} \times \text{U}(1)_\text{Y}$ transformations. In the following, the electromagnetic and weak formalisms are described separately; the unification mechanism is summarized at the end.

The *electromagnetic interaction* is described by Quantum Electrodynamics (QED). It explains phenomena involving electrically charged particles interacting by means of exchange of a photon. It is an abelian gauge theory with the symmetry group $\text{U}(1)_\text{em}$. Following the same rationale as in QCD, the QED Lagrangian is expressed as

$$\mathcal{L}_{\text{QED}} = i\bar{\psi}\gamma^{\mu}\partial_{\mu}\psi - m\bar{\psi}\psi - eQ\bar{\psi}\gamma^{\mu}\psi A_{\mu} - \frac{1}{4}F_{\mu\nu}F^{\mu\nu} , \tag{2.7}$$

where ψ is a fermionic field, involving both quarks and leptons. The first term corresponds to the free Lagrangian of a massive spin-1/2 field, whereas the second term is the mass term. The third term arises from the introduction of the $\text{U}(1)_\text{em}$ covariant derivative, namely

$$D_{\mu} = \partial_{\mu} + ieQA_{\mu} , \tag{2.8}$$

and corresponds to the interaction between the photon, represented by the gauge potential A_{μ}, and the fermion. The strength of the interaction is proportional to the charge eQ of the fermion, Q being the quantum number associated to this interaction and e the electron charge. The last term in Eq. 2.7 corresponds to the free propagation of the photon, where $F_{\mu\nu}$ is the electromagnetic or Maxwell tensor defined as $F_{\mu\nu} = \partial_{\mu}A_{\nu} - \partial_{\nu}A_{\mu}$. As in the case of the gluons, the photon is massless, but it does not interact with itself, since QED is an abelian theory.

The *weak interaction* is described with the non-abelian gauge group $\text{SU}(2)_\text{L}$ group. It is a unique theory: unlike other interactions, it has the peculiarity that it violates parity. This is accounted for in the theoretical description by the property of the *chi-*

rality of a fermion field, which introduces a vector-axial structure in the Lagrangian of the weak force. The chirality is a Lorentz-invariant quantity corresponding to the eigenvalues of the operator $\gamma^5 = i\gamma^0\gamma^1\gamma^2\gamma^3$, which can be -1 or $+1$, giving rise to the so-called left (ψ_L) and right (ψ_R) chirality fields, represented as $SU(2)_L$ doublets and $SU(2)_L$ singlets, respectively. The left and right components of a fermion field ψ are obtained by applying the P_L and P_R projectors

$$\begin{aligned}\psi &= \psi_L + \psi_R\,;\\ \psi_L &= P_L\psi = \frac{1}{2}(1-\gamma^5)\psi\,,\\ \psi_R &= P_R\psi = \frac{1}{2}(1+\gamma^5)\psi\,.\end{aligned} \tag{2.9}$$

Each fermionic field of the SM is represented as one left chirality doublet (Ψ_L) and two right chirality singlets (ψ_R, ψ'_R). For the first family of fermions, the fields of the electron-neutrino pair are expressed as

$$\Psi_L(x) = \begin{pmatrix}\nu_{eL}\\ e_L\end{pmatrix}\;;\quad \psi_R(x) = \nu_{eR}\,,\quad \psi'_R(x) = e_R\,; \tag{2.10}$$

the up-down quark pair is expressed as

$$\Psi_L(x) = \begin{pmatrix}u_L\\ d_L\end{pmatrix}\;;\quad \psi_R(x) = u_R\,,\quad \psi'_R(x) = d_R\,. \tag{2.11}$$

The same holds for the other two families. Under this notation, the weak Lagrangian for a spin-1/2 field can be written as

$$\mathcal{L}_{\text{weak}} = i\bar{\Psi}_L\gamma^\mu D_\mu\Psi_L + i\bar{\psi}_R\gamma^\mu D_\mu\psi_R + i\bar{\psi}'_R\gamma^\mu D_\mu\psi'_R - \frac{1}{4}W^i_{\mu\nu}W^{\mu\nu}_i\,, \tag{2.12}$$

with the omision of the mass term. The covariant derivative of the $SU(2)_L$ group is

$$D_\mu = \partial_\mu + ig_w T^i_\mu W^i_\mu \tag{2.13}$$

and the tensor field $W^i_{\mu\nu}$ is defined as

$$W^i_{\mu\nu} = \partial_\mu W^i_\nu - \partial_\nu W^i_\mu - g_w\epsilon_{ijk}W^j_\mu W^k_\nu\,, \tag{2.14}$$

with $i = 1, 2, 3$. Here W^i_μ are three gauge fields and $T^i_\mu = \sigma_i/2$ are the generators of the $SU(2)_L$ group, proportional to the Pauli matrices σ_i. The g_w is the coupling constant and ϵ_{ijk} are the structure constants of $SU(2)_L$, defined by the commutation rules $[\sigma_i/2, \sigma_j/2] = i\epsilon_{ijk}\sigma_k/2$. Again, the term $-g_w\epsilon_{ijk}W^j_\mu W^k_\nu$ in Eq. 2.14 adds the cubic and quartic self-interactions among the gauge fields, expected in non-abelian gauge theories.

The quantum number associated to the $SU(2)_L$ group is the *weak isospin*, which has three components $I_{1,2,3}$. The right chirality fields ψ_R and ψ'_R have a third isospin component of $I_3 = 0$ as they are singlets under $SU(2)_L$; the left-handed field Ψ_L has $I_3 = +1/2$ and $I_3 = -1/2$ for the upper and lower components, since they form a doublet under $SU(2)_L$. Consequently, the $SU(2)_L$ group acts only on the left component of the chiral fields, as indicated with the subscript L, with an interaction term $-g_w \bar{\Psi}_L \gamma^\mu T_i \Psi_L W^i_\mu$, which couples the left chirality field to the gauge fields.

The gauge fields W^i_μ in Eq. 2.14 do not correspond to the physical $W^\pm_\mu$ and Z_μ fields associated to the $W^\pm$ and Z bosons. The first can be related via the linear combinations

$$W^\pm_\mu = \frac{1}{\sqrt{2}}(W^1_\mu \mp i W^2_\mu) . \tag{2.15}$$

An additional $U(1)_Y$ group has to be introduced in order to relate W^3_μ to the physical Z_μ boson field. It is an abelian group with an associated gauge field B_μ, described by a Lagrangian similar to the QED case, except that instead of the charge, the quantum number associated is the *weak hypercharge* Y, hence the subscript in $U(1)_Y$. The Lagrangian of this group is

$$\mathcal{L}_Y = i\bar{\psi}\gamma^\mu D_\mu \psi - \frac{1}{4} B_{\mu\nu} B^{\mu\nu} , \tag{2.16}$$

omitting the mass term. The covariant derivative of $U(1)_Y$ takes the form

$$D_\mu = \partial_\mu + i g_Y B_\mu , \tag{2.17}$$

with a coupling constant g_Y. The tensor field $B_{\mu\nu}$ is defined as

$$B_{\mu\nu} = \partial_\mu B_\nu - \partial_\nu B_\mu . \tag{2.18}$$

The $U(1)_Y$ group acts on the fermionic field ψ, which can have left or right chirality. The interaction term arising from the covariant derivative is $-g_Y \bar{\psi}\gamma^\mu (Y/2)\psi B_\mu$, which couples the left or right handed field ψ to the gauge field B_μ.

Combining the Lagrangians in Eqs. 2.12 and 2.16, it is possible to unify the electromagnetic and weak interactions into the electroweak interaction, represented by the $SU(2)_L \times U(1)_Y$ symmetry group. The mixing of the B_μ and W^3_μ gauge bosons to obtain the electromagnetic field A_μ and the field Z_μ associated to the Z boson is done via the rotation

$$\begin{pmatrix} A_\mu \\ Z_\mu \end{pmatrix} = \begin{pmatrix} \cos\theta_W & \sin\theta_W \\ -\sin\theta_W & \cos\theta_W \end{pmatrix} \begin{pmatrix} B_\mu \\ W^3_\mu \end{pmatrix} , \tag{2.19}$$

where θ_W is the so-called Weinberg or weak angle. In order to recover the electromagnetic coupling, the charge Q and the hypercharge Y are related via the Gell-Mann-Nishijima formula

Table 2.1 Electroweak quantum numbers of the fermion fields in the $SU(2)_L$ representation, categorized under quarks and leptons. The subscripts L and R correspond to the left and right chiralities. I_3 is the weak isospin, Y is the weak hypercharge and Q is the electric charge

Fermions	1st gen. 2nd gen. 3rd gen.	I_3	Y	Q
Quarks	$\begin{pmatrix} u_L \\ d_L \end{pmatrix}$ $\begin{pmatrix} c_L \\ s_L \end{pmatrix}$ $\begin{pmatrix} t_L \\ v_L \end{pmatrix}$	$\begin{pmatrix} +1/2 \\ -1/2 \end{pmatrix}$	$+1/3$	$\begin{pmatrix} +2/3 \\ -1/3 \end{pmatrix}$
	u_R c_R t_R	0	$+4/3$	$+2/3$
	d_R s_R b_R	0	$-2/3$	$-1/3$
Leptons	$\begin{pmatrix} \nu_{e,L} \\ e_L \end{pmatrix}$ $\begin{pmatrix} \nu_{\mu,L} \\ \mu_L \end{pmatrix}$ $\begin{pmatrix} \nu_{\tau,L} \\ \tau_L \end{pmatrix}$	$\begin{pmatrix} +1/2 \\ -1/2 \end{pmatrix}$	-1	$\begin{pmatrix} 0 \\ -1 \end{pmatrix}$
	e_R μ_R τ_R	0	-2	$+1$
	$\nu_{e,R}$ $\nu_{\mu,R}$ $\nu_{\tau,R}$	0	0	0

$$Y = 2(Q - I_3) . \tag{2.20}$$

Moreover, the Weinberg angle θ_W is related to the coupling constants e, g_w and g_Y by

$$e = g_w \cos\theta_W = g_Y \sin\theta_W . \tag{2.21}$$

Table 2.1 shows the values of th electroweak quantum numbers for all the fermions, depending on their chiralities. The overall unified electroweak Lagrangian is

$$\mathcal{L}_{\mathrm{EWK}} = i\bar{\Psi}_L\gamma^\mu D_\mu \Psi_L + i\bar{\psi}_R\gamma^\mu D_\mu\psi_R + i\bar{\psi}'_R\gamma^\mu D_\mu\psi'_R - \frac{1}{4}W^i_{\mu\nu}W^{\mu\nu}_i - \frac{1}{4}B^i_{\mu\nu}B^{\mu\nu}_i , \tag{2.22}$$

where the expressions of $W^i_{\mu\nu}$ and $B^i_{\mu\nu}$ remain the same as in Eqs. 2.14 and 2.18. The covariant derivative is

$$D_\mu = \partial_\mu + ig_w W^i_\mu \frac{\sigma_i}{2} + ig_Y B_\mu \frac{Y}{2} . \tag{2.23}$$

Upon expansion of the first three terms of the Lagrangian in Eq. 2.22, one finds the different interaction terms among the fermions and the electroweak bosons. These can be grouped into those involving the $W^\pm$ bosons and those involving the Z boson. The former is denoted as the *charged current* interaction, where the coupling only involves left chirality fields, and therefore parity is maximally violated. The latter is denoted as *neutral current* interactions, and can involve both left and right chirality fields.

Both the charged and neutral currents occur with different strengths due to the mixing of the gauge fields via the Weinberg angle. The electromagnetic force, mediated by the photon γ, is not sensitive to the chirality of the fermion fields. The last two terms of Eq. 2.22 are the kinetic terms of the electroweak bosons; when developed, they show a large spectrum of self-interactions among them. These interactions can be cubic (ZWW, γWW) or quartic (ZZWW, $\gamma\gamma$WW, γZWW, WWWW), as expected the electroweak interaction being a non-abelian gauge theory.

It is noteworthy that the charged weak interaction, mediated by the $W^{\pm}$ bosons, is the only type of interaction which can change the flavour of the fermions. In the case of the quarks, the transition probability within the three families is encoded in the Cabibbo-Kobayashi-Maskawa (CKM) matrix [7, 8], which relates the weak eigenstates of the down-type quarks (d', s', b') to their mass eigenstates (d, s, b) via

$$\begin{pmatrix} d' \\ s' \\ b' \end{pmatrix} = \begin{pmatrix} V_{ud} & V_{us} & V_{ub} \\ V_{cd} & V_{cs} & V_{cb} \\ V_{td} & V_{ts} & V_{tb} \end{pmatrix} \begin{pmatrix} d \\ s \\ b \end{pmatrix} . \tag{2.24}$$

The diagonal elements of the CKM matrix are close to unity, meaning that transitions between quarks of the same generation are favoured. In the case of the leptons, the probability of transition is encoded in the Pontecorvo-Maki-Nakagawa-Sakata (PMNS) matrix [9], relating the weak eigenstates of the neutrinos (ν_e, ν_μ and ν_τ) to their mass eigenstates (ν_1, ν_2 and ν_3) through the expression

$$\begin{pmatrix} \nu_e \\ \nu_\mu \\ \nu_\tau \end{pmatrix} = \begin{pmatrix} U_{e1} & U_{e2} & U_{e3} \\ U_{\mu 1} & U_{\mu 2} & U_{\mu 3} \\ U_{\tau 1} & U_{\tau 2} & U_{\tau 3} \end{pmatrix} \begin{pmatrix} \nu_1 \\ \nu_2 \\ \nu_3 \end{pmatrix} . \tag{2.25}$$

According to the SM, the PMNS matrix should be a unit matrix with null off-diagonal elements, whereas the mixing between the three families has been observed [6].

Up to this point, the $W^{\pm}$ and Z bosons have been assumed to be massless, since adding explicit mass terms of the gauge fields to Eq. 2.22 would break the gauge invariance. The same holds for the left and right chirality fermions: a direct mass term of a fermion $m\bar{\psi}\psi = m\bar{\psi}_R\Psi_L + m\bar{\Psi}_L\psi_R$ would mix components having different weak isospin and weak hypercharge and is therefore not allowed. This limitation is in clear contradiction with the experimental observation of massive weak bosons and fermions. The mechanism through which fermions and bosons acquire mass, known as the Brout-Englert-Higgs mechanism [10–12], is explained in the following section.

2.1.2 The Brout-Englert-Higgs Mechanism

The Brout-Englert-Higgs (BEH) mechanism, also known as *electroweak symmetry breaking* (EWSB), was introduced as a solution to generate the gauge boson masses and explain the fermion masses, which were not accounted for in the electroweak gauge formalism. It was proposed in 1964 independently by Englert and Brout [10], by Higgs [11] and by Guralnik et al. [12]. It is based on the concept of spontaneous symmetry breaking, a phenomenon that is often observed in Nature in which a physical system in a symmetric state ends up in an asymmetric state. In particular, it describes systems where the Lagrangian obeys certain symmetries, but an individual ground state of the system does not exhibit the symmetries of the system itself.

The BEH mechanism is realized through the addition of two complex scalar fields (ϕ^+, ϕ^0) grouped in a weak isospin doublet ϕ known as the Higgs field. It is expressed as

$$\phi = \begin{pmatrix} \phi^+ \\ \phi^0 \end{pmatrix} = \frac{1}{\sqrt{2}} \begin{pmatrix} \phi_1 + i\phi_2 \\ \phi_3 + i\phi_4 \end{pmatrix} , \tag{2.26}$$

where ϕ_i $(i = 1, 2, 3, 4)$ are real scalar fields. The Lagrangian describing the gauge interaction of the Higgs field is

$$\mathcal{L}_{\mathrm{BEH}} = (D_\mu \phi)^\dagger (D_\mu \phi) - V(\phi) , \tag{2.27}$$

where D_μ is the covariant derivative defined in Eq. 2.12. The potential $V(\phi)$ has the form

$$V(\phi) = \mu^2 \phi^\dagger \phi + \lambda (\phi^\dagger \phi)^2 , \tag{2.28}$$

where μ and λ are constants. The shape of the Higgs potential $V(\phi)$ depends on the sign of μ; λ must be positive for the potential to be bounded from below. The case where $\mu^2 > 0$ corresponds to a potential well around a trivial minimum state $\phi_{\mathrm{ground}} = 0$, and the Lagrangian in Eq. 2.27 would represent a scalar particle with mass μ and a self-interaction of coupling λ. The case where $\mu^2 < 0$ leads to the so-called *Mexican hat* potential, whose shape is illustrated in Fig. 2.2.

In the configuration where $\mu^2 < 0$, the ground state with minimal energy is given for the infinite set of doublets that minimize this potential by fulfilling the condition

$$\phi^\dagger \phi = \frac{-\mu^2}{2\lambda} \equiv \frac{v^2}{2} , \tag{2.29}$$

where v is the *vacuum expectation value* (VEV) of the scalar potential. This makes the system unstable and forces it to choose a particular value for its ground state among all possible states. Upon selection of a specific ground state ϕ_{ground}, the $\mathrm{SU}(2)_{\mathrm{L}} \times \mathrm{U}(1)_{\mathrm{Y}}$ symmetry is explicitly broken. Without loss of generality, this ground state can be expressed as

$$\phi_{\mathrm{ground}} = \frac{1}{\sqrt{2}} \begin{pmatrix} 0 \\ v \end{pmatrix} . \tag{2.30}$$

Although the ground state does not respect the initial $\mathrm{SU}(2)_{\mathrm{L}} \times \mathrm{U}(1)_{\mathrm{Y}}$ symmetry, it is invariant under the $\mathrm{U}(1)_{\mathrm{em}}$ symmetry group, as it is parallel to the ϕ^0 component of the doublet. Hence, the Higgs field can be expanded around its minimum as

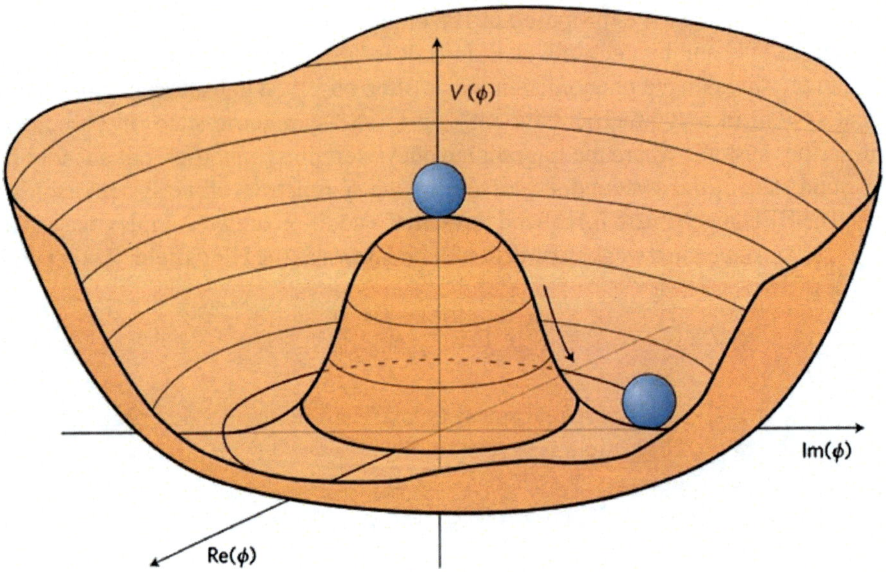

Fig. 2.2 Illustration of the Higgs potential in Eq. 2.28 with $\mu^2 < 0$, also known as the *Mexican hat* potential

$$\phi_{\text{ground}} \rightarrow \phi(x) = \frac{1}{\sqrt{2}} e^{i\sigma_a \theta^a(x)} \begin{pmatrix} 0 \\ v + h(x) \end{pmatrix}, \tag{2.31}$$

with $a = 1, 2, 3$. The field $h(x)$ corresponds to a massive field and the $\theta_1(x)$, $\theta_2(x)$ and $\theta_3(x)$ fields correspond to three massless scalar bosons. The latter are expected as consequence of the Goldstone theorem [13], that states that the spontaneous breaking of a continuous symmetry gives rise to as many massless bosons (i.e. *Goldstone bosons*) as broken generators of the symmetry. These Goldstone bosons are unphysical; they can be absorbed through an $SU(2)_L$ transformation of the Higgs field, known as the *unitary gauge*, leading to

$$\phi(x) \rightarrow \phi'(x) = \frac{1}{\sqrt{2}} \begin{pmatrix} 0 \\ v + h(x) \end{pmatrix}. \tag{2.32}$$

The Goldstone bosons have been translated to additional degrees of freedom of the $W^{\pm}$ and Z bosons, which correspond to their longitudinal polarizations: the mechanism confers mass to the weak bosons. The fourth real scalar field $h(x)$ remaining corresponds to a new physical massive particle, the Higgs boson (H).

Upon substitution of Eq. 2.32 in the Lagrangian in Eq. 2.27, and adding the kinetic terms of the physical fields $W^{\pm}$ and Z, the BEH Lagrangian becomes

$$
\begin{aligned}
\mathcal{L}_{\mathrm{BEH}} =& \frac{1}{2}\partial_\mu h \partial^\mu h - \frac{1}{2}2\mu^2 h^2 \\
&+ \frac{g_w^2 v^2}{4} W^{\mu+} W_\mu^- + \frac{1}{2}\frac{(g_w^2 + g_Y^2)v^2}{4} Z^\mu Z_\mu \\
&+ \frac{g_w^2 v}{2} h W^{\mu+} W_\mu^- + \frac{g_Y^2 v}{2} h Z^\mu Z_\mu \\
&+ \frac{g_w^2}{4} h^2 W^{\mu+} W_\mu^- + \frac{g_Y^2}{4} h^2 Z^\mu Z_\mu \\
&+ \frac{\mu^2}{v} h^3 + \frac{\mu^2}{4v^2} h^4 .
\end{aligned}
\tag{2.33}
$$

The first line of the Lagrangian corresponds to the evolution of the scalar Higgs field h, with mass

$$
m_{\mathrm{H}} = \sqrt{2}|\mu| = \sqrt{2\lambda} v , \tag{2.34}
$$

where m_{H} is a free parameter of the theory. The second line corresponds to the mass terms of the weak bosons $\mathrm{W}^{\pm}$ and Z, of masses

$$
\begin{aligned}
m_{\mathrm{W}} &= \frac{g_w v}{2} , \\
m_{\mathrm{Z}} &= \frac{\sqrt{g_w^2 + g_Y^2}}{2} v = \frac{m_W}{\cos\theta_W} .
\end{aligned}
\tag{2.35}
$$

Hence, the masses of the weak bosons depend on v. The cubic and quartic interactions between the Higgs field and the weak bosons are described in the third and forth lines, respectively. The corresponding couplings, denoted as λ_{HVV} and λ_{HHVV}, respectively, are proportional to the squared masses of the bosons via

$$
\begin{aligned}
\lambda_{\mathrm{HVV}} &= \frac{2}{v} m_V^2 , \\
\lambda_{\mathrm{HHVV}} &= \frac{1}{v^2} m_V^2 ,
\end{aligned}
\tag{2.36}
$$

where V denotes a $\mathrm{W}^{\pm}$ boson or a Z boson. Finally, the cubic and quartic self-interactions of the Higgs boson are represented in the fifth line. On this basis, the BEH potential $V(\phi)$ can be expressed more conveniently as

$$
\begin{aligned}
V(h) &= \frac{1}{2}(2\mu)^2 h^2 + \frac{\mu^2}{v} h^3 + \frac{\mu^2}{4v^2} h^4 \\
&= \frac{1}{2} m_{\mathrm{H}}^2 h^2 + \lambda_{\mathrm{HHH}} v h^3 + \lambda_{\mathrm{HHHH}} h^4 .
\end{aligned}
\tag{2.37}
$$

The Higgs boson self-coupling is thus related to its mass via the expression

$$\lambda_{\rm HHH} = \lambda_{\rm HHHH} = \frac{m_{\rm H}^2}{v^2} . \tag{2.38}$$

The Higgs boson self-interactions are purely related to the scalar sector, as they are entirely determined from the parameters of the scalar potential, contrary to the weak bosons, whose self-interaction have a gauge nature.

Yukawa Interaction

Thus far, the spontaneous symmetry breaking of the local gauge invariance of $SU(2)_L \times U(1)_Y$ gives rise to the mass terms of the weak bosons through their interaction with the Higgs field. The same field is responsible of conferring mass to the fermions in a gauge-invariant way through the so-called *Yukawa interaction.* The interaction couples a fermionic field ψ ($s = 1/2$) to a scalar field ϕ ($s = 0$), as illustrated in Fig. 2.3. In the context of the electroweak theory, this translates into the coupling of the left chirality doublets (Ψ_L) and the right chirality singlets (ψ_R, ψ'_R) to the Higgs scalar field (ϕ) via the Lagrangian

$$\mathcal{L}_{\rm Yukawa} = -y_{\rm f'}(\bar{\Psi}_L \phi \psi'_R + \bar{\psi}'_R \phi^\dagger \Psi_L) - y_{\rm f}(\bar{\Psi}_L \phi^c \psi_R + \bar{\psi}_R \phi^{c\dagger} \Psi_L) . \tag{2.39}$$

The values $y_{\rm f}$ and $y_{\rm f'}$ are known as the *Yukawa couplings*, where f corresponds to the up-type fermions and f' corresponds the down-type fermions. The symbol ϕ^c represents the charge conjugate of the Higgs field, defined as $\phi^c = i\sigma^2\phi^*$. Upon electroweak symmetry breaking, the Lagrangian becomes

$$\mathcal{L}_{\rm Yukawa} = \sum_f -m_{\rm f}\bar{\psi}\psi - \frac{m_{\rm f}}{v} h\bar{\psi}\psi , \tag{2.40}$$

where the sum runs over all fermions. The first term in Eq. 2.40 represents the mass term of the fermion and the second term corresponds to the coupling of the fermion to the Higgs field. The value of the masses of the fermions are

$$m_{\rm f} = \frac{y_{\rm f}}{\sqrt{2}} v , \tag{2.41}$$

i.e. the fermions couple to the Higgs boson with a coupling strength proportional to their mass. Being the fermion masses free parameters of the SM, the couplings

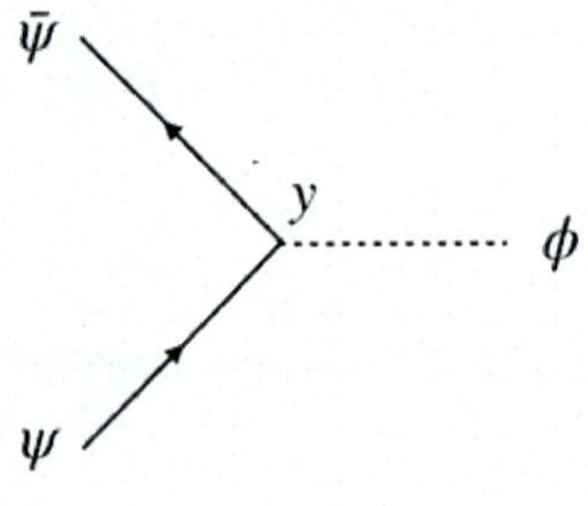

Fig. 2.3 Feynman diagram representing a Yukawa interaction of strength y between a fermion field $\psi(x)$ and a scalar field $\phi(x)$

can be inferred from mass measurements. In the SM, neutrinos are massless and do not exist as right-handed particles, hence they do not have a Yukawa coupling. As the top quark is the most massive fermion of the SM, its coupling to the Higgs boson (y_t) is the highest, with a value close to unity. The precise measurement of this crucial parameter, conducted in this thesis, constitutes an essential probe of the SM; deviations from the prediction would unambiguously reveal physics manifestations beyond the SM.

In conclusion, the BEH mechanism introduces a scalar Higgs field which permeates the entire universe and has a non-zero vacuum expectation value. The interaction of gauge bosons and fermions with this field generates their masses in a gauge-invariant way. The mechanism thus proposes an elegant solution to an important shortcoming of the gauge theory, leading to new verifiable predictions. The phenomenology and latest measurements related to the Higgs boson are summarized in the following section.

2.1.3 Phenomenology of the Higgs Boson at the LHC

On the 4th July 2012, the ATLAS and CMS experiments at the LHC at CERN announced independently the observation of a new boson in the mass region of around 125 GeV compatible with the one predicted by the SM [3, 4]. The discovery was performed with the data collected at a centre-of-mass energy $\sqrt{s} = 7$ and 8 TeV during the first two years of the LHC Run 1 (2010–2013). The importance of this discovery was emphasized a year later, when F. Englert and P. Higgs were jointly awarded the Nobel prize in physics "for the theoretical discovery of a mechanism that contributes to our understanding of the origin of mass of subatomic particles, and which recently was confirmed through the discovery of the predicted fundamental particle, by the ATLAS and CMS experiments at CERN's Large Hadron Collider". With the observation of the Higgs boson and the measurement of its mass m_H, the last important parameter of the SM had been determined.

Increasingly sensitive measurements of the Higgs boson mass were conducted by both the ATLAS and CMS collaborations after the discovery as more data were gathered throughout the Run 2 (2016–2018) of the LHC. To date, the most precise measurement was achieved with the combination of the data recorded in 2011, 2012 and 2016 by the CMS experiment at centre-of-mass energies of $\sqrt{s} = 7, 8$ and 13 TeV, respectively. The value of the mass was measured with 0.1% accuracy to [14]

$$m_\mathrm{H} = 125.38 \pm 0.14\ \mathrm{GeV}\ . \tag{2.42}$$

Alongside the determination of m_H, other important properties of the Higgs boson have been established since the discovery. The Higgs boson was confirmed to be electrically neutral, to have spin $s = 0$ and to be even under charge conjugation parity (CP) transformations [15], in agreement with the SM predictions.

Once a value of the Higgs boson mass is assumed, predictions for its production and decay rates are possible. In proton collisions, four main production modes are identified; the Feynman diagrams are shown in Fig. 2.4. Their expected cross sections for a Higgs boson of mass of 125 GeV are shown in Fig. 2.5a.

- The **gluon-gluon fusion** (ggH) constitutes the dominant production mode, with a cross section of 49 pb, because of the large Parton Density Function for gluons. The two gluons of the protons interact to produce a Higgs boson via an intermediate loop of virtual heavy quarks; the loop is necessary since the Higgs boson does not couple to the massless gluons. Being the Higgs boson couplings to fermions proportional to their mass, the process is dominated at leading order by the top quark loop, and therefore is adequate to probe the top Yukawa coupling. However, any particle beyond the SM potentially sensitive to QCD could in principle contribute to the loop: the top Yukawa coupling can only be inferred in a model-dependent way with this process. As he Higgs boson is produced alone in this process, gluon fusion can lead to rather clean final states with few particles.
- The **vector boson fusion** (VBF) process has a cross section of 3.8 pb, about 10 times smaller than ggH. The two quarks from the protons radiate massive vector bosons, which interact to produce a Higgs boson: the coupling of the Higgs boson to the vector bosons can be measured with this process. At the LHC, the fusion of two W bosons is three times more probable than the fusion of two Z bosons as a result of the different couplings of the quarks to the W and Z bosons. The experimental signature of this production mode is very clean and distinct, as the process is purely

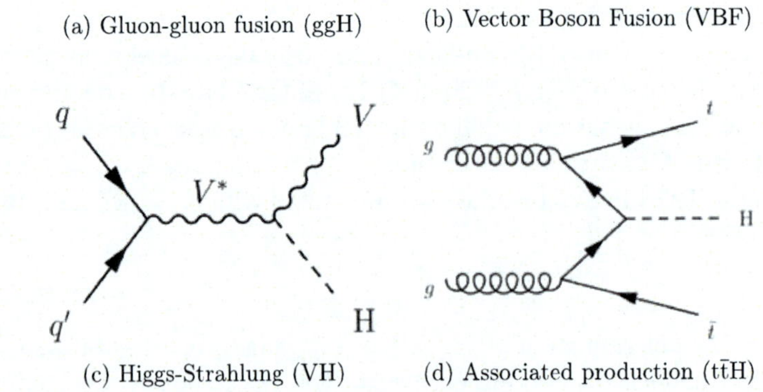

Fig. 2.4 Leading-order Feynman diagrams of the four main Higgs boson production modes at hadron colliders. Their cross sections at the LHC are given in Fig. 2.5a

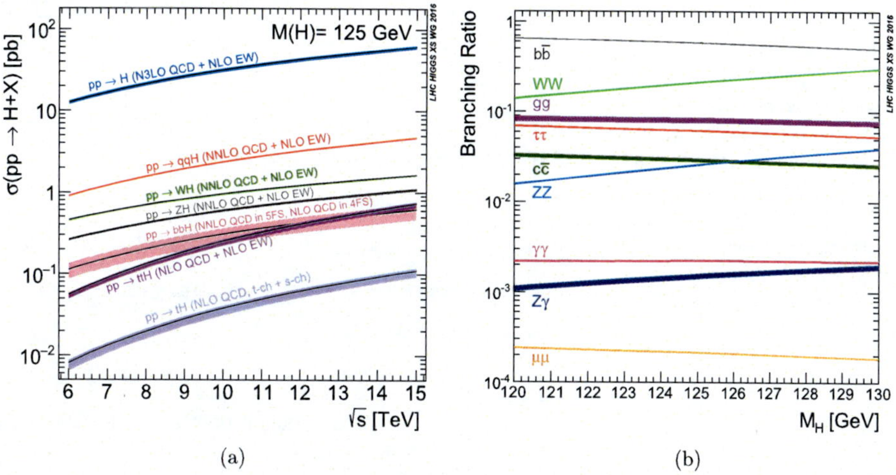

Fig. 2.5 **a** Cross sections of the main Higgs boson production modes at hadron colliders as a function of the centre-of-mass energy for a mass of $m_{\mathrm{H}} = 125$ GeV. **b** Branching ratios of the main decay modes of the Higgs boson as a function of the Higgs boson mass. The acronym NLO refers to next-to-leading order computation in perturbation theory, and similarly for NNLO and N3LO in higher-order expansions [16]

electroweak and the QCD activity is concentrated around the outgoing forward quarks, that present high dijet invariant mass and large pseudorapidity separation.

- The **Higgs-Strahlung** (VH) is the third most frequent mechanism, with a cross section 2.3 pb. In this process, a quark interacts with an antiquark to produce a massive vector boson which radiates a Higgs boson. As in the VBF case, the VH process is instrumental to probe the Higgs boson coupling to vector bosons. At the LHC, the WH cross section is approximately two times higher than the ZH cross section; the precision in the reconstruction of the system is nonetheless limited due to the decays of the W boson to neutrinos, which are undetectable.
- The **associated production of the Higgs boson with heavy quarks** is the last of the four main production modes. The heavy quarks can be either top quarks ($\mathrm{t\bar{t}H}$ and tH), or bottom quarks ($\mathrm{b\bar{b}H}$). Both can be initiated by either two incoming gluons or a quark and an antiquark, with similar cross sections at $\sqrt{s} = 13$ TeV, but the latter presents final states harder to distinguish from the backgrounds. The tH process is initiated by a b-quark together with another quark or a gluon, with a cross section around 5 times smaller than in the $\mathrm{t\bar{t}H}$ case. Unlike in the ggH case, there are no loops involved in these processes, and thus the top and bottom Yukawa couplings can be measured directly at tree-level.

The latest combination of the Higgs boson production cross sections, conducted by the ATLAS experiment with the data collected during Run 2 [17], are shown in Fig. 2.6. All main Higgs production modes have cross sections in agreement with the SM predictions within uncertainties. Up to date, all but the tH and $\mathrm{b\bar{b}H}$ processes have been observed at the LHC [18, 19]. In particular, the observation of the $\mathrm{t\bar{t}H}$ process

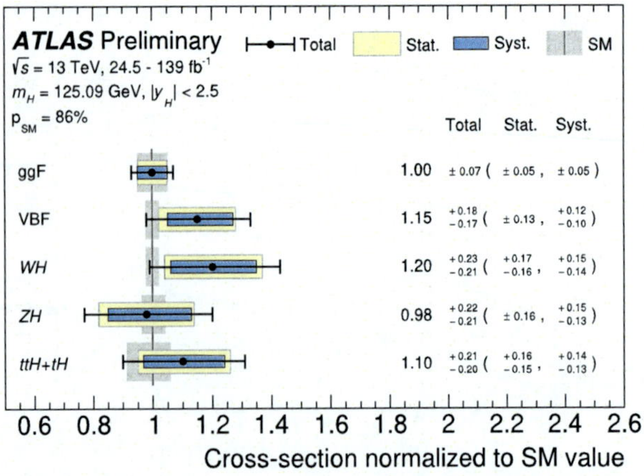

Fig. 2.6 Cross sections of the main Higgs boson production modes normalized to their SM predictions, obtained from the data collected by the ATLAS experiment in Run 2 at $\sqrt{s} = 13$ TeV. The cross sections are shown assuming SM values of the Higgs boson branching ratios [17]

Table 2.2 Branching ratios (BR) of the main Higgs boson decay modes assuming a mass m_{H} = 125.09 GeV. The uncertainties are due to missing higher-order corrections to the partial widths, uncertainties on the quark masses and on the value of the strong coupling constant α_s [16]

Decay mode	BR [%]
$\mathrm{H} \rightarrow \mathrm{b\bar{b}}$	$58.09^{+0.72}_{-0.73}$
$\mathrm{H} \rightarrow \mathrm{WW}^*$	21.52 ± 0.33
$\mathrm{H} \rightarrow \mathrm{gg}$	8.18 ± 0.42
$\mathrm{H} \rightarrow \tau^+\tau^-$	6.27 ± 0.10
$\mathrm{H} \rightarrow \mathrm{c\bar{c}}$	$2.88^{+0.16}_{-0.06}$
$\mathrm{H} \rightarrow \mathrm{ZZ}^*$	2.641 ± 0.040
$\mathrm{H} \rightarrow \gamma\gamma$	0.2270 ± 0.0047
$\mathrm{H} \rightarrow \mathrm{Z}\gamma$	0.1541 ± 0.0090
$\mathrm{H} \rightarrow \mu^+\mu^-$	$0.02171^{+0.00036}_{-0.00037}$

by the ATLAS and CMS collaborations in 2018 [20, 21] constituted a milestone in Run 2, as the measurement established the tree-level coupling of the Higgs boson to the top quark, and hence to an up-type quark. An extensive introduction to the tH and $\mathrm{t\bar{t}H}$ processes is given in Sect. 2.2.

With a lifetime of 1.56×10^{-22} s, the Higgs boson can not reach the detector: it must be studied via its decay products. The possible Higgs boson decay modes and the expected branching ratios (BR) can be found in Fig. 2.5b as a function of the Higgs boson mass; the values of the BR for a mass of $m_{\mathrm{H}} = 125.09$ GeV are shown in Table 2.2. The Higgs boson couples to massive electroweak bosons, and can therefore decay as $\mathrm{H} \rightarrow \mathrm{WW}^*$ (BR $\approx 22\%$) and $\mathrm{H} \rightarrow \mathrm{ZZ}^*$ (BR $\approx 2.6\%$). Note that the mass difference between the Higgs boson and the weak bosons ($m_{\mathrm{H}} < 2m_{\mathrm{W,Z}}$) implies that one of the two vector bosons must be off mass shell[1] in both cases. The Higgs boson

[1] An on mass shell particle satisfies the energy-momentum relation of $E^2 = p^2 + m_0^2c^4$, while an off mass shell particle does not.

can also decay via loops of fermions (mostly top quarks) or W bosons to massless particles like in H → $\gamma\gamma$ (BR ≈ 0.23%), H → Zγ (BR ≈ 0.15%) and H → gg (BR ≈8.2%). At the LHC, the H → WW*, H → ZZ* and H → $\gamma\gamma$ decays have been probed. Interestingly, even if the last two have some of the lowest branching ratios, they were the main drivers of the Higgs boson discovery in 2012. The former benefits from a high signal-to-background ratio and an excellent invariant mass resolution in leptonic decays; the latter profits from very high resolution in the reconstruction of the invariant diphoton mass. With the amount of data gathered in Run 2, these decay modes lead to the most stringent measurements of the Higgs boson properties and the first differential cross section studies in the ATLAS and CMS collaborations [14, 22–27].

The Higgs boson also couples to fermions via the Yukawa interaction, with a strength proportional to the fermion mass. Since a pair of b-quarks is the heaviest possible final state of the Higgs boson, the H → b$\bar{\text{b}}$ channel is the most dominant (BR ≈ 58%). This decay suffers from a high QCD background, which can be partially reduced by using dedicated algorithms to identify the jets which originate from b-quarks. These algorithms are nonetheless insufficient to remove the $b\bar{b}$ irreducible background in the gluon fusion production mode, hence searches of the H → b$\bar{\text{b}}$ process are mostly focused on the VH production mode. The decay H → $\tau^+\tau^-$ is the second most dominant decay to fermions (BR ≈ 6.3%), but in this case the invariant mass of the $\tau^+\tau^-$ pair cannot be reconstructed precisely due to the decay of the τ into neutrinos, and multivariate reconstruction techniques are used instead. Other decays of the Higgs boson into fermions are potentially accessible, but they suffer from a very large QCD background (H → c$\bar{\text{c}}$, BR ≈ 2.9%) or from a very low branching ratio (H → $\mu^+\mu^-$, BR ≈ 0.022%). The decay to a pair of top quarks (with one quark off-shell) is heavily suppressed and is not expected to be observable. The existence of the Higgs boson decays to fermions was established via the observation of the H → $\tau^+\tau^-$ [28, 29] and H → b$\bar{\text{b}}$ [19, 30] processes by the ATLAS and CMS experiments during Run 2. The coupling of the Higgs to the second generation of fermions was probed with the evidence of the H → $\mu^+\mu^-$ process found by the CMS collaboration, with a significance of 3.0σ [31].

Upon measurement of the Higgs boson production and decay rates, the proportionality of the Higgs boson couplings and the mass of the fermions ($y_{\mathrm{f}} \propto m_{\mathrm{f}}$) and the square mass of the weak bosons ($\lambda_{\mathrm{HVV}} \propto m_{\mathrm{V}}^2$) can be verified. This is illustrated in Fig. 2.7 with the full Run 2 data collected by the CMS experiment. The couplings of the Higgs boson are probed over about 3 orders of magnitude, and the dependence of their strength on the boson and fermion masses is established, confirming that the Higgs boson breaks the degeneracy between the fermion families.

The unparalleled amount of data collected at the LHC during Run 1 and Run 2 granted an extensive set of combined ATLAS and CMS measurements of the Higgs boson production and decay rates, resulting in a number of constraints on the Higgs couplings to vector bosons and fermions. Thur far, all the measurements suggest that the scalar particle discovered in 2012 presents properties consistent with the ones postulated by the SM. However, more stringent constraints on the couplings are needed. In particular, the self-coupling of the Higgs field, at the heart of the Higgs

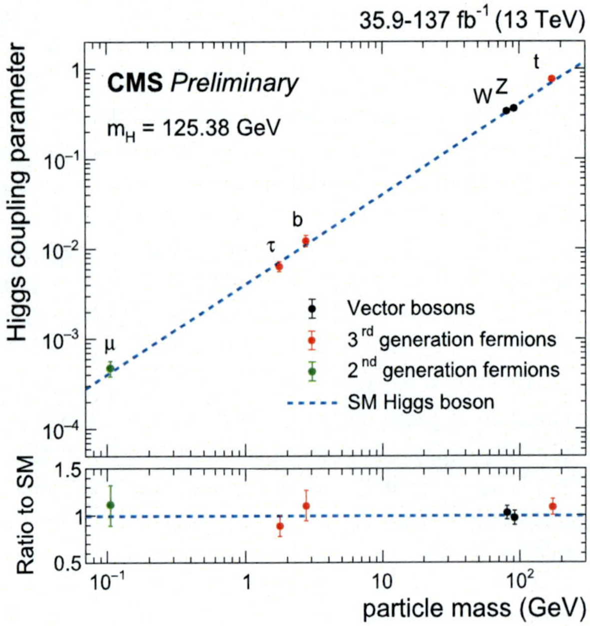

Fig. 2.7 Normalized Higgs boson coupling constants as a function of the boson or fermion mass, obtained from the combination of the data collected by the CMS experiment in Run 2 at $\sqrt{s} = 13$ TeV [31]

mechanism, needs to be established. In spite of the SM providing a coherent picture, many open questions about our universe still remain unanswered. Specifically, the Higgs sector faces the *naturalness* problem, resulting from the fact that the mass of the Higgs boson is highly sensitive to loop corrections but has been nonetheless measured to a value of 125 GeV. This and other shortcomings of the SM are summarized in the next section.

2.1.4 Shortcomings of the Standard Model

The SM theory provides a consistent description of all the elementary particles known to date and their interactions at the considered energy scales. Its predictive power makes it one of the most rigidly tested models in physics. Nonetheless, it still fails to explain various experimental observations; a non-exhaustive list is given in the following.

- *Neutrinos* in the SM are regarded as massless particles, while non-zero masses are required as a consequence of the neutrino flavour oscillations observed by several neutrino experiments [6].
- The measurement of the rotational speed of galaxies [32] suggests that there is a large amount of undetected mass in the universe, or *dark matter*, which can not be related to any SM particle. Experiments designed for the direct detection of dark matter candidates, such as the Weakly Interacting Massive Particles (WIMP), have had no success yet.

- The accelerated expansion of the universe requires the presence of a repulsive action, denoted as *dark energy*, that compensates for the gravitational attractive forces. It has been corroborated by cosmological observations [33] but is not contemplated in the SM.
- The *matter-antimatter asymmetry* observed in the universe is not accounted for in the SM. Such mechanism manifests, for instance, in the CP-violation in the SM weak interaction, but its magnitude is not sufficiently large to account for the observed imbalance [34].
- The SM does not account for the existence of the *gravitational force*, accurately described in the theory of general relativity but not perturbatively renormalisable in its quantized version. This prevents theoretical physics from reaching the much acclaimed unification of the four interactions at high energy scales.
- In the Higgs sector, the *naturalness* problem [35] arises from the mass of the Higgs boson being highly sensitive to loop corrections: if the theory was considered valid up to the Planck scale (10^{19} GeV), the magnitude of these corrections would blow. Since a tiny value of 125 GeV has been measured, it means that the mass parameter must be extremely fine-tuned to cancel out the divergences.
- There is an *arbitrariness* in the construction of the SM which has no obvious theoretical explanation, such as the 19 free parameters only determined experimentally and the large range of orders of magnitude among the fermion masses and force strengths.

These experimental observations suggest that the SM works only at the electroweak scale, and a more fundamental theory beyond the SM is needed to address these shortcomings. Numerous BSM theories have been postulated as possible extensions of the SM, typically expressed in terms of new interactions, quantum numbers or parameters describing additional degrees of freedom or symmetries. Among these, *supersymmetry* (SUSY) [36] is regarded as the most promising theory. It relies on the introduction of a new symmetry between fermions and bosons, under which each fermion is associated to a supersymmetric boson, and vice-versa. Extensions of the SM based on this symmetry solve the naturalness problem and provide candidates for dark matter, for instance. In spite of the central role of SUSY searches in the ATLAS and CMS physics program, no evidence for sypersymmetry has been found up to date.

2.2 Associated Production of the Higgs Boson with Top Quarks

Fermions in the SM couple to the Higgs boson through the Yukawa interaction with a coupling strength y_f proportional to the mass of the fermion m_f. The proportionality is expressed as $y_\mathrm{f} = \sqrt{2}m_\mathrm{f}/v$, where v is the VEV of the Higgs field, with a measured value of 245.22 GeV [37]. The values of the fermion masses are not directly predicted by the SM and can only be constrained by experimental observations. In the case of the top quark, the combined measured value of its mass by the Tevatron and LHC experiments [5] amounts to

$$m_{\mathrm{t}} = 172.9 \pm 0.4\ \mathrm{GeV}\ , \tag{2.43}$$

Being the heaviest fermion of the SM, the top quarks presents the largest Yukawa coupling, with a value of $y_{\mathrm{t}} \approx 0.995$. It is yet not known if the fact that the top Yukawa coupling has a value very close to unity is by chance or can be explained by an unknown underlying theory.

Given the large value of the top Yukawa coupling, its experimental characterization is a crucial step towards a deeper understanding of the electroweak symmetry breaking and the underlying physics. Many physical processes are sensitive to the value of this coupling: the top Yukawa interaction dominates the main Higgs boson production mechanism (gluon fusion) and contributes significantly to one of its most relevant decay channels ($\mathrm{H} \rightarrow \gamma\gamma$). In turn, deviations of the coupling from the SM predictions can have a strong impact on the theory. In the absence of observed BSM manifestations, the top Yukawa coupling is at the moment the only measurable quantity that can give an insight on the possible existence and the energy scale of new physics above the electroweak scale, which are extremely sensitive to y_{t}. Small changes of the value of y_{t} can result in corrections to the Higgs boson self-coupling that can destabilize the vacuum [38] and corrections to the value of the Higgs mass m_{H}^2 that can destabilize the weak scale. Additionally, a large variety of models beyond the SM predict modified top-Higgs couplings. This is for instance the case of the two Higgs doublet model (2HDM) extensions of the SM [39]. These can arise in several BSM theories such as the minimal supersymmetric extension of the SM (MSSM) [40], twin Higgs models [41] and certain composite Higgs models [42]. The possible existence of charge-parity violation in the top Yukawa sector would likewise manifest via CP-violating couplings that would affect the production rate of the physics processes involved.

Even though the top Yukawa coupling is involved in the Higgs boson production via the gluon fusion and in the $\mathrm{H} \rightarrow \gamma\gamma$ decay, its value can only be determined indirectly through the rate of these processes, as unknown BSM particles could potentially appear in the loops provided they are sensitive to the QCD interaction. The only direct way to probe the top Yukawa is from the associated production of the Higgs boson with top quarks, where the Higgs boson couples to the top quark at tree-level. In this context, the Higgs boson can be produced in association with a pair of top quarks ($\mathrm{t\bar{t}H}$) or with a single top quark (tH). In the case of the $\mathrm{t\bar{t}H}$ production, the cross section varies as y_{t}^2, meaning the process is sensitive to the magnitude of the coupling. The tH process is sensitive to the relative sign of the top-Higgs coupling and the W-Higgs couplings via the interference of its LO Feynman diagrams. The combined study of both processes serves to test the BSM models that predict modified top Yukawa couplings without making any assumption regarding the new physics. An overview of the $\mathrm{t\bar{t}H}$ and tH production at hadron colliders is given in the following. The latest measurements in the top Yukawa sector are given at the end.

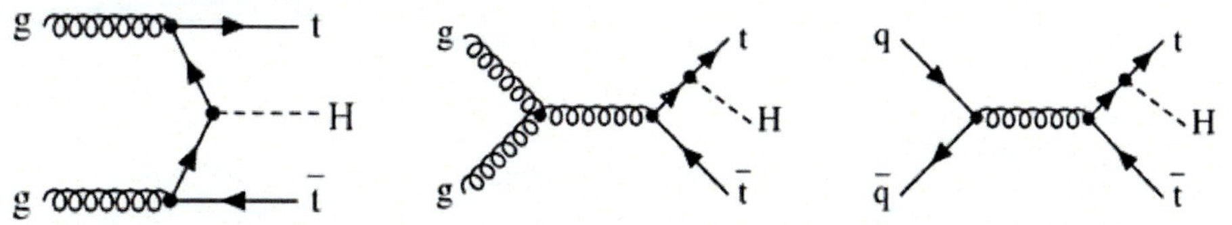

Fig. 2.8 Leading-order Feynman diagrams of the $t\bar{t}H$ production at hadron colliders

2.2.1 Standard Model Production

At hadron colliders, the $t\bar{t}H$ process can be initiated either by a pair of gluons or by a quark-antiquark pair. Some LO Feynman diagrams for the $t\bar{t}H$ process are shown in Fig. 2.8. The Higgs boson couples at tree-level to a top or antitop quark, which is produced either as an intermediate particle or as an outgoing particle, and hence the cross section is proportional to y_t^2. As the process plays a prominent role in the LHC physics program, the theoretical predictions of its cross section have benefited from significant efforts during the last years. The most recent improvements in this area are presented in Ref. [16]. In particular, the $t\bar{t}H$ signal modelling has been significantly improved by the addition of the next-to-leading order electroweak (NLO EWK) corrections to the NLO QCD corrections that have been available for several years. As an illustration, Fig. 2.9 shows the $t\bar{t}H$ cross section as a function of the Higgs boson mass at $\sqrt{s} = 13$ TeV including only QCD and both QCD+EW NLO corrections. These NLO EW corrections are of special relevance in the context of the measurement of the cross section, as they introduce additional corrections to the Higgs boson coupling to vector bosons and to itself, which slightly modify the dependence of the cross section on y_t^2. The most up-to-date prediction of the $t\bar{t}H$ cross section at $\sqrt{s} = 13$ TeV amounts to 506.5 fb, as shown in Table 2.3 along with the associated uncertainties.

The measurement of the cross section of the $t\bar{t}H$ process serves to determine the magnitude $|y_t|$, but not the sign of the coupling. The sign of y_t can be inferred by studying the tH and $\bar{t}$H processes, referred to jointly as tH. The tH process can occur via several channels, the corresponding LO diagrams shown in Fig. 2.10. The dominant channel is the t-channel (Fig. 2.9a, b), denoted as tHq, where a virtual W boson is exchanged. After this follows the W-channel (Fig. 2.9a, b), or tHW, which features a real W boson in the final state. There is an additional contribution to tH in the s-channel, denoted as tHb, via the exchange of a virtual W boson, but its cross section is negligible at the LHC energies, since the initial state requires a quark and an antiquark. In both the tHq and tHW processes, the Higgs boson can either couple to the top quark (Fig. 2.9a, c, denoted as Higgs-top diagrams) with a coupling strength of y_t, or to the W boson (Fig. 2.9c, d, denoted as Higgs-W diagrams) with a coupling strength of λ_{HWW}. The values of the tHq and tHW cross sections depend strongly on the interference between the contributions of the Higgs-top and the Higgs-W diagrams, as explained in Sect. 2.2.2. Under the SM assumption, the diagrams interfere destructively: the cross section acquires the lowest possible values.

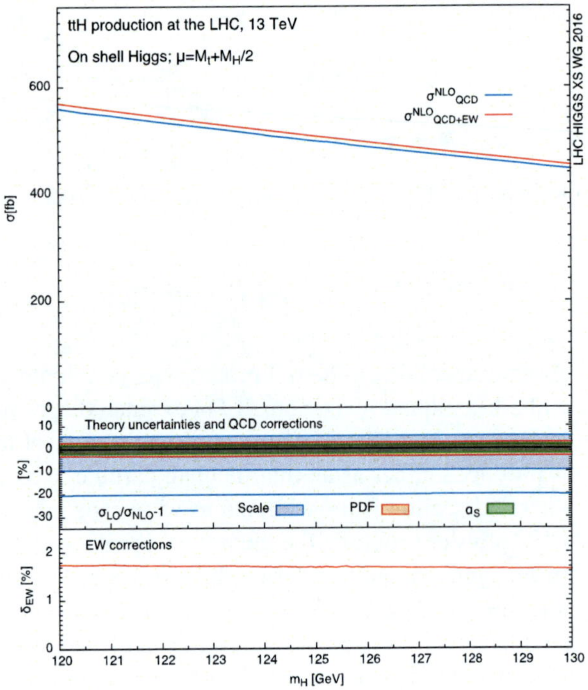

Fig. 2.9 Cross section of the $t\bar{t}H$ process at $\sqrt{s} = 13$ TeV as a function of the mass of the Higgs boson, including NLO QCD corrections (in blue) and NLO QCD+EW corrections (in red) [16]

Table 2.3 Cross sections at NLO QCD and NLO QCD+EWK of the $t\bar{t}H$ and tH processes for $\sqrt{s} = 13$ TeV and $m_H = 125.09$ GeV. K_{QCD} is the ratio of the NLO QCD and LO cross sections, δ_{EW} is the relative difference between the NLO QCD+EW and NLO cross section, Scale is the relative uncertainty due to the renormalization and factorization scale (including uncertainties on the flavour scheme in the tH t-channel), and PDF+α_s is the relative uncertainty arising from the PDF and the value of α_s [16, 43]. The values are computed with the generator MADGRAPH5_AMC@NLO [44]

Process	σ^{NLO}_{QCD} [fb]	σ^{NLO}_{QCD+EW} [fb]	K_{QCD}	δ_{EW} [%]	Scale [%]	PDF+α_s [%]
$t\bar{t}H$	498.0	506.5	1.25	1.7	+5.8/ − 9.2	± 3.6
tH t-channel	74.25	–	1.20	–	+6.5/ − 14.9	± 3.7
tH W-channel	15.17	–	1.38	–	+4.9/ − 6.7	± 6.3
tH s-channel	2.879	–	1.20	–	+2.4/ − 1.8	± 2.2

In turn, assuming a negative sign of y_t, as contemplated in several BSM models, the interference turns constructive and the cross section is enhanced significantly.

In the SM framework, the cross sections of the tHq and tHW processes amount to $\sigma_{tHq} = 74.3$ fb and $\sigma_{tHW} = 15.1$ fb, respectively, as shown in Table 2.3 together with the associated uncertainties. The values of these cross sections are to the state-of-

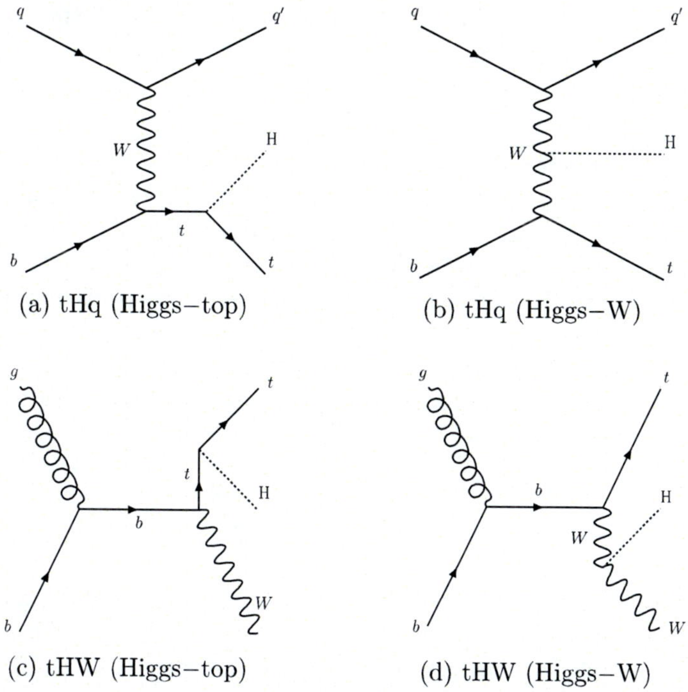

Fig. 2.10 Leading-order Feynman diagrams of the tHq (top row) and tHW (bottom row) processes at hadron colliders. The Higgs boson can couple to the top quark with a coupling strength y_t (left column) or to the W boson with a coupling strength λ_{HVV} (right column).

the-art predictions, computed at NLO accuracy in perturbative QCD [16]. They are evaluated in the so-called *five-flavour scheme* (5FS), where the initial bottom quarks are considered as sea quarks of the proton and are modelled by the parton distribution functions (PDFs) of the proton. The alternative approach is the *four-flavour scheme* (4FS), where the bottom quarks are produced by gluon splitting, described as part of the hard scattering process. The kinematic properties of the bottom quark are described more accurately with the 4FS, but the 5FS results in easier calculations, as there is one quark less in the final state. In the 4FS, the cross section of the tHq process amounts to approximately 71 fb [45]. The tHW process interferes with the $t\bar{t}H$ process in the 4FS, whereas the two processes result in different final states in the 5FS; the latter is used for the simulation of the tHW process in this analysis.

2.2.2 Interpretations Beyond the Standard Model

Besides giving direct access to the measurement of the top Yukawa coupling in the SM framework, the $t\bar{t}H$ and tH processes are windows to new physics beyond the SM. In what follows, two possible BSM interpretations are described, affecting both the kinematics and cross sections of the $t\bar{t}H$ and tH processes. The first one is related to the relative sign of the top Yukawa coupling and the coupling of the Higgs boson to the W bosons, while the second one addresses the CP-violation in the Yukawa sector.

Relative sign of the top Yukawa coupling

As mentioned in Sect. 2.2.1, if the value of the top Yukawa coupling y_t were to deviate from the SM expectation of $y_t \approx 1$, the cross section of the tH process would change. To understand this effect, it is convenient to express the tH cross section as a function of the Higgs boson coupling modifiers. This is done in the so-called *kappa framework* [16], where the coupling modifiers κ are defined as the ratio of the measured value of a given coupling to the one predicted by the SM. The relevant coupling modifiers in the tH process are the ones corresponding to the top quark (κ_t) and vector bosons (κ_V), defined as

$$\kappa_t = \frac{y_t}{y_t^{SM}}, \quad \kappa_V = \frac{\lambda_{HVV}}{\lambda_{HVV}^{SM}}, \tag{2.44}$$

where V represents a $W^{\pm}$ boson or a Z boson. Anomalous couplings leading to values $\kappa_t \neq 1$ or $\kappa_V \neq 1$ would affect the interference between the Higgs-top and Higgs-W diagrams, thus inducing changes in the tHq and tHW cross sections. The cross sections of these processes with non-SM couplings can be approximately parametrized with [46]

$$\begin{aligned} \sigma_{tHq} &= (2.63\kappa_t^2 + 3.58\kappa_V^2 - 5.21\kappa_t\kappa_V) \times \sigma_{tHq}^{SM}, \\ \sigma_{tHW} &= (2.91\kappa_t^2 + 2.31\kappa_V^2 - 4.22\kappa_t\kappa_V) \times \sigma_{tHW}^{SM}. \end{aligned} \tag{2.45}$$

This dependence is illustrated in Fig. 2.11 in units of the SM cross sections. In the case where $\kappa_t = -\kappa_V = -1$,[2] referred to as the *Inverted Top Coupling* (ITC) scenario, the tHq (tHW) cross section is enhanced by a factor of 11.4 (9.4) with respect to the SM prediction. In this case, the tHq cross section, with a value of ~793 fb, would exceed that of the $t\bar{t}H$ production, and be well within the reach of the current experimental sensitivity. The precise measurement of the rate at which the tH process occurs, conducted in this thesis as explained in Chap. 7, can therefore provide an insight to new physics manifesting via anomalous top-Higgs couplings. As the $t\bar{t}H$ process is insensitive to the sign of the top Yukawa coupling, the cross sections of this process in the SM and the ITC scenarios are equal.

Charge-parity Violation in the Top Yukawa Sector

The tH and $t\bar{t}H$ productions can be used to study the possible CP-violating components in the top Yukawa sector. It has already been established from the study of its decay to vector boson pairs that the Higgs boson is a pure scalar particle, invariant under charge-parity (CP) transformations. However, these measurements only probe the Higgs boson couplings to vector bosons: if the Higgs boson is a mixture of CP

[2] Since κ_V is more precisely determined at the LHC to be SM-like, by convention the negative sign is allowed for κ_t.

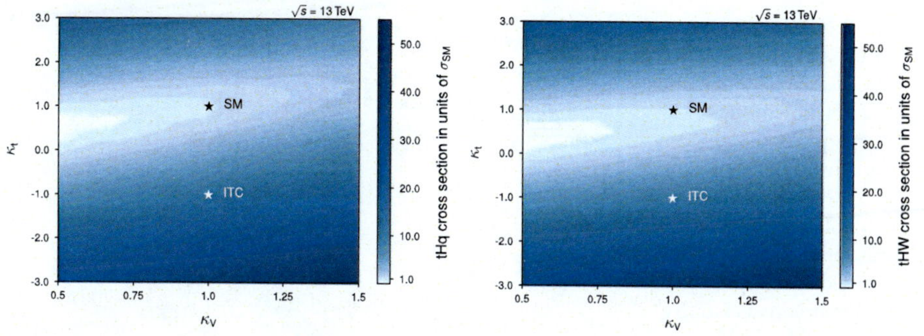

Fig. 2.11 Cross sections of the tHq (left) and tHW (right) processes in hadron colliders at $\sqrt{s} =$ 13 TeV as a function of the coupling modifiers $\kappa_t = y_t/y_t^{SM}$ and $\kappa_V = \lambda_{HVV}/\lambda_{HVV}^{SM}$. The points corresponding to the Standard Model and the Inverted Top Coupling scenarios are indicated with a black and a white star, respectively [47]

eigenstates where only the CP-even component couples to the vector bosons, then CP-violation in the Yukawa interaction might not have been detected. The effective Lagrangian density for this coupling can be written as [45]

$$\mathcal{L} = \bar{\psi}_t \left(\cos(\alpha)\kappa_{Htt}g_{Htt} + i\sin(\alpha)\kappa_{Att}g_{Att}\gamma_5\right)\psi_t X_0 \, , \tag{2.46}$$

where ψ_t is the top quark field, X_0 represents the Higgs field and α is the CP-mixing phase. The scalar component (CP-even) is described by Htt, while the pseudoscalar component (CP-odd) is described by Att. The values $g_{itt} = m_i/v$ are the coupling strengths and κ_{itt} are the coupling modifiers (with κ_{Htt} being equal to the above defined κ_t). The case $\alpha = 0$ corresponds to a pure CP-even coupling and recovers the SM cross section when $\kappa_{Htt} = 1$. The case $\alpha = 90°$ holds for a pure CP-odd coupling. Any intermediate value indicates CP violation, being $\alpha = 45°$ the maximally CP-violating angle. The case where $\alpha = 180°$ corresponds to the ITC scenario when $\kappa_{Htt} = 1$. The cross sections of the $t\bar{t}H$ and tH productions for different mixing angles are shown in Fig. 2.12. The tH process is sensitive to the whole range of mixing angles, while the $t\bar{t}H$ process is degenerate in α. Although briefly discussed here, the CP-violation interpretation is not contemplated in this manuscript.

2.2.3 *Experimental Searches*

The precise measurement of the $t\bar{t}H$ and tH productions requires the experimental capability to identify and reconstruct these signals in the detector. A joint search of the tH and $t\bar{t}H$ processes is convenient for this purpose, as they present a high degree of overlap in their experimental signatures, with either one or two top quarks and a Higgs boson, and more precise physics interpretations can be extracted from the combined analysis. With respect to other Higgs boson production mechanisms, the

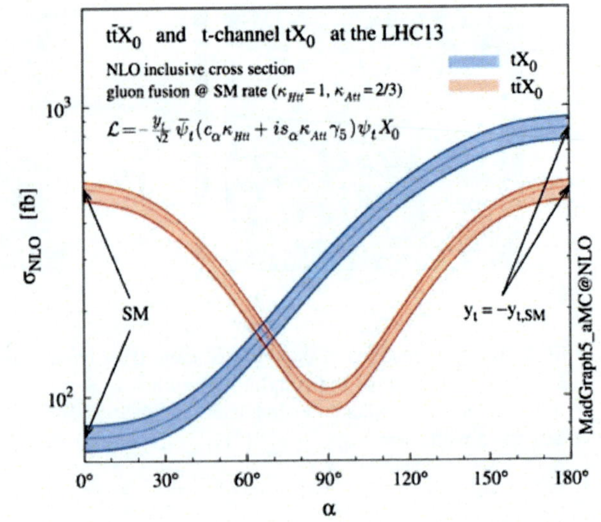

Fig. 2.12 Cross sections at NLO for the tX$_0$ and ttX$_0$ processes at hadron colliders as a function of the charge-parity (CP) mixing angle α. X_0 labels a generic spin $s = 0$ particle with CP-violating couplings. The Standard Model and the Inverted Top Coupling scenarios correspond to $\alpha = 0°$ and $\alpha = 180°$, respectively [45]

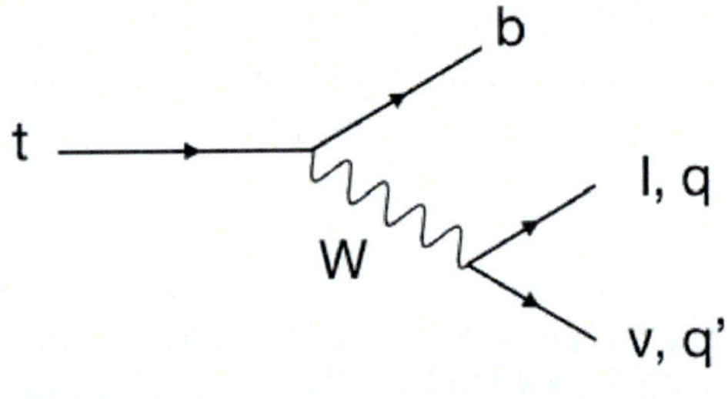

Fig. 2.13 Leading-order Feynman diagram of the top quark decay to a W boson and a bottom quark. The W boson can decay to a lepton ($\ell = e, \mu, \tau$) plus a neutrino (ν_ℓ), or to two quarks (q and q')

associated production of the Higgs boson with top quarks benefits from the special properties of the top quark compared to the other quarks in the SM. In addition, the various decay modes available to the Higgs boson result in a very rich and varied search that, upon combination, enhances the sensitivity to these rare production mechanisms.

Besides being the most massive particle, the top quark has a decay width of $\Gamma_t = 1.42^{+0.19}_{-0.15}$ GeV [5], and hence presents a very short lifetime of $\tau_t = \hbar/\Gamma_t \approx 5 \times 10^{-25}$ s, much shorter than the typical timescale of the formation of hadronic bound states ($\tau_{QCD} \approx 3 \times 10^{-24}$ s). This means that the top quark decays before it hadronizes, and therefore it is the only quark that exists as a free particle. Because of the large matrix element $|V_{tb}|^2$ in the CKM matrix, the top quark decays almost exclusively to a bottom quark (b) and an on-shell W boson, as depicted in Fig. 2.13, with a measured branching ratio of $(95.7 \pm 3.4)\%$ [5]. This lone decay mode t $\rightarrow$ W + b gives the top quark a distinctive signature and leads to a limited number of possible final states, constraining the searches considerably.

The W boson produced in the top decay can decay either to a quark-antiquark pair (BR $\sim$ 67%), creating light jets, or to a charged lepton and its associated neutrino (BR $\sim$ 33%). The former are called hadronic decays and the latter are called leptonic decays. For top quark pairs, this yields to three different categories (fully-

Table 2.4 Branching ratios (BR) of the decay modes of a top-antitop quark pair [5]. The symbol ℓ represents a lepton (e, μ or τ)

Category	Decay mode	BR [%]
Fully-leptonic	$\mathrm{t\bar{t}} \rightarrow (\mathrm{W^+b})\ (\mathrm{W^-\bar{b}}) \rightarrow (\ell^+\nu_\ell \mathrm{b})\ (\ell'^-\bar{\nu}_{\ell'}\mathrm{\bar{b}})$	10.5
Semileptonic	$\mathrm{t\bar{t}} \rightarrow (\mathrm{W^+b})\ (\mathrm{W^-\bar{b}}) \rightarrow (\mathrm{q\bar{q}'b})\ (\ell'^-\bar{\nu}_{\ell'}\mathrm{\bar{b}})$	43.8
	$\mathrm{t\bar{t}} \rightarrow (\mathrm{W^+b})\ (\mathrm{W^-\bar{b}}) \rightarrow (\ell^+\nu_\ell \mathrm{b})\ (\mathrm{q\bar{q}'\bar{b}})$	
Fully-hadronic	$\mathrm{t\bar{t}} \rightarrow (\mathrm{W^+b})\ (\mathrm{W^-\bar{b}}) \rightarrow (\mathrm{q\bar{q}'b})\ (\mathrm{q\bar{q}'\bar{b}})$	45.7

leptonic, semileptonic and fully-hadronic), whose branching ratios and decay chains are shown in Table 2.4. At least one (two) b-quark(s) are always present in the tH (t$\bar{\mathrm{t}}$H) final state. The b-quark is easily identifiable in the detector, as the B mesons which originate from the its hadronization have very specific features which are exploited with sophisticated reconstruction algorithms. They have a long lifetime of about 1.5 ps ($c\tau \sim 450$ mm), which leads to the presence of a secondary vertex that can be measured with high resolution tracking detectors. The masses of B mesons are large ($m \sim 5$ GeV) with respect to their decay products, which in turn receive a large momentum transfer that results in a wider opening angle. The jets originated from b-quarks have larger daughter particle multiplicity compared other quarks. In approximately 40% of the b-quark decays, an electron or a muon is produced inside the jet, which gives an additional handle to identify the jets originating from b-quarks, but which is also a source of lepton misidentification for processes involving leptonic decays from other mother particles.

In turn, the Higgs boson exhibits a large spectrum of decay modes (see Fig. 2.5b), which gives rise to a wide variety of possible final states. The most popular t$\bar{\mathrm{t}}$H and tH searches include the decays $\mathrm{H} \rightarrow \mathrm{b\bar{b}}$, $\mathrm{H} \rightarrow \gamma\gamma$, $\mathrm{H} \rightarrow \tau^+\tau^-$, $\mathrm{H} \rightarrow \mathrm{WW}^*$ and $\mathrm{H} \rightarrow \mathrm{ZZ}^*$. A summary of the experimental strategies for these final states in the context of the t$\bar{\mathrm{t}}$H search is given in the following; similar rationale applies to the corresponding tH search.

t$\bar{\mathrm{t}}$H with $\mathrm{H} \rightarrow \mathrm{b\bar{b}}$

The $\mathrm{H} \rightarrow \mathrm{b\bar{b}}$ final state profits from the largest branching ratio and is therefore very attractive experimentally. Nonetheless, it is a very complex final state, as it presents large jet multiplicities from light quarks and b-quarks, which gives rise to a large combinatorial background. Hence, the event reconstruction requires sophisticated multivariate techniques, as well as dedicated algorithms to identify the jets arising from b-quarks, or b-jets. This analysis suffers from an overwhelming background arising from the t$\bar{\mathrm{t}}$+jets process and, to a lesser extend, multijet QCD production; the latter is typically estimated from the data. The uncertainties on the measurement of the t$\bar{\mathrm{t}}$H cross section are systematic-dominated, the largest contributions resulting from the imprecise modelling of the t$\bar{\mathrm{t}}$+b$\bar{\mathrm{b}}$ and t$\bar{\mathrm{t}}$H backgrounds in the simulation.

t$\bar{\mathrm{t}}$H with $\mathrm{H} \rightarrow$ Multileptons

The multileptons final state is the channel targeted in this thesis; the analysis techniques are described in detail in Chap. 6. The search is sensitive to the $\mathrm{H} \rightarrow \mathrm{WW}^*$,

$H \rightarrow ZZ^*$ and $H \rightarrow \tau^+\tau^-$ decays, which present large particle multiplicities and, in particular, several light leptons (electrons and muons) and hadronically decaying τ leptons in the final state. The targeted Higgs boson decays have intermediate branching ratios; however, the presence of leptons reduces the QCD background contributions and profits from a higher resolution in the detector. The sensitivity of the analysis is enhanced by dedicated light lepton reconstruction techniques which discriminate the prompt leptons from the signal against the non-prompt leptons from B meson decays. The latter constitutes one of the main backgrounds, together with the top quark pair production in association with weak bosons ($t\bar{t}V$). The uncertainty on the measurement of the $t\bar{t}H$ cross section is equally dominated by statistical and systematic effects; the main systematic uncertainties result from the modelling of the signal and the $t\bar{t}V$ background in the simulation, and estimation of the non-prompt lepton background from the data.

$t\bar{t}H$ with $H \rightarrow \gamma\gamma$
The $H \rightarrow \gamma\gamma$ final state has the lowest branching ratio; it profits, nonetheless, from a low associated background and a clean final state. Thus, the diphoton invariant mass $m_{\gamma\gamma}$ can be reconstructed with high precision. This enables the signal to be extracted directly from a fit to the $m_{\gamma\gamma}$ distributions, which shows a characteristic narrow peak on a slowly falling background. The main backgrounds to this process arise from photon pairs produced in association with jets or leptons, and from γ+jet or jet+jet processes in which one or two jets are misidentified as photons. The uncertainties on the measurement of this process are dominated by statistical effects, being the leading systematic uncertainty the theoretical estimation of the $t\bar{t}H$ cross section.

2.2.4 *Latest Measurements*

Searches for the Higgs boson produced in association with either one or two top quarks have been performed at the LHC in proton-proton collisions at $\sqrt{s} = 7$, 8 TeV (Run 1) and $\sqrt{s} = 13$ TeV (Run 2) by both the ATLAS and the CMS collaborations. The most relevant milestone of these searches is the observation of the $t\bar{t}H$ production by both collaborations in 2018 [20, 21]. After confirming the existence of this rare production mechanism, efforts now centre on increasing the precision of the measurements and establishing the individual Higgs boson decay modes within the $t\bar{t}H$ process. In parallel, searches for the tH process have been undertaken by the ATLAS and CMS collaborations, establishing increasingly stringent upper limits to the value of the cross section. Several BSM interpretations in the top Yukawa sector have been explored upon combination with the $t\bar{t}H$ process, in particular concerning modified top-Higgs coupling models and the presence of CP-violation.

Associated Production of the Higgs Boson with Two Top Quarks
The observation of the Higgs boson production in association with a top quark pair was reported independently by the ATLAS and CMS collaborations in 2018 [20, 21]. The observation was based on a combined analysis of proton-proton collision

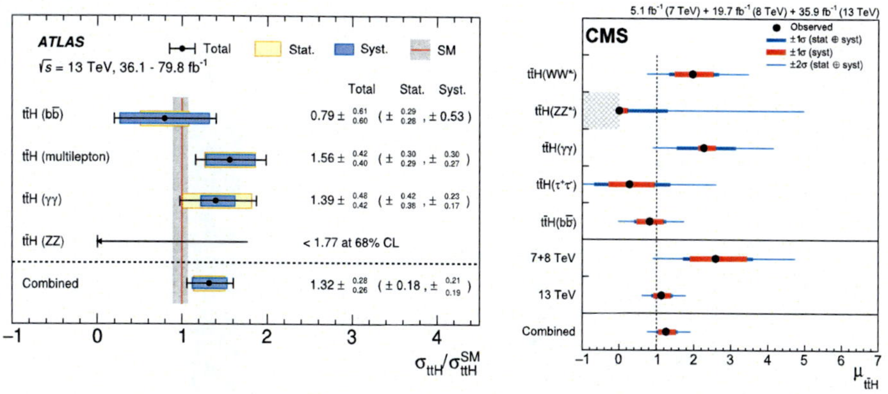

Fig. 2.14 Ratio of the measured cross section to the SM prediction of the $t\bar{t}H$ process in different final states measured in the **a** ATLAS and **b** CMS experiments with data collected at $\sqrt{s} = 7$, 8 and 13 TeV. The combination of the different final states lead to the observation of the $t\bar{t}H$ process by both collaborations independently [20, 21]

data collected at centre-of-mass energies of $\sqrt{s} = 7$, 8 and 13 TeV, amounting to integrated luminosities of 61 fb^{-1} for CMS and 105 fb^{-1} for ATLAS. The decay modes $H \to b\bar{b}$, $H \to WW^*$, $H \to ZZ^*$, $H \to \tau^+\tau^-$ and $H \to \gamma\gamma$ were combined in the search. The ratio between the measured cross section of the $t\bar{t}H$ process and its SM expectation, known as the signal strength modifier $\mu = \sigma/\sigma_{SM}$, is shown in Fig. 2.14 for ATLAS and CMS. The results are presented separately for each Higgs boson decay mode and for the combination of all of them. The combined signal strength amounts to $\mu = 1.32^{+0.28}_{-0.26}$ for ATLAS and $\mu = 1.26^{+0.31}_{-0.26}$ for CMS; it corresponds to an observed significance over the background-only hypothesis of 6.3σ for ATLAS and 5.2σ for CMS. The measured cross sections are consistent with the SM prediction: the tree-level coupling of the Higgs boson to the an up-type quark was demonstrated for the first time.

After the observation, the ATLAS and CMS collaborations oriented their searches towards establishing the $t\bar{t}H$ processes in the individual Higgs boson decay modes. The first evidence of the $t\bar{t}H$ process in the $H \to b\bar{b}$ final states was found by the CMS collaboration with the analysis of 77 fb^{-1} of data at $\sqrt{s}$ =13 TeV; the measured excess over the background-only hypothesis was reported with a significance of 3.9σ [48]. The observation of the $t\bar{t}H$ process with $H \to \gamma\gamma$ arrived with the analysis of the full Run 2 dataset at $\sqrt{s}$ =13 TeV: a significance of 6.6σ was reported by CMS [22] and 4.9σ by ATLAS [49]. The same analysis was used to probe the CP structure of the top Yukawa coupling in the $t\bar{t}H$ process [22, 50]. Both collaborations found no indication of CP-violation in the Yukawa sector. A CP-mixing angle α greater than 43° was excluded at 95% CL by the ATLAS collaboration; the fractional contribution of the CP-odd component was measured to $f_{CP} = 0.00 \pm 0.33$ by the CMS experiment, meaning the pure CP-odd structure was excluded at 3.2σ.

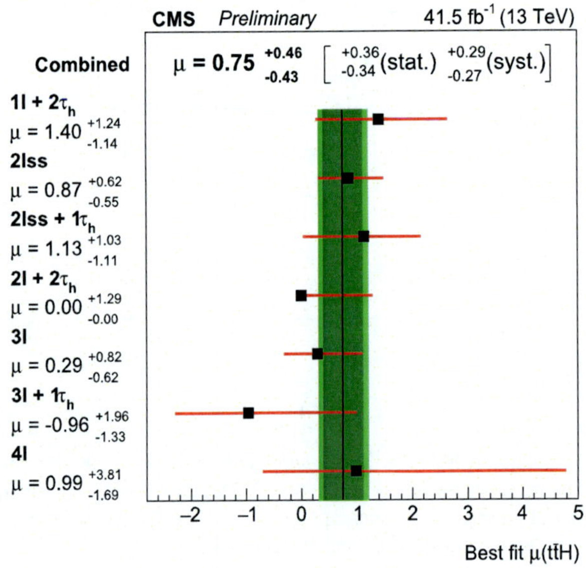

Fig. 2.15 Ratio of the measured cross section to the SM prediction of the $\mathrm{t\bar{t}H}$ process in multileptonic final states with 41.5 fb^{-1} of data collected at $\sqrt{s} = 13$ TeV by the CMS experiment [51]. The integrated luminosity corresponds to a partial dataset of the full result reported in Ref. [51], which includes an additional 39.5 fb^{-1} dataset previously analyzed [52]

Figure 2.15 shows the most precise result of the $\mathrm{t\bar{t}H}$ signal strength in multileptons final states prior to this thesis [51]. The result is obtained with 42 fb^{-1} of data recorded by the CMS experiment. The combination with the additional 40 fb^{-1} previously analyzed [52] lead to a measured signal strength modifier of

$$\mu_{\mathrm{t\bar{t}H}} = 0.96^{+0.34}_{-0.31} \,. \tag{2.47}$$

An observed significance of 3.2σ over the background-only hypothesis was reported, corresponding to the first evidence of the $\mathrm{t\bar{t}H}$ process in the multileptons final state. The result was derived in seven mutually exclusive categories according to the number of light leptons ℓ (electrons, muons) and hadronic τ leptons (τ_h) in the final state, where dedicated Boosted Decision Trees were developed to enhance the separation of the $\mathrm{t\bar{t}H}$ signal from the relevant backgrounds. The sensitivity of the measurement is driven by the categories containing no hadronic τ_h, which have larger statistical power and a cleaner experimental signature.

The analysis presented in this manuscript is inspired by the latest $\mathrm{t\bar{t}H}$ multilepton cross section measurement [51] with several improvements. The analyzed dataset is enlarged to the integrated luminosity of 137 fb^{-1} corresponding to the full Run 2, decreasing the statistical uncertainties significantly. The tH process is included in the search for the first time, and the measured $\mathrm{t\bar{t}H}$ and tH production rates are used to constrain the magnitude and sign of the top Yukawa coupling. The search is enriched by the inclusion of three additional categories. The identification algorithms for the τ_h and the b-jets have been optimized with Deep Neural Network (DNN) approaches with respect to the previous BDT-based techniques. The separation of the signal from the backgrounds is enhanced with machine learning techniques which include dedicated DNNs besides the optimized BDTs of the previous search. A multiclassification approach is implemented in the most sensitive categories, resulting in an improved disambiguity of the signals and the backgrounds, and also between the tH and $t\bar{t}H$ signals themselves. A complementary analysis approach is developed based on the multivariate Matrix Element Method, granting additional robustness to the results.

Associated Production of the Higgs Boson with a Top Quark
Following a similar approach to the $t\bar{t}H$ analyses, dedicated searches for the tHq and tHW processes were carried out by the ATLAS and CMS collaborations using data collected at $\sqrt{s} = 13$ TeV during Run 2 [46, 50]. In the context of the SM, the latest search conduced by the CMS collaboration includes an integrated luminosity of 36 fb^{-1}. It combines the decay mode $H \rightarrow b\bar{b}$ with at least one lepton in the final state and the decay $H \rightarrow$ multileptons, which provide the highest experimental sensitivity, along with a reinterpretation of the measurement in the $H \rightarrow \gamma\gamma$ channel in the context of tHq production. The main results are shown in Fig. 2.16 (left). For a SM-like tH process, an upper limit on the cross section times branching ratio was set to 25 times the SM expectation at 95% CL. The same measurement was conducted by the ATLAS collaboration, in this case restricted to the $H \rightarrow \gamma\gamma$ decay but including the full Run 2 dataset of 139 fb^{-1} [50]. In this case, the upper limit on the tH cross section is constrained to 12 times the value predicted by the SM at 95% CL; it constitutes the most stringent measurement to date.

Upon combination of the $t\bar{t}H$ and tH production modes, BSM interpretations of the results were conducted by the CMS experiment, shown in Fig. 2.16 (right). These showed that the data slightly favour a positive value of $\kappa_t = y_t/y_t^{SM}$ over a negative value by about 1.5σ, while excluding values outside the ranges [−0.9, −0.5] and [1.0, 2.1] at 95% CL, assuming $\kappa_V = 1.0$. The work presented in this

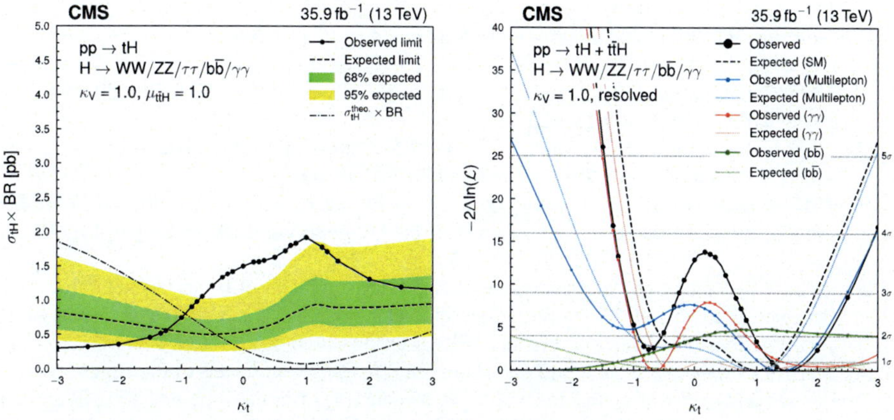

(a) (b)

Fig. 2.16 **a** Upper limits on the tH cross section times branching ratio as a function of the coupling modifier κ_t, assuming the SM value of the $t\bar{t}H$ cross section. **b** Likelihood scan as a function of the coupling modifier κ_t combining the $t\bar{t}H$ and tH processes. Both results correspond to 35.9 fb^{-1} of data collected by the CMS experiment at $\sqrt{s} = 13$ TeV, assuming $\kappa_V = 1.0$ [46]

thesis extends these searches with the analysis of the full Run 2 dataset, providing a more precise measurement of the tH production rate in the SM framework and posing more stringent constraints to the value of κ_t.

References

1. Paganini P (2020) Physique des particules avancée: an introduction to the Standard Model of Particle Physics. Lectures presented at Ecole Polytechnique, Palaiseau (France)
2. Herrero MJ (1998) The standard model. Lectures presented at the NATO ASI 98 School, Techniques and Concepts of High Energy Physics, St Croix, Virgin Islands (USA). https://arxiv.org/abs/hep-ph/9812242
3. ATLAS Collaboration (2012) Observation of a new particle in the search for the Standard Model Higgs boson with the ATLAS detector at the LHC. Phys Lett B 716:1–29
4. CMS Collaboration (2012) Observation of a new boson at a mass of 125 GeV with the CMS experiment at the LHC. Phys Lett B 716:30–61
5. Tanabashi M et al (2018) Review of particle physics. Phys Rev Lett D 98:030001
6. Fukuda Y et al (1998) Evidence for oscillation of atmospheric neutrinos. Phys Rev Lett 81:1562–1567
7. Cabibbo N (1963) Unitary symmetry and leptonic decays. Phys Rev Lett 10:531–533
8. Kobayashi M, Maskawa T (1973) CP-Violation in the renormalizable theory of weak interaction. Prog Theor Phys 49(2):652–657, 02 1973
9. Maki Z, Nakagawa M, Sakata S (1962) Remarks on the unified model of elementary particles. Prog Theor Phys 28(5):870–880, 11 1962
10. Englert F, Brout R (1964) Broken symmetry and the mass of gauge vector mesons. Phys Rev Lett 13:321–323
11. Higgs PW (1964) Broken symmetries and the masses of Gauge Bosons. Phys Rev Lett 13:508–509
12. Guralnik GS, Hagen CR, Kibble TWB (1964) Global conservation laws and massless particles. Phys Rev Lett 13:585–587
13. Goldstone J (1961) Field theories with "Superconductor" solutions. In: Nuovo Cimento 1955–1965(19):154–164
14. CMS Collaboration (2020) A measurement of the Higgs boson mass in the diphoton decay channel. Phys Lett B 805:135425
15. CMS Collaboration (2013) Study of the mass and spin-parity of the Higgs Boson candidate via its decays to Z Boson Pairs. Phys Rev Lett 110:081803
16. LHC Higgs Cross Section Working Group, D de Florian et al (2016) Handbook of LHC higgs cross sections: 4. Deciphering the nature of the Higgs sector. Technical report FERMILAB-FN-1025-T, CERN-2017-002-M, 10 2016
17. ATLAS Collaboration (2020) A combination of measurements of Higgs boson production and decay using up to 139 fb^{-1} of proton-proton collision data at $\sqrt{s} = 13$ TeV collected with the ATLAS experiment. Technical report ATLAS-CONF-2020-027, CERN, Geneva
18. ATLAS Collaboration and CMS Collaboration (2016) Measurements of the Higgs boson production and decay rates and constraints on its couplings from a combined ATLAS and CMS analysis of the LHC pp collision data at $\sqrt{s} = 7$ and 8 TeV. J High Energy Phys 2016:081803
19. ATLAS Collaboration (2018) Observation of Higgs to bb decays and VH production with the ATLAS detector. Phys Lett B 786:59–86
20. ATLAS Collaboration (2018) Observation of Higgs boson production in association with a top quark pair at the LHC with the ATLAS detector. Phys Lett B 784:173–191
21. CMS Collaboration (2018) Observation of $t\bar{t}$H production. Phys Rev Lett 120:231801

22. CMS Collaboration (2020) Measurements of $t\bar{t}$H production and the CP structure of the Yukawa interaction between the Higgs boson and top quark in the diphoton decay channel. https://arxiv.org/abs/2003.10866
23. ATLAS Collaboration (2018) Measurements of Higgs boson properties in the diphoton decay channel with 36 fb^{-1} of pp collision data at $\sqrt{s} = 13$TeV with the ATLAS detector. Phys Rev D 98:052005
24. ATLAS Collaboration (2018) Combined measurement of differential and total cross sections in the H $\rightarrow \gamma\gamma$ and the H $\rightarrow ZZ^* \rightarrow 4l$ decay channels at $\sqrt{s} = 13$TeV with the ATLAS detector. Phys Lett B 786:114–133
25. CMS Collaboration (2019) Measurements of properties of the Higgs boson in the four-lepton final state in proton-proton collisions at $\sqrt{s} = 13$ TeV. Technical Report CMS-PAS-HIG-19-001, CERN, Geneva
26. ATLAS Collaboration (2019) Measurements of the Higgs boson inclusive and differential fiducial cross sections in the 4ℓ decay channel at $\sqrt{s} = 13$ TeV. Technical report ATLAS-CONF-2019-025, CERN, Geneva, Jul 2019
27. ATLAS Collaboration (2020) Higgs boson production cross-section measurements and their EFT interpretation in the 4ℓ decay channel at $\sqrt{s} = 13$ TeV with the ATLAS detector. Technical report CERN-EP-2020-034, CERN, Geneva
28. CMS Collaboration (2018) Observation of the Higgs boson decay to a pair of tau leptons with the CMS detector. Phys Lett B 779:283–316
29. ATLAS Collaboration (2019) Cross-section measurements of the Higgs boson decaying into a pair of τ-leptons in proton-proton collisions at $\sqrt{s} = 13$TeV with the ATLAS detector. Phys Rev D 99:072001
30. CMS Collaboration (2018) Observation of higgs boson decay to bottom quarks. Phys Rev Lett 121:121801
31. CMS Collaboration (2020) Measurement of Higgs boson decay to a pair of muons in proton-proton collisions at $\sqrt{s} = 13$ TeV. Technical report CMS-PAS-HIG-19-006, CERN, Geneva
32. Sofue Yoshiaki, Rubin Vera (2001) Rotation curves of spiral galaxies. Ann Rev Astron Astrophys 39(1):137–174
33. Planck Collaboration (2020) Planck 2018 results. VI. Cosmological parameters. Astron Astrophys 641:A6
34. Ligeti Z (2005) The CKM matrix and CP violation. Int J Mod Phys A 20:5105–5118
35. Wells JD (2017) Higgs naturalness and the scalar boson proliferation instability problem. Synthese 194(2):477–490
36. Shadmi Y (2016) Introduction to supersymmetry. In: 2014 European school of high-energy physics, pp 95–123
37. MuLan Collaboration (2011) Measurement of the positive muon lifetime and determination of the fermi constant to part-per-million precision. Phys Rev Lett 106:041803
38. Bezrukov F, Shaposhnikov M (2015) Why should we care about the top quark Yukawa coupling? J Exp Theor Phys 120(3):335–343
39. Branco GC, Ferreira PM, Lavoura L, Rebelo MN, Sher M, Silva JP (2012) Theory and phenomenology of two-Higgs-doublet models. Phys Rep 516(1–2):1–102
40. Djouadi A (2008) The anatomy of electroweak symmetry breaking Tome II: the Higgs bosons in the Minimal Supersymmetric Model. Phys Rep 459(1–6):1–241
41. Chacko Z, Nomura Y, Papucci M, Perez G (2006) Natural little hierarchy from a partially goldstone twin Higgs. J High Energy Phys 2006(01):126
42. Mrazek J, Pomarol A, Rattazzi R, Redi M, Serra J, Wulzer A (2011) The other natural two Higgs doublet model. Nucl Phys B 853(1):1–48
43. Demartin F, Maier B, Maltoni F, Mawatari K, Zaro M (2017) tWH associated production at the LHC. Eur Phys J C 77(1)
44. Alwall J et al (2014) The automated computation of tree-level and next-to-leading order differential cross sections, and their matching to parton shower simulations. J High Energy Phys 07:079

45. Demartin F et al (2015) Higgs production in association with a single top quark at the LHC. Eur Phys J C 75
46. CMS Collaboration (May 2019) Search for associated production of a Higgs boson and a single top quark in proton-proton collisions at $\sqrt{s} = 13$TeV. Phys Rev D 99:092005
47. CMS Collaboration (2018) Search for the tH(H→ b$\bar{\text{b}}$) process in pp collisions at $\sqrt{s} = 13$ TeV and study of Higgs boson couplings. Technical report CMS-PAS-HIG-17-016, CERN, Geneva
48. CMS Collaboration (2019) Measurement of t$\bar{t}$H production in the H → $b\bar{b}$ decay channel in 41.5 fb^{-1} of proton-proton collision data at $\sqrt{s} = 13$ TeV. Technical report CMS-PAS-HIG-18-030, CERN, Geneva
49. ATLAS Collaboration (2019) Measurement of Higgs boson production in association with a $t\bar{t}$ pair in the diphoton decay channel using 139 fb^{-1} of LHC data collected at $\sqrt{s} = 13$ TeV by the ATLAS experiment. Technical report ATLAS-CONF-2019-004, CERN, Geneva
50. ATLAS Collaboration (2020) CP properties of Higgs Boson interactions with top quarks in the t$\bar{t}$H and tH processes Using $H \rightarrow \gamma\gamma$ with the ATLAS detector. Phys Rev Lett 125:061802
51. CMS Collaboration (2018) Measurement of the associated production of a Higgs boson with a top quark pair in final states with electrons, muons and hadronically decaying τ leptons in data recorded in 2017 at $\sqrt{s} = 13$ TeV. Technical report CMS-PAS-HIG-18-019, CERN, Geneva
52. CMS Collaboration (2018) Evidence for associated production of a Higgs boson with a top quark pair in final states with electrons, muons, and hadronically decaying τ leptons at $\sqrt{s} =$ 13TeV. J High Energy Phys 2018(8):66

Chapter 3
The CMS Experiment at the LHC

CERN, or the *European Laboratory for Particle Physics*, is an international research centre that operates the largest particle physics laboratory in the world. It sits astride the Franco-Swiss border west of Geneva and was founded in 1954 by twelve European countries. It was initially dedicated to the fields of nuclear and particle physics: its original name stands for *Conseil Européen pour la Recherche Nucléaire*, or European Council for Nuclear Research. Today, as our understanding of matter goes much deeper than the nucleus, the laboratory is oriented towards particle physics research. It has become an example of international scientific collaboration, with more than 13000 collaborators of over 100 nationalities representing more than 500 universities and institutes. As particle physics demands the ultimate in performance, CERN is at the forefront of technology development and knowledge transfer, and most notably served as the birthplace of the World Wide Web (WWW) in 1989.

CERN's current major facility is the *Large Hadron Collider* (LHC), the largest and most powerful particle accelerator ever built. It is a circular proton accelerator designed to reach a centre-of-mass energy of 14 TeV. Built between 1998 and 2008, the design of the LHC was largely driven to profit from the pre-existing CERN infrastructures: the LHC is installed in a 26.7 km long tunnel that was built to host its predecessor, the Large Electron Positron (LEP) collider, located between 45 and 170 m below ground level. The LHC hosts two beam-pipes where protons circulate in opposite directions, and which are brought to collision at four interaction points, where four particle detectors are installed. At one of this points sits the *Compact Muon Solenoid* (CMS) experiment, a general-purpose detector designed to explore a broad range of physics processes, from precision electroweak measurements to searches of supersymmetric particles. It is with the proton collision data collected by this detector that this thesis was conducted.

This chapter gives an overview of the LHC accelerator and the CMS detector. Section 3.1 reviews the design, parameters and operations of the different phases of the LHC. The CMS subdetector structure is presented in Sect. 3.2. A special focus is put on the CMS trigger system in Sect. 3.3: an important fraction of the work

C. Martin Perez, *Search for the Higgs Boson Produced in Association with Top Quarks with the CMS Detector at the LHC*, Springer Theses,
https://doi.org/10.1007/978-3-030-90206-3_3

of this thesis was dedicated to its optimization and the subsequent evaluation of its performance. The algorithms used to reconstruct and identify the physics objects in the CMS detector, later used in the physics analyses, are described in Sect. 3.4.

3.1 The Large Hadron Collider

The LHC was designed to deliver proton-proton (pp) collisions at an unprecedented maximum centre-of-mass energy of $\sqrt{s} = 14$ TeV with a very high instantaneous luminosity of 1×10^{34} cm^{-2}s^{-1}. It was conceived to investigate the nature of the spontaneous symmetry breaking through the search of the Higgs boson, which was observed by the ATLAS and CMS collaborations in 2012 [1, 2]. Additionally, it was intended to scan the accessible phase space in the search of new phenomena beyond the SM, aiming at favouring or ruling out the postulated scenarios. Complementary to the proton runs, a physics program of heavy ion collisions (Pb–Pb) is also carried out with the goal of studying the collective behaviour of quarks and gluons in plasma.

The realization of the LHC constituted a two decade-long international journey. Its first proposal dates back to 1984 with the official recognition of the project, subsequently approved in 1994 and inaugurated in 2008. Two eras of physics operations have already been conducted: Run 1, which lasted from 2009 to 2013, and Run 2, from 2015 to 2018. The LHC is currently in a phase of maintenance and upgrade in preparation for Run 3, the third data-taking era that will take place from 2022 to 2024. After that, the LHC and the accelerator complex will undergo a profound upgrade towards the High Luminosity LHC (HL–LHC), scheduled to start in 2027. A description of the accelerator complex and operations is given in the following.

3.1.1 Accelerator Complex

The complete accelerator complex is illustrated in Fig. 3.1. The LHC is the last ring in a chain of particle accelerators, built well before the LHC and upgraded to meet its stringent requirements. The first step of the chain consists in the extraction of protons from a bottle of hydrogen gas making use of a strong electric field. The protons are then sent to a *Radio Frequency Quadrupole* (RFQ), where they are grouped into bunches and accelerated until they reach an energy of 750 keV. After that, the protons are supplied to the *Linear Accelerator* (LINAC 2), which brings the proton beam to an energy of about 50 MeV. The particles then arrive to the first circular collider, the *Proton Synchrotron Booster* (PSB), a 150 m ring that accelerates the beam up to an energy of 1.4 GeV and increases the intensity of the proton bunches. Next, the beam enters the *Proton Synchrotron* (PS) and then the *Super Proton Synchrotron*, two circular accelerators of 620 m and 6912 m in length which raise the energy of the beam to 26 and 450 GeV, respectively.

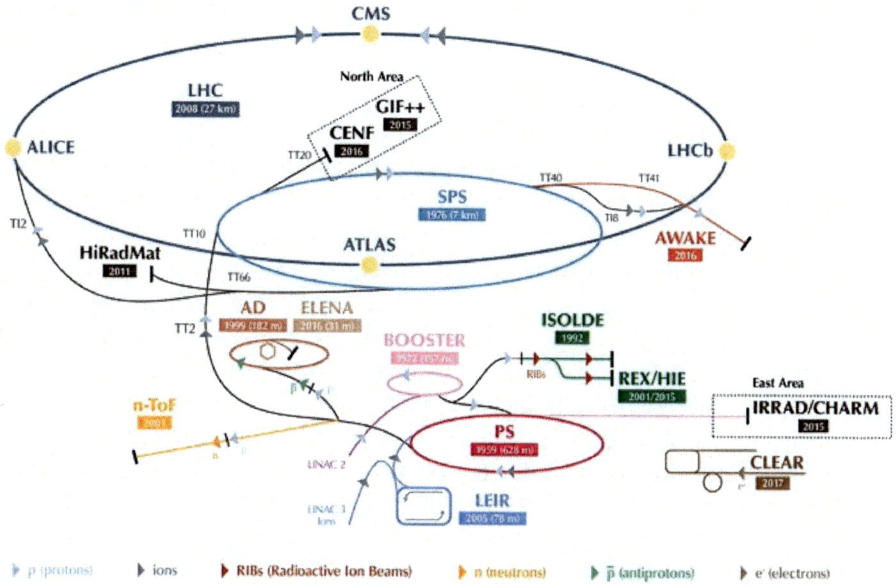

Fig. 3.1 Illustration of the accelerator complex at CERN. Protons are accelerated to increasing energies at the LINAC 2 (in light pink), Booster (in light pink), PS (in dark pink), SPS (in light blue) and LHC (in grey) accelerators. The counter-circulating proton beams at the LHC collide in the centre of the CMS, ATLAS, LHCb and ALICE detectors [3]

The proton bunches are fed into the LHC with fast kicker magnets, which split the beam into two parallel beamlines that travel in opposite directions in the LHC tunnel. Once in the LHC, the beams are further accelerated to their maximal energy. The acceleration is performed in the *high frequency accelerating cavities*, placed in eight 545 m long straight sections along the ring. The trajectory of the beam is bent with 1232 superconducting *dipole* magnets placed throughout eight 2.45 km long arcs. These magnets generate a field of 8.3 T and need to be cooled down to a temperature of 1.9 K (−271.25 °C) with superfluid helium-4. Additional magnets, namely *quadrupoles*, are used to keep the particles focused in narrow beams.

Once the proton beam reaches the nominal energy and the beam is stabilized, protons are brought to collide at four different points along the LHC instrumented with particle detectors. ATLAS (*A Toroidal LHC ApparatuS*) [4] and CMS (*Compact Muon Solenoid*) [5] are multipurpose detectors which can measure the products of both proton and heavy-ion collisions. They are installed in the diametrically opposite points of the LHC, where the highest instantaneous luminosity is achieved. The LHCb (*LHC beauty*) [6] experiment consists of an asymmetric single-arm detector devoted to heavy flavour quarks physics,; its primary goal is to searching for evidence of new physics in charge-parity (CP) violation and rare decays. The last experiment, ALICE (*A Large Ion Collider Experiment*) [7], was designed to cope with very high particle multiplicities and is mainly devoted to the study of quark-gluon plasma in heavy-ion collisions.

3.1.2 Nominal Design Parameters

The LHC accelerates protons, which are charged, composite and stable particles. Being fundamental particles, electrons would be more appropriate for precision measurements, but protons have the advantage that they suffer much smaller synchrotron radiation losses due to their higher mass. This type of radiation is emitted when a charged particle is accelerated radially; it produces energy losses which limit the maximum reachable energy in a circular collider. Thus a proton collider can achieve much higher energy than an electron collider, being the maximum value limited by the capacity of the magnets to maintain the protons in the circular trajectory. The nominal *centre-of-mass energy* of the proton-proton collisions at the LHC is $\sqrt{s} = 14$ TeV, meaning each beam has an energy of 7 TeV. An important fraction of the momentum of the proton is carried by the sea quarks and gluons that composes them; it is therefore possible to generate interesting physics without colliding protons with their antiparticles, which are much more difficult to produce.

Alongside the beam energy, a key parameter of the LHC machine is the *instantaneous luminosity* $\mathcal{L}$, which characterizes the collision rate and serves as an indicator of its performance. It relates the number of events per unit time $\partial N/\partial t$ produced for a given process with its cross section σ via

$$\frac{\partial N}{\partial t} = \mathcal{L} \times \sigma\,. \tag{3.1}$$

A large instantaneous luminosity is essential to produce low probability processes such as $\mathrm{t\bar{t}H}$ and tH, but it also represents a challenge for the data acquisition system. Upon integration of $\mathcal{L}$ over time, one obtains the *integrated luminosity*, $L = \int \mathcal{L}\, dt$, which characterizes the amount of data produced. The instantaneous luminosity is usually expressed in units of $\mathrm{cm^{-2}s^{-1}}$, while the integrated luminosity is expressed in units of inverse picobarns ($\mathrm{pb^{-1}}$) or femtobarns ($\mathrm{fb^{-1}}$).[1]

Under the assumption that the two counter-rotating beams are identical, the instantaneous luminosity relates to the beam properties as

$$\mathcal{L} = \frac{N_p^2 n_b f \gamma_r}{4\pi \epsilon_n \beta^*} F\,, \tag{3.2}$$

where N_p is the number of protons per bunch and n_b is the number of bunches. The symbol f represents the revolution frequency of the bunches and γ_r is the relativistic factor. The transverse emittance ϵ_n characterizes the confinement of the beam in space and momentum, whereas the beta function β^* represents its focus at the interaction point. Finally, F is a geometric factor which accounts for the luminosity reduction due to the crossing-angle of the beams at the interaction point ($F \leq 1$). The values of these parameters in the LHC nominal design are given in Table 3.1.

[1] The conversion is 1 barn = 10^{-28} m^2.

Table 3.1 Nominal parameters of the LHC accelerator in proton-proton collisions [8]

Symbol	Parameter	Nominal value
$\sqrt{s}$	Centre-of-mass energy	14 TeV
Δt	Bunch spacing	25 ns
$\mathcal{L}$	Instantaneous luminosity	1×10^{34} cm^{-2}s^{-1}
n_b	Number of bunches per beam	2808
N_p	Number of protons per bunch	1.15×10^{11}
f	Revolution frequency	11245 Hz
ϵ_n	Transverse emittance	3.75 mm mm
β^*	Beta function	0.55 m

In its nominal design, the LHC accelerates and collides as many as 2808 proton bunches per beam, each bunch containing about 115 billion protons. The bunches are grouped in trains of 48 bunches ("48b" scheme) spaced in intervals of 25 ns each, and circulate around the ring about 11 000 times per second, only 3.1 m/s slower than the speed of light. Of these, 2544 bunches collide at the CMS interaction point at a bunch collision rate of 40 MHz. This configuration yields a luminosity of $\sim 1 \times 10^{34}$ cm^{-2}s^{-1} at the beginning of the fill, defined as the point when the proton injection is complete at the LHC cannot accommodate any more bunches.

Any collider with high instantaneous luminosity faces an important drawback: the *pileup* (PU), defined as the number of simultaneous interactions taking place in each bunch crossing. The average PU is directly proportional to the instantaneous luminosity and relates to the beam properties as

$$\langle PU \rangle = \frac{\mathcal{L}\, \sigma_{pp}^{inel}}{n_b\, f}\,, \tag{3.3}$$

where σ_{pp}^{inel} is the inelastic pp cross section, which amounts to 69 mb at $\sqrt{s} = 13$ TeV [9], leading to a nominal average pileup of $\sim$22 interactions per bunch crossing at the LHC, frequently exceeded during Run 2 operations. High pileup values result in a very high detector occupancy that degrades the efficiency and resolution of the particle reconstruction.

3.1.3 Operational Runs

A summary of the past operational runs and the baseline future schedule for the LHC is shown in Fig. 3.2. The LHC operations are expected to cover a period of around 30 years, divided in two main operational phases: *Phase 1* (2011–2024) and *Phase 2* (2027–2037). During Phase 1, the centre-of-mass energy is gradually increased from 7 to 14 TeV, and the delivered instantaneous luminosity is expected to reach twice

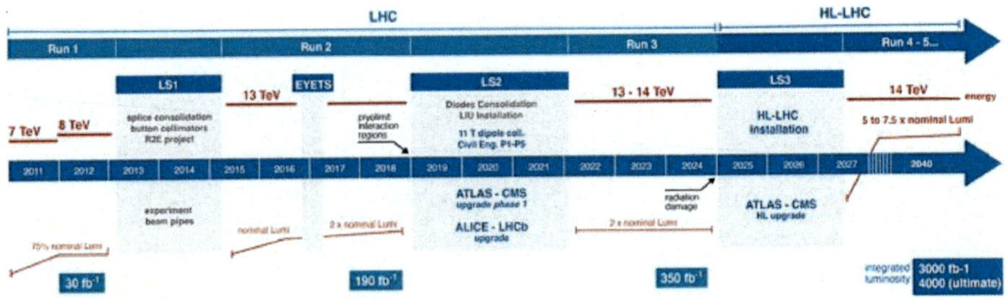

Fig. 3.2 Baseline schedule of the LHC and HL-LHC operations [10]

the nominal value at the end of this phase. In practice, the nominal design values have already been exceeded, as shown in Fig. 3.3, with the corresponding average pileup distributions shown in Fig. 3.4. The machine reached a record instantaneous luminosity of 2.1×10^{34} cm^{-2}s^{-1} in 2018, with a peak average pileup of around 60. A profound upgrade of the accelerator and experiments will take place in 2024, after which the Phase-2 will start in 2027, comprising 10 years of operations. The centre-of-mass energy will be maintained at its nominal value but an instantaneous luminosity of at least 5 times the design value is foreseen to be delivered.

After more than a decade of construction and installation, the first proton beams circulated through the LHC on the 10th September 2008. The inaugural tests were stopped a few days later, when a faulty electrical connection between two magnets resulted in a considerable loss of liquid helium and an important mechanical damage. Due to the repairs required, the collider was not fully operational again until November 2009, when low-energy beams circulated for the first time since the incident. The beginning of 2010 saw the continuous ramp-up of the beam energy, and on the 30th March of the same year the LHC reached the record for high-energy collisions with proton beams at a centre-of-mass of 7 TeV. This milestone marked the beginning of the so-called *Run 1*. By the end of 2011, an integrated luminosity of 6.1 fb^{-1} was delivered to CMS at $\sqrt{s} = 7$ TeV. In 2012, the LHC centre-of-mass energy was increased to $\sqrt{s} = 8$ TeV, and a larger set of 23.3 fb^{-1} was delivered that year to CMS. A peak instantaneous luminosity of 0.8×10^{34} cm^{-2}s^{-1} was achieved in Run 1, corresponding to a peak average pileup of around 45. The amount of data collected allowed for the prompt discovery of the Higgs boson in 2012 and the measurement of its properties.

The LHC was shut down on 13th February 2013 for its 2-year upgrade, period referred to as *Long Shutdown 1* (LS1). A series of renovation work took place to push the LHC towards its design parameters, being the key driver the consolidation of the high-current in the superconducting magnets. The experiments also took advantage of the LS1 to conduct important detector upgrades to cope with increased luminosity. In particular, the CMS trigger electronics underwent a significant upgrade, discussed in Sect. 3.3.1.

The *Run 2* of the LHC started on the 5th April 2015, and lasted until the end of 2018. One month after restarting, protons collided in the LHC at the record-breaking

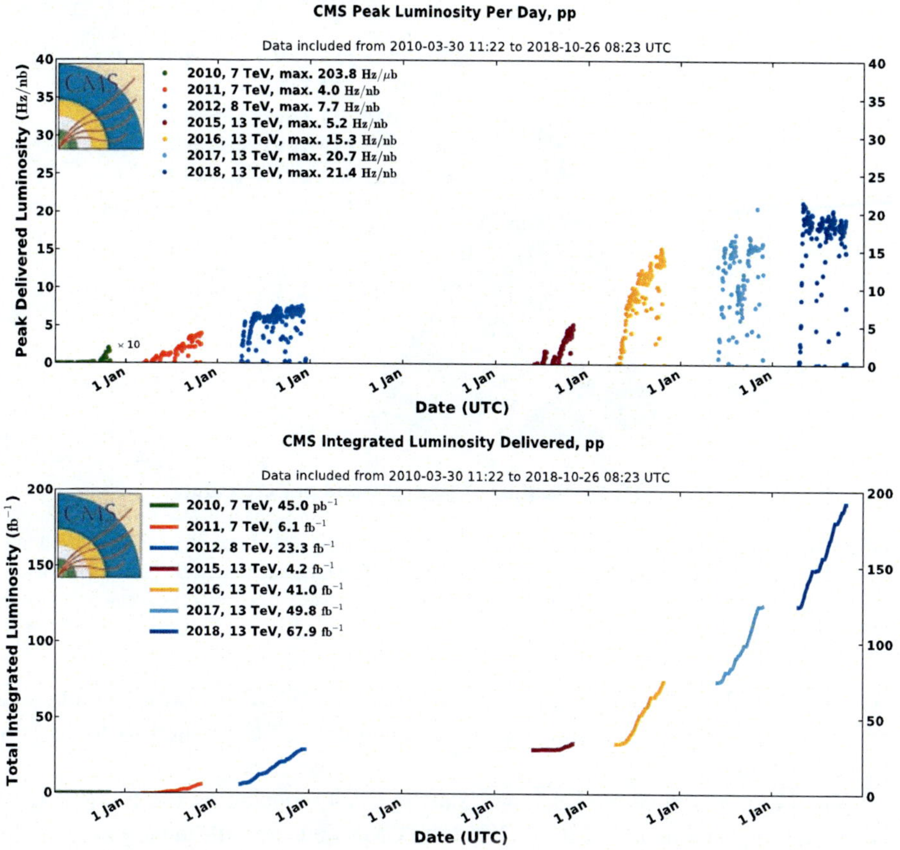

Fig. 3.3 Peak luminosity (top) and total integrated luminosity (bottom) delivered per day to the CMS experiment during the LHC Run 1 and Run 2 [11]

centre-of-mass energy of 13 TeV. Operations in 2015 were oriented to the commissioning of the LHC at this new energy, but starting from 2016 the instantaneous luminosity was brought beyond the original LHC design value. Integrated luminosities of 4.2 fb^{-1}, 41.0 fb^{-1}, 49.8 fb^{-1} and 67.9 fb^{-1} were delivered to CMS in the years from 2015 to 2018. The data collected during Run 2 allowed CMS to learn more about the Higgs boson, notably how it couples to the third generation of quarks and leptons, which was not expected to be within the reach of the LHC experiments until much more data had been recorded. The data delivered in Run 2 are used to derive the results presented in this thesis.

The *Long Shutdown 2* (LS2) started on the 10th December 2018; it is ongoing as this thesis is being written. The LHC and the whole CERN accelerator complex are being maintained and upgraded as this thesis is being completed. As in LS1, this is an occasion for the experiments to upgrade their detectors in preparation for *Run 3*, expected to start in 2022, comprising three years of operations at a new record centre-of-mass energy of $\sqrt{s} = 14$ TeV. After Run 3, the experiments are foreseen

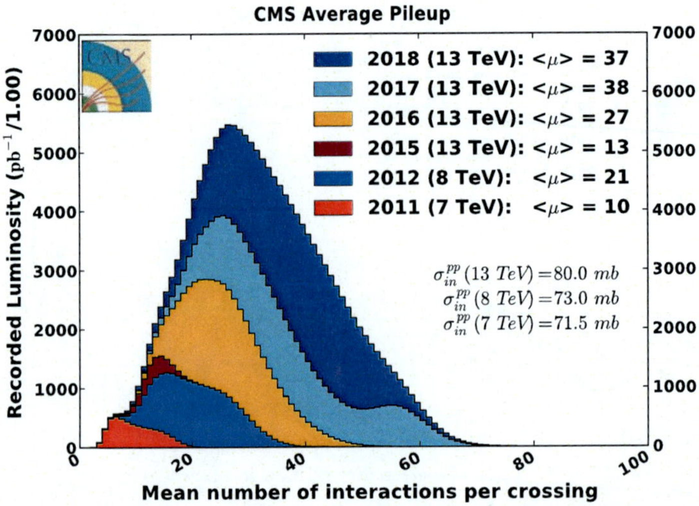

Fig. 3.4 Average pileup profile represented in stacked histograms for each year of data-taking of the CMS experiment during the LHC Run 1 and Run 2. The measured inelastic proton-proton cross section for each centre-of-mass energy is also shown [11]

to have recorded an integrated luminosity of about 300 fb^{-1}, instrumental to study new rare SM phenomena and reduce the uncertainties associated to the existing measurements.

The *Long Shutdown 3* (LS3) will start in 2024, concluding the Phase 1 of the LHC. During this period, the LHC and the CMS detector will undergo a profound upgrade towards the *High Luminosity LHC* (HL-LHC), which will run in *Phase 2*. The goal of the upgraded machine is to reach a peak instantaneous luminosity of 5×10^{34} $\text{cm}^{-2}\text{s}^{-1}$, which yields to a total integrated luminosity of about 3000 fb^{-1} after a decade of operations, enhancing significantly the sensitivity to rare phenomena. The unprecedented collision rate will produce an average pileup of $\sim$140. In its ultimate configuration, the machine could be pushed to an instantaneous luminosity of 7.5×10^{34} $\text{cm}^{-2}\text{s}^{-1}$, corresponding to an average pileup of $\sim$200. The extreme conditions require important upgrades in trigger system, described in Chap. 5.

3.2 The CMS Detector

The CMS (*Compact Muon Solenoid*) [5] detector is a cylindrical detector with a length of 21.6 m and a diameter of 14.6 m. The label "compact" in its name reflects the fact that it weights 14500 tons, compared to the ATLAS detector, which is approximately twice as big but has half the weight. It is located 100 m underground in a cavern at the LHC point 5, near the village of Cessy in France. CMS is a general purpose detector originally designed to reconstruct the decay products of the Higgs boson precisely. It is structured in several concentric subdetectors which complement each other in the characterization of the different particles resulting from the *pp* interactions. One of the key features of CMS is the intense magnetic field induced by the solenoid magnet (see Fig. 3.7). Traversing charged particles are bent under the effect of the field, hence their momenta, tracks and interaction vertex can be precisely measured with the pixel and strip trackers close to the interaction point. The electromagnetic and hadronic calorimeters are located around the tracking system; they are designed to absorb and measure the energy of electrons, photons and hadrons. Muons traverse the calorimeters and are measured in the muon tracking systems that sit in the outermost part of CMS.

Collisions take place at the centre of CMS every 25 ns in the *interaction point* (IP). This means that new proton bunches are colliding at the CMS core before the decay products of the previous collision have reached the active volumes of the detectors, hence the experimental signatures overlap (*out-of-time pileup*). Additionally, multiple interactions can take place within the same bunch crossing (*in-time pileup*). Both effects generate the so-called *underlying events*, corresponding to the hadronic activity that does not originate from the hard scattering process. Underlying events are typically softer; they can be disentangled from the signals of interest provided the high granularity, fast response and wide solid angle of CMS, instrumented with radiation-hard detectors and electronics. An overview of the CMS design and performance is given in the following.

3.2.1 Coordinate System and Conventions

The CMS experiment uses a right-handed coordinate system centred in the nominal interaction point, illustrated in Fig. 3.5. The z-axis is the longitudinal coordinate that matches the beam axis and points anticlockwise. The y-axis points upwards in the direction perpendicular to the LHC plane, which is 1.41% tilted with respect to the horizontal plane. The x-axis points radially towards the geometrical centre of the LHC ring. The xy-plane is denoted as the *transverse* plane. Given the cylindrical shape of the detector, polar coordinates are also used. The azimuthal angle ϕ is defined in the xy-plane by the angle with respect to the x-axis, while the radial coordinate in this plane corresponds to r. In the yz-plane, the polar angle θ is measured from the y-axis.

In the *pp* collisions at the LHC, the interactions occur at the level of the fundamental constituents of the proton, i.e. the quarks and the gluons, also referred to as *partons*. The fraction of the proton momentum carried by these partons is unknown, and the longitudinal boost of rest frame of the event differs from one the next. Hence, observables that are not distorted by the centre-of-mass boost are favoured in CMS. The transverse plane is of particular interest, as the projections of observables on this plane are invariant under Lorentz boosts along the z-axis. This is the case of the *transverse momentum* (p_{T}) and the *transverse mass* (m_{T}), often used to describe the kinematics of a particle, defined as

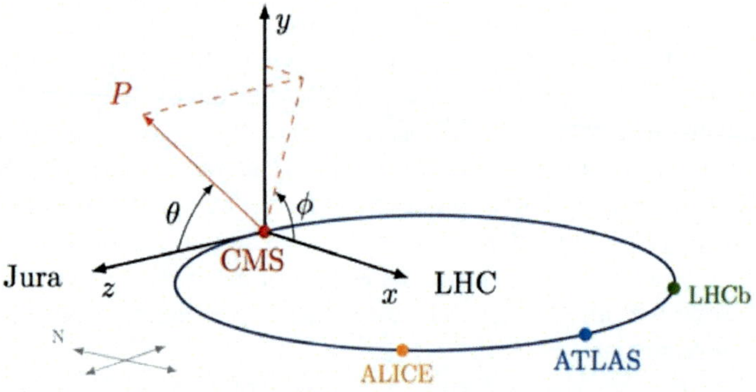

Fig. 3.5 Illustration of the coordinate system at the CMS detector [12]

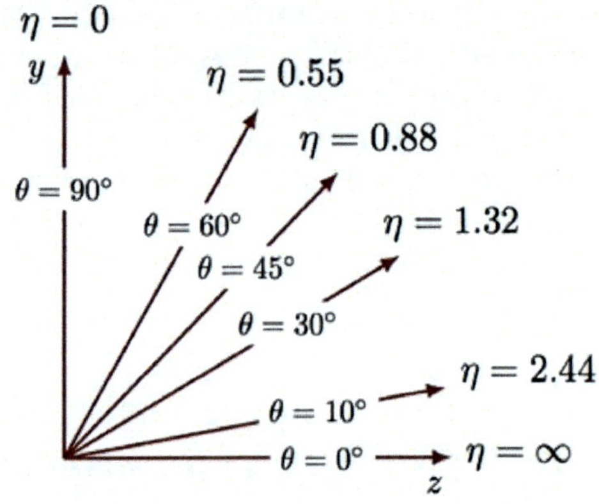

Fig. 3.6 Illustration of the relation between the pseudorapidity η and the polar angle θ at the CMS detector [12]

$$
\begin{aligned}
p_{\mathrm{T}}^2 &= p_{\mathrm{x}}^2 + p_{\mathrm{y}}^2 \,, \\
m_{\mathrm{T}}^2 &= m^2 + p_{\mathrm{x}}^2 + p_{\mathrm{y}}^2 = E^2 - p_{\mathrm{z}}^2 \,.
\end{aligned}
\tag{3.4}
$$

The transverse momentum is instrumental when estimated the presence of neutrinos in the events. These manifest as an imbalance of the summed p_{T} of all the reconstructed particles, expressed via the missing transverse energy ($E_{\mathrm{T}}^{\mathrm{miss}}$) observable (see Sect. 3.4.7).

To describe the angle of a particle relative to the beam axis, a useful spacial coordinate is the *pseudorapidity*, defined as

$$
\eta = -\ln\, \tan(\theta/2) \,. \tag{3.5}
$$

The value of η spans from 0 when perpendicular to the beam line ($\theta = \pi/2$) to infinity when parallel to the beamline ($\theta = 0$), as illustrated in Fig. 3.6. The regions of the detector of large η are often referred to as *forward* direction. The difference in pseudorapidity $\Delta\eta$ between two particles is also an invariant observable under Lorentz boosts along the z-axis. Related to this, another instrumental quantity is the

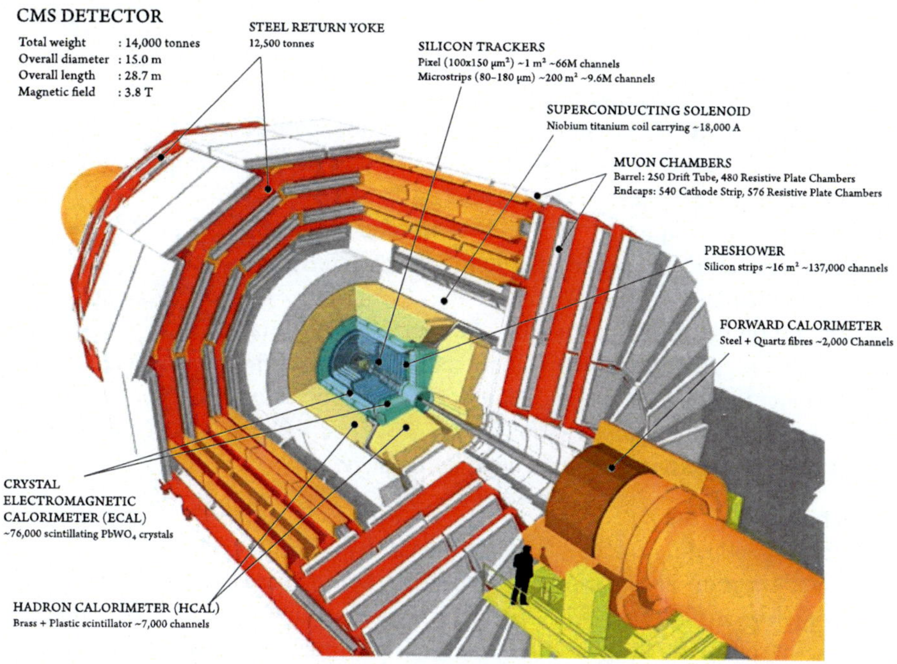

Fig. 3.7 Schematic view of the CMS detector and its subcomponents [13]

angular distance between two particles, defined as

$$\Delta R^2 = \Delta\phi^2 + \Delta\eta^2 , \tag{3.6}$$

with $\Delta\phi = |\phi_1 - \phi_2|$ and $\Delta\eta = |\eta_1 - \eta_2|$. This observable serves to characterize solid angles; it is typically used to evaluate the isolation of a particle with respect to the neighbouring particles.

3.2.2 Subdetectors

The layout of the CMS detector and its subcomponents is shown in Fig. 3.7. The detector is instrumented with multiple concentric layers of subdetectors which use different types of technologies to probe various properties of the particles produced in the collisions. The CMS detector is commonly divided into two sections: the central section, or *barrel*, and the two forward regions, or *endcaps*. The specific η boundaries between those two regions are defined differently for each subsystem. Ordered by vicinity to the interaction point, the subcomponents of the CMS detector are given in the following.

- The **silicon tracker** reconstructs the charged particle trajectories as well as the coordinate of their origin. It is composed of an *inner pixel tracker* and an *outer strip tracker*, with decreasing granularity.
- The **calorimeters** measure the energy of the particles. The *electromagnetic calorimeter* targets electrons and photons, whereas the *hadronic calorimeter* is primarily aimed at measuring the energy deposited by jets.
- The **solenoidal magnet** provides an intense magnetic field so that the charge and momentum of charged particles can be inferred from the bending of their trajectories.
- The **muon chambers** are additional tracking layers placed at the outermost part of CMS to measure the momentum and trajectory of the muons, which cross all the inner layers. Depending on the η-region, the detector technology employed is either *drift tubes*, *cathode strip chambers* or *resistive plate chambers*, overlaid to provide the maximum reconstruction efficiency.

Solenoidal Magnet

The superconducting solenoidal magnet of CMS [14] is the central component under which the rest of the experiment was designed. It encompasses the tracker and the calorimeters, so that these two subdetectors can be in direct contact and the amount of non-active material placed in front of these is minimized. With a weight of 220 tons, a width of 6 m and a length of 12.5 m, the CMS magnet is the largest of its kind. It generates an axial magnetic field of 3.8 T mostly uniform within its volume. As other dipole magnets, the cabling of the solenoid is made of niobium-titanium alloy, cooled down to 4.7 K (−268.5 °C) by a cryostat system using liquid helium. The muon system surrounds the magnet and is embedded in a steel return yoke, which serves as support of its mechanical structure. The yoke is responsible for the return of the magnetic flux to reduce the stray field, hence the muon system is immersed in a 2 T field.

The magnetic field $\vec{B}$ provokes the bending of the paths of the particles of non-zero charge q and speed $\vec{v}$ in the transverse plane via the Lorentz force $\vec{F_L} = q(\vec{v} \times \vec{B})$; the charge and momentum of a particle can be inferred from this bending, alongside the measurement performed by the tracker. To achieve the highest precision, the magnetic field must be accurately characterized over the entire volume of the experiment. A detailed map of the CMS magnet magnetic field is illustrated in Fig. 3.8; it is measured with a precision of less than than 0.1% in the tracker volume [15].

Silicon Tracker

The CMS tracker [16, 17] is the subcomponent located the closest to the interaction point. It has a length of 5.8 m and a diameter of 2.5 m, with an active area of around 200 m^2 and a geometrical acceptance up to $|\eta| = 2.5$. It is composed of two main parts: the inner pixel tracker and the outer strip tracker. The subdivision is justified by the fact that the detector occupancy rapidly decreases with the radial distance, hence higher spatial precision is required close to the beam pipe. The entire tracker employs silicon sensors, as they allow for fine granularity measurements and fast readout. The charged particles traversing the tracker volume deposit their energy through the ionization of the silicon semiconductors, creating electron-hole pairs that induce a

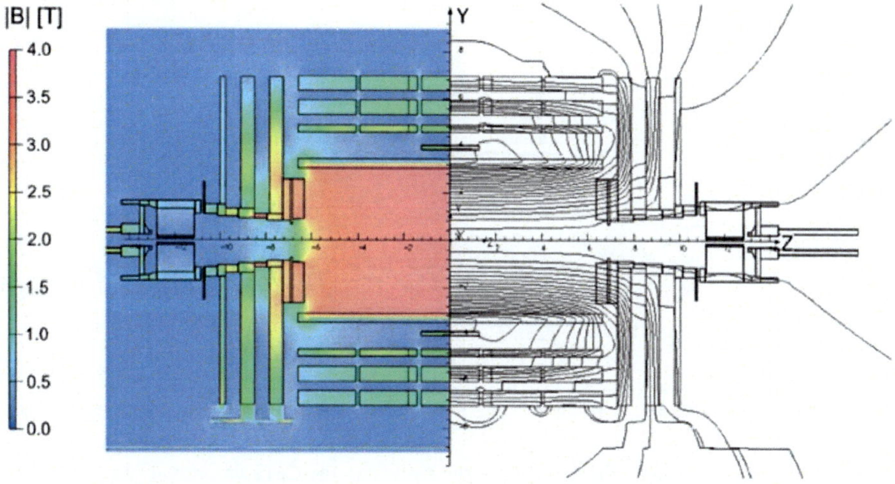

Fig. 3.8 Map of the magnitude of the magnetic field $|B|$ (left) and the field lines (right) in a longitudinal layout of the CMS detector [15]

signal when drifted towards the electrodes. These *hits* in the consecutive silicon layers are reconstructed to determine the particle trajectory. The high segmentation of the sensors allows for a precise spatial measurement; it is used for the determination of the hard scatter interaction point, or *primary vertex*, and its discrimination against pileup interactions. It also enables the identification of heavy-flavour quarks or τ leptons, as their production is characterized by a vertex displaced by a few millimetres from the primary vertex, also known as the *secondary vertex*.

The silicon tracker is mounted with cylindrical layers in the barrel and disks perpendicular to the beam in the endcaps. The general layout of the tracker is shown in Fig. 3.9, where the lines represent detector modules. One can see that, for a given layer, each module is shifted slightly with respect to its neighbour, which allows them to overlap, thereby avoiding gaps in the acceptance. The *pixel tracker* surrounds the interaction point; the *strip tracker* surrounds the pixel detector. The latter comprises the *tracker inner barrel* (TIB) and the *tracker inner disks* (TID) in its innermost section, and the *tracker outer barrel* (TOB) and the *tracker endcaps* (TEC) in its outermost section. The closely-spaced double line elements in the strip modules indicate that one module is rotated with respect to the other in order to permit reconstruction of the hit positions in three dimensions.The system is cooled down to around $-20\,^{\circ}$C so that the detector components can withstand the extremely high level of radiation present in the region closest to the beam pipe.

The pixel tracker is subject to a very high particle flux and employs silicon pixel technology, which profits from excellent spatial and time resolutions even in harsh environments. It plays an essential role in CMS, as it has to provide optimal vertex reconstruction and also serve as seed to most of the tracking algorithms. To cope with the increased luminosity, a pixel detector upgrade was installed in March 2017 [19, 20], featuring an improved faster readout chip, as well as additional tracking

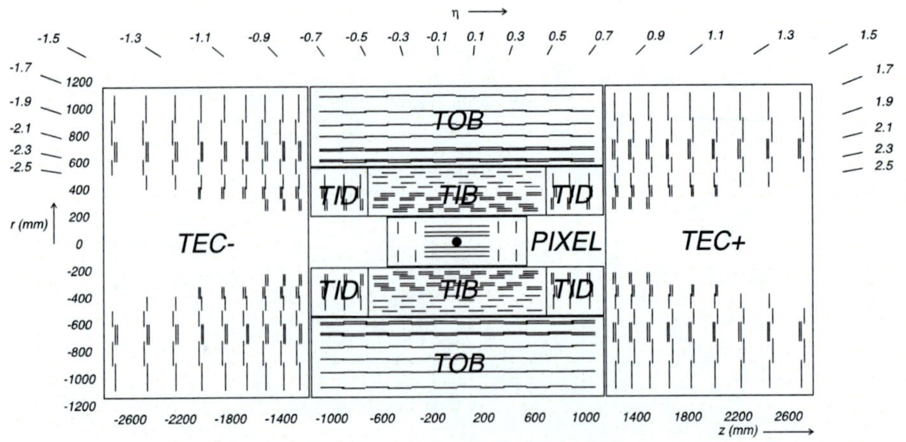

Fig. 3.9 Schematic longitudinal layout of the tracker in the CMS detector. The position of the pixel tracker and the different parts of the strip tracker are shown [18]

layers both in the barrel and in the forward regions. With the new design, the material budget was reduced, and the innermost layer was moved closer to the IP, while the outermost layer was moved further away from it. The upgraded detector contains 4 layers in the barrel, covering the radial range of 3 to 16 cm, while the endcaps contain 3 disks located from $|z| = 29.1$ to $|z| = 51.6$ cm. Comprising a total of 125 million silicon pixels of size 100×150 mm^2, the pixel tracker has a tracking efficiency of 99.95% and an intrinsic spatial resolution of 5–10 mm [21].

After the pixels and on their way out of the tracker stands the silicon strip tracker, with a rather complex configuration. It is composed of about 9.3 millions of silicon strips with 15 different geometries depending on the distance to the IP. The innermost strips have a smaller size, as they have to withstand higher particle fluxes and occupancies. The distance between the strips (*pitch*) and the strip length ranges from 80 mm and 8.5 cm in the inner regions to 205 mm and 20 cm in the outer regions, respectively. The strip tracker is subdivided in four different regions. In the barrel, the TIB covers the region $20 < r < 55$ cm, while the TOB after it reaches up to $r = 116$ cm. Their analogues in the endcaps are the TID and TEC, which extend in the regions $58 < |z| < 124$ cm and $124 < |z| < 282$ cm respectively. The spatial resolution in the transverse plane is 23-34 mm in the TIB and 35–52 mm in the TOB, whereas it is around 10 times larger in the longitudinal direction. Given its optimal configuration, the overall hit efficiency of the strip tracker is 99.8% [18].

The design of the tracker is a result of a compromise between providing the best detector performance and keeping the amount of inactive material as low as possible. The latter is a critical feature of the tracker, as a higher amount of passive material generates multiple scattering, bremsstrahlung, photon conversion and nuclear interactions. These distort the measurement of the trajectory in the tracker and the measurement of the energy in the calorimeters just after. Figure 3.10 shows the simulated

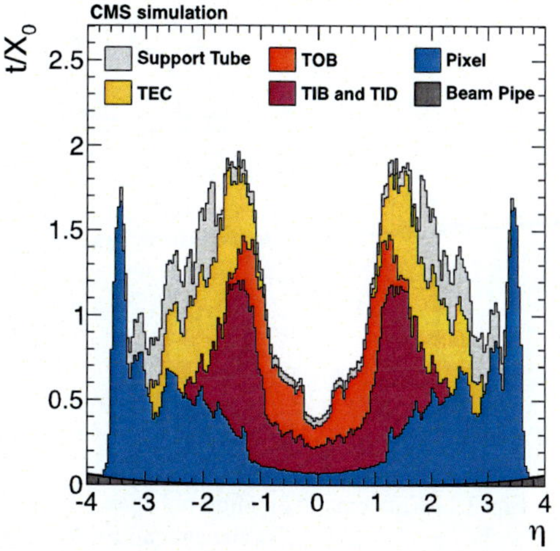

Fig. 3.10 Total thickness t of the tracker material traversed by a particle produced in the interaction point as a function of the pseudorapidity η, expressed in units of hadronic interaction length λ_i. The contribution to the total material budget is shown for each subsystem, the beam pipe and the support tube of the tracker [18]

material budget of the detector in units of hadronic interaction length (λ_i)[2] before the pixel upgrade. The tracking material and the relative services (cables, support, cooling system) represented up to $1.6\lambda_0$; this value was reduced by about 40% in the endcaps and 10% in the barrel with the upgrade, improving the IP resolution by a factor 1.5 in the z-direction [21].

Electromagnetic Calorimeter

The CMS electromagnetic calorimeter (ECAL) [22] is an hermetic array of lead tungstate ($PbWO_4$) scintillating crystals, whose purpose is to measure precisely the energy of incident electrons and photons. The measurement is based on the conversion of the incident electron or photon to an electromagnetic shower, that interacts with the crystal material producing scintillating light. The ECAL is an homogeneous scintillator, meaning it is composed of a single material, as $PbWO_4$ acts both as absorber and active material. This feature grants a compact design that allows both the electromagnetic and hadronic calorimeter to fit inside the volume of the magnet. Owing to its high granularity, the ECAL presents excellent energy and position resolution, instrumental in the Higgs boson discovery via the $\mathrm{H} \rightarrow \gamma\gamma$ and $\mathrm{H} \rightarrow \mathrm{ZZ}^* \rightarrow 4\ell$ decays.

Compared to other crystals, $PbWO_4$ profits from a number of attractive properties. It has a high density ($\rho = 8.28$ g/cm^3), a short radiation length[3] ($X_0 = 0.89$ cm) and small Molière radius[4] ($R = 2.19$ cm). These parameters ensure an excellent shower containment within the crystals, which have a length of approximately $25X_0$. The

---

[2] The nuclear interaction length λ_i is defined as the mean distance travelled by a hadronic particle before undergoing an inelastic nuclear interaction.

[3] The radiation length X_0 characterizes the longitudinal extension of the shower and defined both as (a) the mean distance over which a high energy electron losses all but 1/e of its energy through bremsstrahlung, and as (b) 7/9 of the mean free path for pair production by a high energy photon.

[4] The Molière radius R characterizes the lateral extension of the shower and is defined as the radius of the cylinder that contains on average 90% of the shower energy.

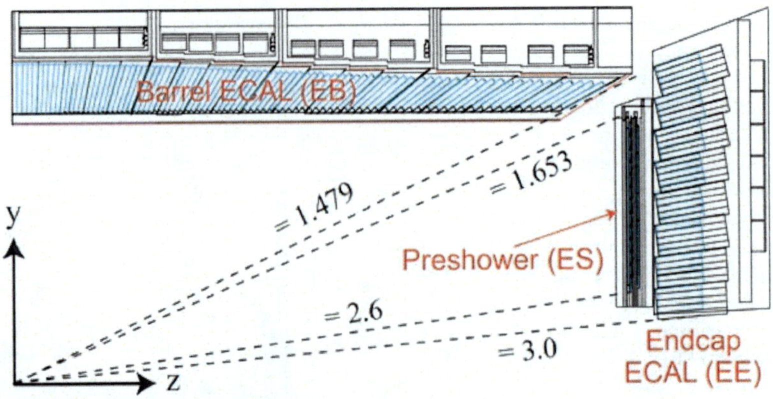

Fig. 3.11 Schematic longitudinal layout of a quadrant of the ECAL in the CMS detector, consisting of the barrel ECAL (EB), the endcap ECAL (EE) and the preshower detector (ES) [23]

crystals also benefit from a fast light yield, with a rise and decay time (5–15 ns) comparable to the bunch crossing time of the LHC, essential for fast triggering. The ECAL has an adequate radiation tolerance, but a modest light yield (~30 photons/MeV), which requires the usage of photodetectors with internal amplification.

The layout of the ECAL is represented in Fig. 3.11. The *barrel ECAL* (EB) extends within the region $|\eta| < 1.479$ and is instrumented with about 61 200 trapezoidal crystals. Each crystal covers a surface of $22 \times 22\ \mathrm{mm}^2$ and has a length of 23 cm. The two *endcap ECAL* (EE) disks have a coverage of $1.479 < |\eta| < 3.0$ and contain 7324 crystals each, with a surface of $28.6 \times 28.6\ \mathrm{mm}^2$ and a length of 22 cm. In both the barrel and the endcaps, crystals are oriented with their axes tilted up to 3° with respect to the direction that points to the nominal IP in order to avoid acceptance gaps between them. The light readout is performed by avalanche photodiodes (APDs) in the EB and vacuum phototriodes (VPTs) in the EE, as the latter can sustain higher particle fluxes.

A much finer electromagnetic *preshower* detector (ES) is installed in front of the two endcaps; it provides an improved spatial resolution in the region $1.65 < |\eta| < 2.6$. It is a sampling calorimeter with two layers of lead absorber followed by 22 mm-long silicon strips to measure the energy deposit and the transverse profile of the shower. In particular, it helps differentiating between single high-energy photons and pairs of collimated low-energy photons which result from $\pi^0 \to \gamma\gamma$ decays. The ES suffers nonetheless from the presence of parasitic signals originated by a large quantity of neutral pions in the tracker material, which significantly affect the identification potential of the preshower layer, only marginally exploited during reconstruction.

The energy measurement in an electromagnetic scintillator is based on the principle that the energy released in the crystals is proportional to the energy of the incident particle. However, the real energy resolution is deteriorated by various factors. In the case of the CMS ECAL, the intrinsic energy resolution σ_E, measured with electron

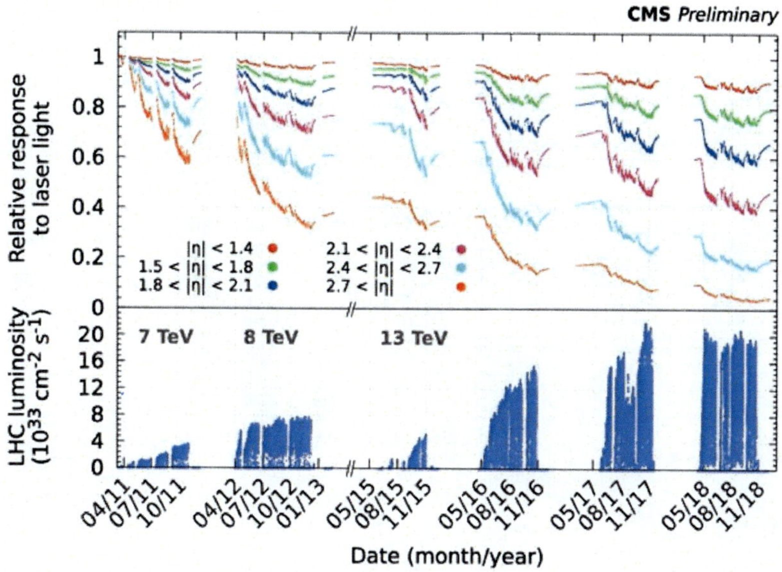

Fig. 3.12 Relative response to laser light injected in the ECAL crystals in bins of pseudorapidity η (top) and instantaneous LHC luminosity delivered by the LHC (bottom) for different periods of data-taking. The crystals transparency is worsened with increasing integrated luminosity [25]

test beam studies [24], is parametrized as

$$\left(\frac{\sigma_E}{E}\right)^2 = \left(\frac{2.8\%}{\sqrt{E}}\right)^2 + \left(\frac{12\%}{E}\right)^2 + (0.3\%)^2 \tag{3.7}$$

The first term on the right-hand side of Eq. 3.7 represents the stochastic fluctuations in the event-by-event measurements; it is small as a result of the homogeneous nature of the calorimeter. The second term corresponds to the noise contributions from the electronics, the digitization chain and the pileup. Finally, the third term is due to non-uniformities in the detector response, miscalibrations and energy leakage.

The resolution of the CMS detector slightly degrades over time, as the large doses of radiation are causing a loss of transparency in the crystals, notably in the forward regions. The loss is monitored by injecting a laser light in each ECAL crystal and measuring its response during each re-fill of the LHC. Figure 3.12 shows the delivered instantaneous luminosities by the LHC and the associated ECAL response; a visible deterioration is found with increasing luminosity. For this reason, the ECAL is constantly re-calibrated with time-dependent corrections based on the measured transparency loss [26]. In view of the large luminosity increase foreseen during the HL-LHC operations, the ECAL will be completely replaced in the endcaps by the *High-Granularity Calorimeter* (HGCAL) [27], based on radiation-hard silicon and scintillation technologies, described in Chap. 5.

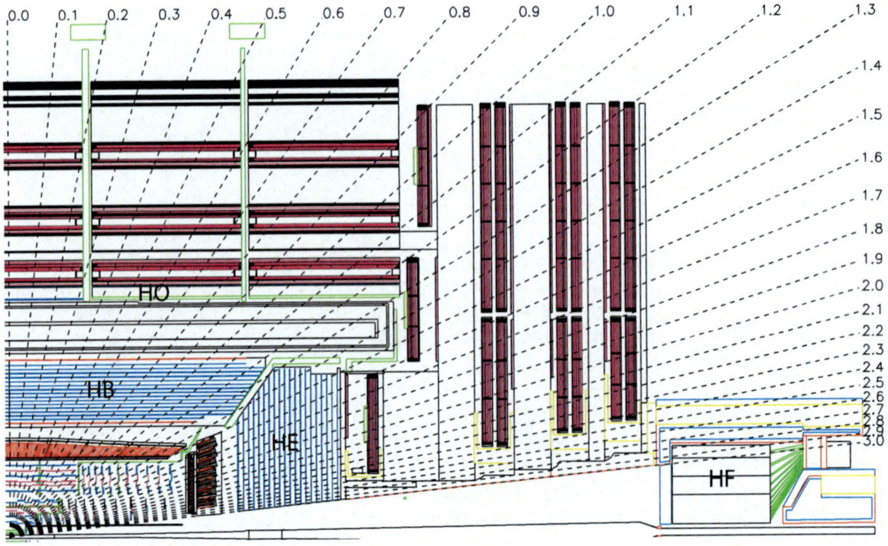

Fig. 3.13 Schematic longitudinal layout of a quadrant of the HCAL in the CMS detector, consisting of the barrel hadronic calorimeter (HB), the endcap hadronic calorimeter (HE), the outer hadronic calorimeter (HO) and the hadronic forward calorimeter (HF) [5]

Hadronic Calorimeter

The hadronic calorimeter (HCAL) [28] measures the energy of the hadrons, which interact mostly via the strong force. These typically traverse the ECAL volume, where they only deposit ~30% of their energy, and are stopped in the dense material of the HCAL, which absorbs and measures the remaining energy. The HCAL is the only CMS component capable of measuring the energy of neutral hadrons, and it is instrumental to infer the energy of non-detectable particles such as neutrinos. Compared to the electrons and photons in the ECAL, hadronic showers are more challenging, as they present large fluctuations in terms of spatial development and energy loss. Additionally, the presence of neutral pions decaying into photons also results in electromagnetic deposits within the shower, which have a different detector response. As the nuclear interaction length is much larger than the electromagnetic radiation length, the HCAL is more massive and larger than the ECAL. The size of the HCAL was significantly limited by the volume constraints of the magnet; the relative size of the material components was optimized for maximal shower containment.

The HCAL is a sampling calorimeter consisting of alternate layers of brass absorbers and plastic scintillator tiles, the latter embedded wavelength shifters that guide the light to hybrid photodiodes. It is divided in four separate calorimeters, represented in Fig. 3.13. The *barrel hadronic calorimeter* (HB) covers the region $|\eta| < 1.4$ and has a thickness of about $7\lambda_i$; the *endcap hadronic calorimeter* (HE) extends the acceptance to $1.3 < |\eta| < 3.0$ with a thickness of $10\lambda_i$. As the material of the electromagnetic and hadronic calorimeters in the barrel may not provide enough stopping power for highly energetic particles in the central region ($|\eta| < 1.4$), the detector is complemented by an *outer hadronic calorimeter* (HO) located outside the solenoid, composed solely of scintillating material that extends the interaction depth to about $11\lambda_i$. The energy measurement in the forward region is conducted by the *hadronic forward calorimeter* (HF), located 11.2 m away from the interaction point and covering the region $2.9 < |\eta| < 5.2$. Exposed to the highest particle fluxes, the HF uses the radiation hard Cerenkov technology of quartz fibers as active elements along with steel absorbers.

The combined energy resolution of the ECAL and the HCAL has been measured to [29]

$$\left(\frac{\sigma_E}{E}\right)^2 = \left(\frac{84.7\%}{\sqrt{E}}\right)^2 + (7.4\%)^2 \,. \tag{3.8}$$

The modest energy resolution degrades the calorimeter-based reconstruction of hadronic objects (jets and hadronically decaying τ leptons); these largely rely on the CMS reconstruction algorithms (see Sect. 3.4), which exploit the whole detector information to correct the energy and angular resolution.

The wide coverage of the HCAL makes it sensitive to most of the collision products; it is pivotal to infer the energy of the undetected particles such as neutrinos. In order to maintain its performance, certain components of the HCAL were upgraded at the end of 2017 [30]. The photodetectors in the HB, HE and HF were replaced and the functionality of the readout electronics was expanded. With the introduction of precision timing measurements and longitudinal depth segmentation, the pattern recognition capabilities in the HCAL were improved, providing an additional handle for background rejection.

Muon Chambers

As the name and the logo of CMS suggest, the precise measurement of the muons momenta is a core specification of the detector design. Muons are produced in many physics processes, and their clean and distinctive signature makes them suitable drivers of trigger decisions. With typical energies ranging from a few to hundreds of GeV, muons lose less energy in their passage through the tracker materials than the electrons, as they are less subject to radiative effects due to their higher mass. For this reason, the muon detector system [31] sits in the outermost part of the CMS detector, outside the magnet volume. The charge and momentum of the muons are measured using the return field of the solenoid (2 T) inside the iron structure where the chambers are embedded, providing a complementary measurement to the one conducted in the silicon tracker.

Since the muon system is located far away from the IP, its detector components must cover a wide surface area. This feature lead to the choice of gas ionization chambers in the muon chambers design, which are reliable and show adequate performance at a reduced cost. The muon chambers are instrumented with three different types of gas detector technologies: *drift tube chambers* (DTs), *cathode strip chambers* (CSCs) and *resistive plate chambers* (RPCs), making a total of 1400 chambers. The DTs are segmented into drift cells, the position of the muons being determined by the measurement of the drift time to the anode wire of the cell. The CSCs operate

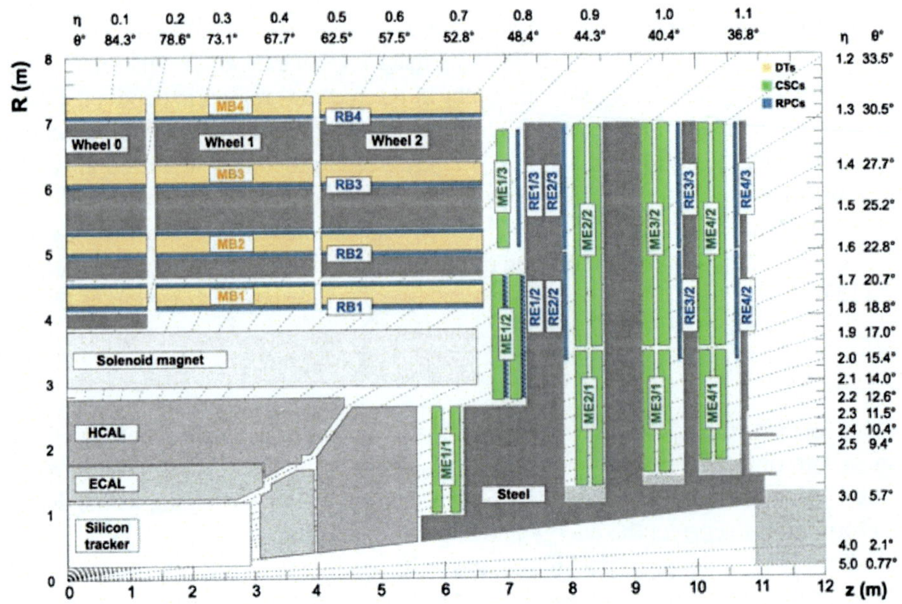

Fig. 3.14 Schematic longitudinal layout of a quadrant of the muon chambers in the CMS detector, consisting of drift tubes (DT), cathode strip chambers (CSC) and resistive plate chambers (RPC) [32]

as standard multiwire proportional counters but add a finely segmented cathode strip readout, which yields an accurate measurement of the coordinate at which the muon crosses the volume. The RPCs are double-gap chambers operated in avalanche mode and are primarily designed to provide timing information for the muon trigger. The different chambers are positioned throughout the muon detector depending on the expected background rates and the uniformity of the magnetic field, as shown in Fig. 3.14. They are arranged to provide a maximal coverage with spatial overlap to maximize the reconstruction efficiency.

The DT chambers are located in the central region ($|\eta| < 1.2$), which is subdivided into four cylindric stations concentric around the beam line. This region is characterized by a relatively low particle rate and a local magnetic field. The basic constituent of the DT detector is a rectangular cell of size 1.3×4.2 cm^2 and length 2.4 m, filled with a mixture of Ar (85%) and CO_2 (15%). The electrodes on the top and bottom of the cell ensure a constant field and a uniform drift velocity of around 55 mm/s, while the cathodes are placed on the sides. When a muon traverses the gas, its position and angle are inferred from the time needed for the knocked-off electrons to drift towards the anode. Each DT cell has a spatial resolution of around 200 mm, yielding a global 80-120 mm position resolution [32].

The endcap regions are instrumented with CSCs in the region $0.9 < |\eta| < 2.4$, where the background rates are higher than in the barrel and the magnetic field is stronger and non-uniform. The CSCs detectors have a trapezoidal shape and contain alternate layers of anode wires and cathode strips, filled with a mixture of Ar (45%),

CO_2 (50%) and CF_4 (10%). The gas is ionized as the muon passes through the chamber, and the induced signals are combined to provide a precise measurement of the muon position. Among other advantages of the CSC technology, it profits from a fast response and fine segmentation, providing a spatial resolution of 40-150 mm.

Finally, RPCs are installed in both the barrel and the endcaps, covering the range $|\eta| < 1.9$. They are formed by two gaps of resistive 2 mm-thick layers interlaid with a mixture of gas (95.2% of $C_2H_2F_4$, 45% of i-C_4H_{10} and 0.3% of SF_6). The detector is operated in avalanche mode: when traversed by a muon, an electron cascade is triggered by the high electric field inside the volume and is read out with strips located in the outer surface. Despite the coarse spatial resolution of the RPCs ($0.8 - 1.2$ cm), these detectors are faster than the DTs and CSCs, with a timing precision better than 3 ns. Hence, they are particularly useful at identifying the bunch crossing associated with a muon track, even in the presence of high pileup, which is an essential feature of the muon trigger system.

Additional detector chambers were introduced in the region $1.6 < |\eta| < 2.2$ at the end of 2017, in order to increase the redundancy of the muon system in the endcaps. The technology used is the *gas electron multiplier* (GEM) [33], well suited for this purpose as it presents a thin profile and excellent rate capability, while it is able to withstand the high particle fluxes of the forward region. The upgrade comprised the pre-installation of 10 prototype chambers; the complete installation of the 144 detectors is ongoing during LS2 [34]. Additionally, the electronics of the CSC subsytem are being upgraded in the chambers closest to beam line, in anticipation of the experimental challenges expected in the HL-LHC.

3.3 The CMS Trigger System

The average bunch crossing rate at the core of CMS is 40 MHz. With the full detector information taking about 1 Mb per event, there is no technology nowadays able to read out and store such huge volumes of data. Nevertheless, most of the collisions result in low-energy pp interactions, which are not contemplated in the physics program of CMS. As shown in Fig. 3.15, even the most frequent SM processes have cross sections of $\sim 10^5$ pb, six orders of magnitude less than the total pp interaction cross section ($\sigma_{pp} \sim 10^{11}$ pb). The goal of the CMS trigger system is to rapidly select events that exhibit a potential physics interest while reducing the acquisition rate by a factor 10^5. It is the bridge between the *online* data-taking and the *offline* data analysis; it must comply with the technical constraints of the former and the high efficiencies and QCD background rejection expected by the latter. The system needs to be flexible to adapt to the different data conditions and robust against the high instantaneous luminosity and pileup conditions delivered by the LHC.

The trigger selection is done based on the kinematic properties of the particles produced. It is implemented in two successive steps with increasing rate reduction and sophistication. The first selection is performed by the *Level-1 trigger*, encoded in custom hardware processors that select up to 100 kHz of the most interesting events

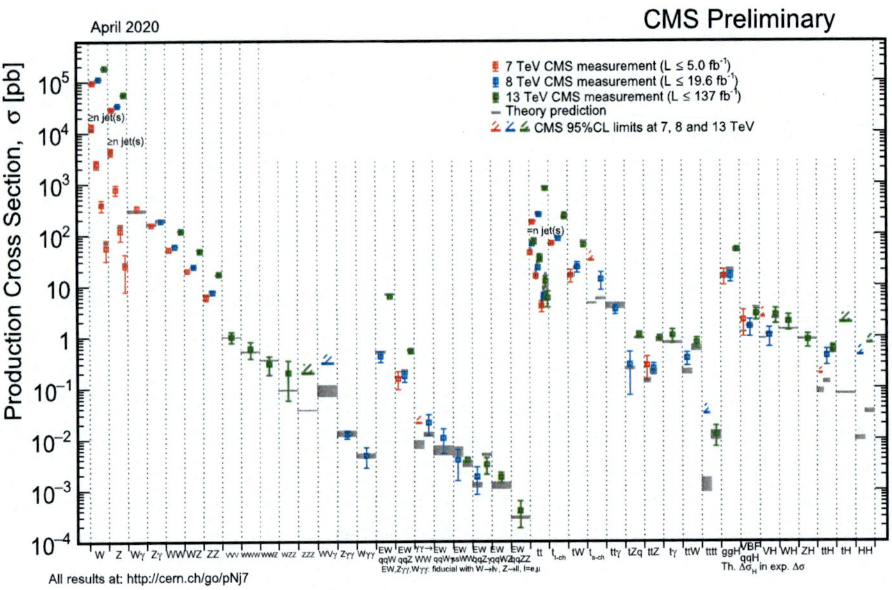

Fig. 3.15 Cross sections of the main SM processes measured with the CMS detector at centre-of-mass energies of $\sqrt{s} = 7$, 8 and 13 TeV, compared to the theoretical predictions [35]

with an available processing time (*latency*) of 3.8 μs. Following this selection, the *High Level Trigger* (HLT) conducts a more detailed scrutiny of the events on a commodity computing processor farm, reducing the rate down to 1 kHz in ~200 ms. More details about the trigger system are provided in the following; a special attention is given to the Level-1 trigger, as it is topic of this thesis. The events selected by the trigger are permanently stored in the tapes of the CERN Tier-0 (the *grid*), where they become available for full reconstruction.

3.3.1 Level-1 Trigger

The input of the Level-1 trigger is the raw data from the front-end readout electronics. As the L1 trigger has only 3.8 μs to make a decision, the full offline reconstruction of the physics objects cannot be applied at this level. Instead, the L1 trigger produces the so-called L1 *candidates*, coarse-granularity and low resolution physics objects. Because of the timing constraints, iterative procedures such as the reconstruction of the particle trajectory in the tracker are not possible; the L1 object reconstruction is performed solely relying on the inputs of the calorimeters and the muon detectors. As the first system in the data acquisition chain, the L1 trigger has to be flexible and scalable to accommodate the increasingly challenging LHC running conditions and the evolving physics searches, while complying with the hardware constraints and the limited bandwidth.

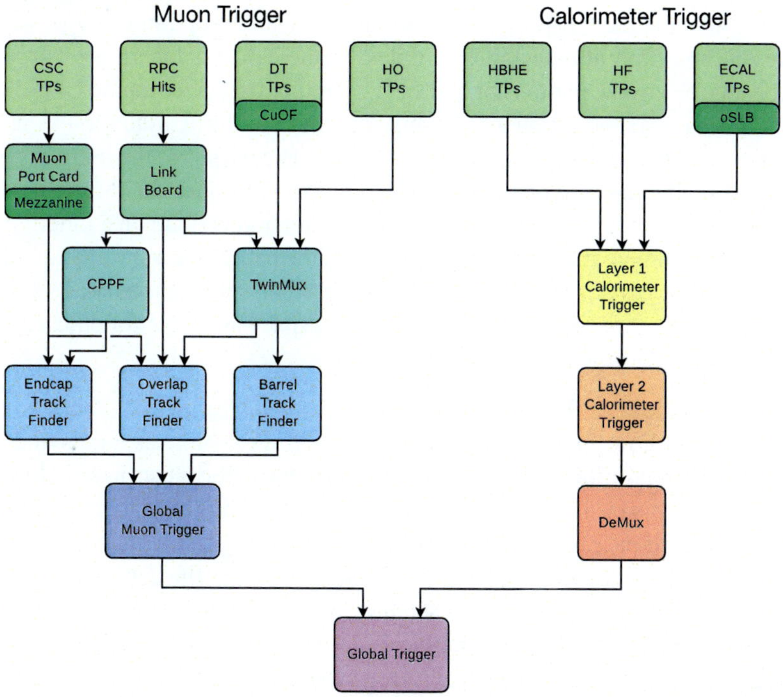

Fig. 3.16 Diagram of the upgraded Level-1 trigger system. The muon trigger builds the muon candidates from the hits in the CSC, RPC and DT. The calorimeter trigger builds the electron/photon, τ_h and jet candidates and computes global quantities from the energy deposits in the HCAL and ECAL. The global trigger collects the output from the muon and calorimeter triggers and performs and event accept or reject decision [36]

The original L1 trigger system was designed to face the nominal LHC operations; in view of the instantaneous luminosities foreseen for Run 2, the L1 trigger could not ensure an adequate physics performance. A major upgrade of the L1 trigger [36, 37] was installed and commissioned between 2015 and 2016. The electronic boards were replaced by advanced mezzanine cards (AMC) equipped with powerful *field-programmable gate arrays* (FPGA). The FPGAs are electronic circuits that can be configured using a hardware description language (HDL), guaranteeing a flexible design of sophisticated and innovative algorithms, able to allocate pattern recognition and pileup mitigation techniques. The communication between subsystems is based on MicroTCA (μTCA) technology to take advantage of additional flexibility and higher bandwidth. The data transfer between the modules and the central CMS data acquisition system is performed with multi-Gb/s serial optical links to allow for a faster and global readout.

The architecture of the upgraded L1 trigger system for Run 2 is sketched in Fig. 3.16. The information of the energy deposits in the HCAL and ECAL subdetectors is collected by the *calorimeter trigger*, while the information of the hits in the DT, RPC and CSC subdetectors is processed in the *muon trigger*. These building bricks constitute the *trigger primitives*, which are combined into trigger candidates representing different detectable particles. In the calorimeter trigger, these objects can be jets, electrons, photons or hadronic τ, further used to compute L1 global event quantities such as the missing transverse energy. Similarly, muon tracks are reconstructed from the hits in the muon subdetectors. The output of the two subsystems is collected by the *micro global trigger* (μGT), that combines the information to perform an event accept or reject decision. This selection is done on a kinematic basis, namely according to the energy and position of the candidate, as well as quality criteria. The energy cutoffs applied at this stage are commonly denoted as L1 thresholds. The lower these thresholds, the wider the phase space available for physics analyses, but also the higher the output rate of the trigger.

Level-1 Calorimeter Trigger

Run 1 showed how an improved granularity of the calorimeter inputs and a global view of the detector information were pivotal for the calorimeter-based object reconstruction. On this basis, the upgraded L1 calorimeter trigger was designed to access the whole detector information in the form of highly-granular trigger towers (TT), with typical sizes $\Delta\eta \times \Delta\phi$ of 0.087×0.087 in most of the detector acceptance. A global view of the detector requires sending the full TT information of a bunch crossing to single electronic board, possible with the usage of a *time-multiplexed trigger* (TMT) architecture, built in a two-layered system, explained in the following.

The upgraded calorimeter trigger architecture is depicted in Fig. 3.17. The inputs of the calorimeter subdetectors arrive to the 18 processor cards (CTP7) of the *Layer-1* system, each processing a 20° region in ϕ. Layer-1 performs a series of preprocessing operations, namely the calibration and sorting of the local energy deposits from the ECAL and the HCAL. The output from Layer-1 is sent to one of the 9 master processor cards (MP7) of the *Layer-2* system. It is done in a time-multiplexed fashion: each processing node of Layer-2 has access to the whole event with full granularity. The identification and reconstruction algorithms of the physics objects (electrons or photons, hadronic τ leptons, jets and energy sums) is implemented at this stage. The usage of multiplexed boards removes the regional boundaries in the object reconstruction and grants additional latency to accommodate for more sophisticated algorithms than in Run 1. The L1 candidates built in the Layer-2 are sent to a demultiplexer node, which reorganizes the reconstructed objects and transmits them to the μGT.

As the L1 trigger does not use the tracker information, the *electrons* and the *photons* are reconstructed jointly in the Layer-2 calorimeter trigger as L1 e/γ objects. The reconstruction algorithm proceeds by clustering energy deposits around the local energy maximum, or *seed*. These clusters are built dynamically, meaning they include surrounding towers with energy deposits above a certain threshold, further trimmed following the typical electron footprint. The trimming results in various candidate shapes categorized and used for identification purposes. A shape veto is applied to reject the clusters least compatible with a genuine e/γ candidate, such as pileup-induced deposits. Additional identification is applied based on the compactness of the shower and the ratio of hadronic-to-electromagnetic deposits. The L1 e/γ candidates

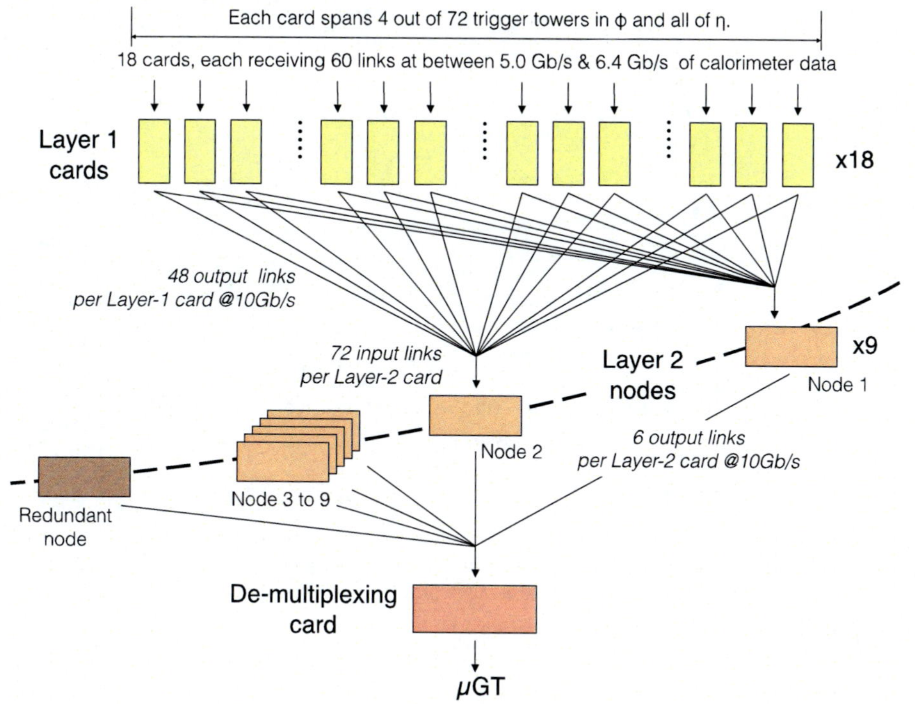

Fig. 3.17 Layout of the upgraded L1 calorimeter trigger. The inputs from the calorimeters are collected by Layer-1, which calibrates and sorts the energy deposits, and sends the full information to a single processor card in Layer-2, which runs the identification and reconstruction algorithms of the electrons or photons, τ_h, jets and energy sums. The output of Layer-2 is sent to the μGT, where the global trigger decision is made [38]

are calibrated according to their energy, η position and shape. As the electron and photon deposits are typically narrower than the ones produced by jets, isolation criteria are applied by setting an upper limit on the presence of calorimeter activity around the candidate. The excellent efficiency achieved for isolated and inclusive L1 e/γ candidates is shown in Fig. 3.18a for typical thresholds used in 2018 data-taking.

The L1 *jet* reconstruction algorithm is based on a square approach: it considers the energy deposit in a 9×9 trigger tower area centred on a local maximum, with similar size to the cone used for the offline reconstruction with the anti-k_T algorithm (see Sect. 3.4.5). The contribution of pileup is estimated and subtracted on a jet-by-jet basis in each bunch crossing according to the energy deposit in the region surrounding this square. The procedure dynamically corrects the fluctuating pileup conditions, which degrade considerably the jet energy resolution, and provides a significant rate reduction. L1 jets are calibrated as a function of their energy and their η position. Besides the jet reconstruction, the full calorimeter granularity is used to estimate *energy sums*. These include the magnitude of the vector sum of the transverse energy over all trigger towers (E_T^{miss}) and the total scalar transverse energy of all jets (H_T). The reconstruction of these objects specially profits from the removal of the

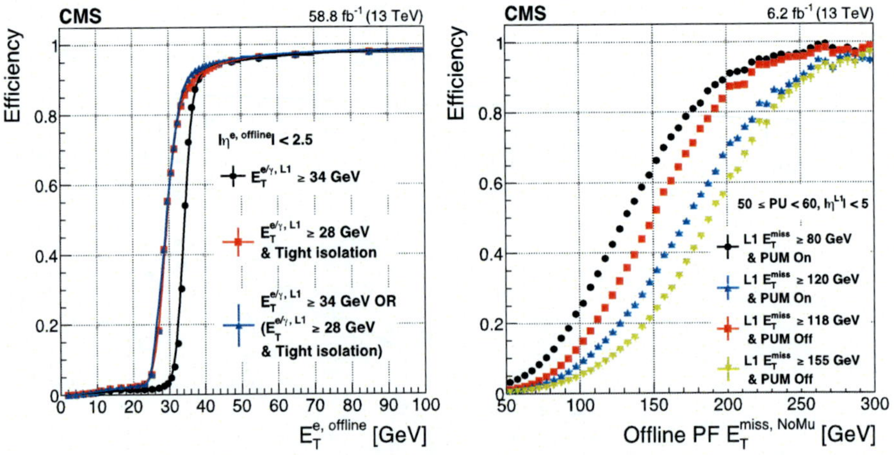

(a) (b)

Fig. 3.18 **a** Level-1 e/γ efficiency as a function of the offline electron transverse energy, for two typical L1 thresholds in 2018 data-taking, with and without isolation requirement [38]. **b** Level-1 $E_{\mathrm{T}}^{\mathrm{miss}}$ efficiency as a function of the offline $E_{\mathrm{T}}^{\mathrm{miss}}$, for L1 $E_{\mathrm{T}}^{\mathrm{miss}}$ thresholds giving the same rate with and without pileup mitigation [38]

regional boundaries resulting from the time multiplexing architecture. A dedicated pileup subtraction is applied on an event-by-event basis to the L1 $E_{\mathrm{T}}^{\mathrm{miss}}$ candidates, specially affected by the pileup deposits. Similarly to the L1 e/γ algorithm, the pileup is estimated based on the low energy deposits in the central part of the detector; it is used along with the η position to derive an energy threshold below which the trigger towers do not enter the $E_{\mathrm{T}}^{\mathrm{miss}}$ computation. Besides a significant rate reduction, this results in a increase of efficiency, illustrated in Fig. 3.18b for a fixed rate.

The τ leptons resulting from the collisions can decay leptonically or hadronically. Leptonic τ leptons are reconstructed at L1 via the electrons or muons produced in the decays. In turn, *hadronic* τ leptons are reconstructed via their decays to one, two or three charged pions, with or without neutral pions. The L1 objects built this way are denoted as L1 τ_h. These objects are narrow, hence the algorithm inherits the individual dynamic clustering of the L1 e/γ, adapted to the L1 τ_h footprint. The charge pions can be spaced out because of the magnetic field and they might result in more than one calorimeter cluster; these are merged into a single candidate under proximity conditions. The L1 τ_h candidates are calibrated and their isolation is evaluated to discriminate them against the QCD-induced jets. It is done in a similar manner as in the e/γ algorithm, except that the isolation cutoff is relaxed in the at higher energies to achieve the maximum efficiency. The L1 τ_h algorithm, object of this thesis, is described in detail in Chap. 4.

Level-1 Muon Trigger

During Run 1, the data from each muon subdetector (CSCs, DTs, RPCs) were used separately to build independent muon tracks. The upgraded muon trigger changed the track reconstruction approach by exploiting the redundancy from the partially over-

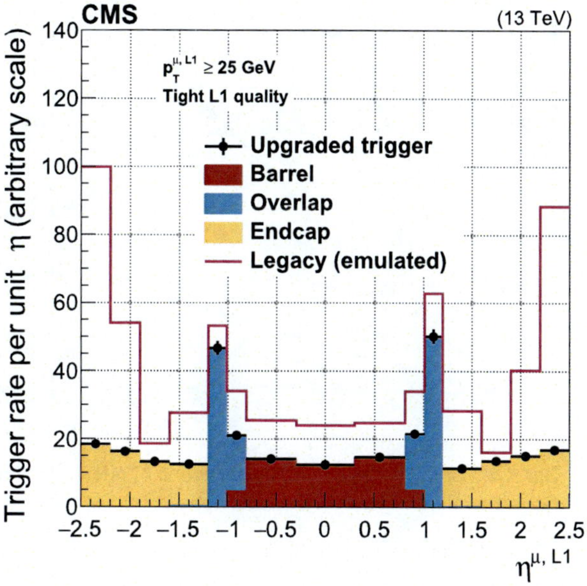

Fig. 3.19 Level-1 muon rate with the legacy Run 1 trigger and the upgraded Run 2 trigger as a function of the L1 muon η. The rate reduction is achieved by combining the information of the muon subsystems to build the trigger primitives [38]

laying detectors (see Fig. 3.16). After the upgrade, the DT, RPC and CSC track finders were replaced by the barrel, overlap and endcap muon track finders, denoted respectively as BMTF ($|\eta| < 0.83$), OMTF ($0.83 < |\eta| < 1.24$) and EMTF ($|\eta| > 1.24$). The signals generated by the three muon subdetectors are used simultaneously to build the trigger primitives, which provide coordinates, timing and quality information from detector hits. The track finders use the muon detector trigger primitives to reconstruct tracks, assign quality to each and measure the muon transverse momenta from the bending in the fringe field of the magnet yoke. The upgraded approach improves the reconstruction efficiency and resolution of the muon candidates with respect to Run 1, while significantly reducing the rate, as illustrated in Fig. 3.19. The outputs of the BMTF, OMTF and EMTF are collected by the *global muon trigger* (μGMT), that sorts the muon candidates, removes duplicates and transmits them to the μGT.

The BMTF in the barrel uses the information from DT and RPC subdetectors, merged into *super-primitives* in the *TwinMux* system; the combination of the spatial resolution in the DT and the precise timing in the RPC improves the efficiency. Each superprimitive is assigned a quality depending on its position and an internal bending angle. The BMTF uses this information to form an acceptance window for the outer layers through extrapolation, within which the superprimitives are grouped to form a muon candidate. The number of hits and the quality of the extrapolation serve as an additional trigger requirement to control the rate. Owing to the large computing power required by the BMTF and Twinmux systems, these are equipped with MP7 cards.

The EMTF uses the data from the CSC and RPC detectors, while the OMTF receives the information of the three muon subsystems. In both cases, tracks are

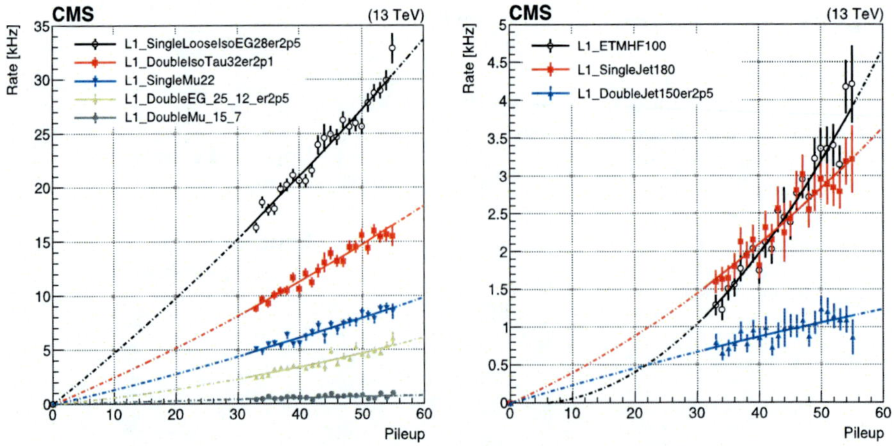

Fig. 3.20 Level-1 trigger rates as a function of pileup for some benchmark seeds, measured using data recorded in 2018 [38]. The description of the seeds is provided in Table 3.2

seeded by a single reference hit, preferably from the inner layers, to which additional hits are associated based on fast *pattern recognition* techniques. These patterns encode information on the hit probability density and the average track propagation between layers for a given energy. The number and topology of the hits are additionally used to establish the muon quality criteria and to reduce the rate. Since the OMTF and EMTF require large memories to store the associative patterns, modular track finders (MTF7) are used in these subsystems.

Level-1 Global Trigger

The final decision of accepting or rejecting an event is taken by the *micro global trigger* (μGT). It is instrumented with MP7 boards which take as inputs the objects reconstructed in the calorimeter and muon trigger systems. The selection is done based on a list of algorithms (*seeds*), collectively referred to as the *trigger menu*. The most basic seeds consist of thresholds applied on kinematic observables (E_T, p_T or η) to one or more L1 objects of one type (L1 e/γ, L1 τ_h, L1 jets, L1 E_T^{miss}, L1 H_T, L1 muon). The menu also comprises more sophisticated seeds (*cross-triggers*), which encode criteria applied on L1 objects of different types; these result in lower thresholds and a larger phase space sensitive to a more diverse range of signals. A list of benchmark seeds during Run 2 can be found in Table 3.2; the corresponding rates as a function of the pileup are shown in Fig. 3.20. The rate of an algorithm can be reduced by applying a *prescale* that determines what fraction of events selected by the seed will pass the trigger: a prescale of N means that only $1/N$ events satisfying the condition are accepted. As the beam intensity of the LHC decreases throughout the fill, decreasing prescale values are applied in view of maximizing the signal efficiency.

With the upgraded μGT, the CMS trigger benefited from extended correlation capabilities at L1, such as the computation of $\Delta\eta$, ΔR, charge comparisons or invariant masses between two L1 objects. As illustration, the L1 dimuon invariant mass

Table 3.2 Description of several benchmark Level-1 triggers used during Run 2 [38]. The rates of these triggers as a function of pileup are shown in Fig. 3.20

L1 seed	Description
L1_SingleLooseIsoEG28er2p5	One loosely isolated e/γ with $E_T > 28$ GeV and $\|\eta\| < 2.5$
L1_DoubleIsoTau32er2p1	Two isolated τ_h with $E_T > 32$ GeV and $\|\eta\| < 2.1$
L1_SingleMu22	One muon with $p_T > 22$ GeV
L1_DoubleEG2512er2p5	Two e/γ with $E_T > 25/12$ GeV and $\|\eta\| < 2.5$
L1_DoubleMu157	Two muon with $p_T > 15/7$ GeV
L1_ETMHF100	Missing transverse energy > 100 GeV
L1_SingleJet180	One jet with $E_T > 180$ GeV
L1_DoubleJet150er2p5	Two jet with $E_T > 150$ GeV and $\|\eta\| < 2.5$

spectrum can be found in Fig. 3.21, showing how the improved resolution in the L1 muon reconstruction yields to visible dimuon mass peaks corresponding to the Φ and Υ resonances. Making use of these novel correlations, analysis-targeted triggers have been designed and implemented during Run 2. These select highly specific signal topologies, such as the Higgs boson produced via vector boson fusion or low-mass dimuon triggers [38]. Considering both the classic and analysis-targeted seeds, the L1 menu comprised a total of 350–400 seeds during Run 2. The ability to keep the L1 thresholds low and the global L1 trigger rate to less than 100 kHz was ensured with the increased sophistication of the algorithms, which enhanced the selectivity of the interesting physics phenomena and granted a rich CMS physics program during Run 2.

3.3.2 High Level Trigger

The High Level Trigger (HLT) is the second step of the CMS trigger; it reduces the output rate of the Level-1 trigger to $\sim$1 kHz, a bandwidth compatible with the data acquisition capability. The HLT is implemented in a software computing farm instrumented with up to 32 000 cores in 2018. It takes as input the full detector data, including tracking information, and conducts a more sophisticated reconstruction than at L1. The most stringent constraint on the HLT is the latency, since the limited number of CPUs imposes a maximum processing time of $\sim$320 ms per event. Therefore, the HLT runs an online object reconstruction that is a streamlined version of the offline reconstruction used in CMS (see Sect. 3.4), based on simplified clustering and tracking algorithms.

The HLT reconstruction is performed only locally around the L1 seeds, reducing the time needed to read the raw detector information. It is designed as a sequence of *reconstruction modules* and *selection filters*, resulting in different *paths*, namely

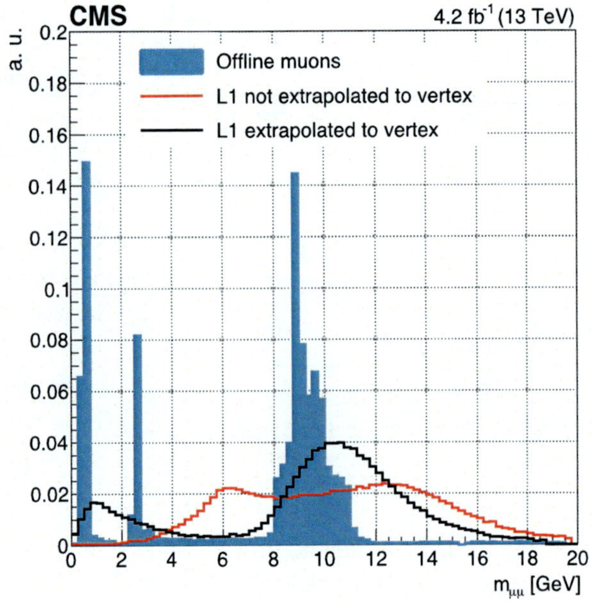

Fig. 3.21 Offline and Level-1 dimuon spectra of oppositely charged muons, with and without extrapolation to vertex. The Level-1 spectrum is shifted due to the p_T offset designed to make the muon trigger 90% efficient at any p_T threshold. The highest mass resonance corresponds to Υ, identifiable at Level-1 after extrapolation to vertex [38]

groups of algorithmic steps with increasing complexity which run in a predefined order. The reconstructions and selections in the HLT rely on the information from the calorimeters and muon detectors in the first place so the rate can be reduced before the CPU-expensive tracking reconstruction is carried out. Then, the tracks are matched to the calorimeter energy deposits or to the muon system tracks. The HLT can additionally derive advanced topologies from the individual inputs; it can reconstruct jets arising from b-quarks and identify displaced vertices, for instance.

The HLT menu consists of around 600 independent algorithms, closely related to the L1 menu seeds. The HLT was regularly updated to cope with the ever changing LHC and detector operation conditions during Run 2, and it benefited significantly from the pixel tracker upgrade in 2017 [19]. Special efforts were devoted to improve the muon and τ_h reconstructions and the E_T^{miss} noise mitigation. In addition, extra data taking strategies were developed to fully exploit the Run 2 physics potential, namely the *scouting* and *parking* techniques [38]. The first one is built on the event content reduction to afford higher rates, whereas the second records extra data on tape to be reconstructed *a posteriori* if deviations from the SM are suspected. The extensive coverage of physics phenomena provided by a typical HLT menu is illustrated in Fig. 3.22, where the bandwidth devoted to the diverse CMS physics analyses is shown.

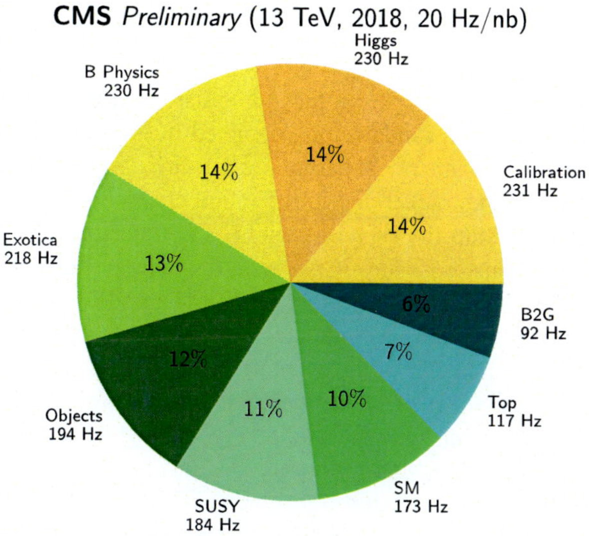

Fig. 3.22 HLT rate usage in the different physics analyses for a typical menu configuration during 2018. *B2G* stands for *Beyond-Two-Generations*, featuring physics models which study the decays of new resonances to heavy SM particles

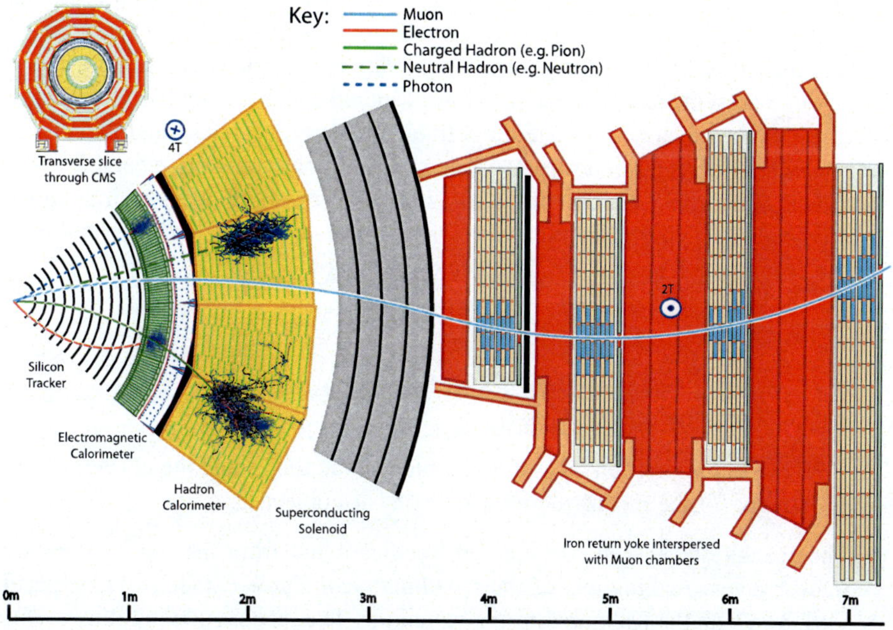

Fig. 3.23 Transverse slice of the CMS detector, showing the experimental signature of the different final-state particles [39]

3.4 Particle Reconstruction and Identification

Each type of particle resulting from the *pp* collisions at the centre of CMS leaves a distinct signature in the subdetectors, depicted in Fig. 3.23. Muons lose a small fraction of energy in the innermost part of the detector and, as charged particles, they are detected both in the tracker and in the muon chambers. Electrons and photons are both absorbed within the ECAL, but only the trajectory of the former reveals in the tracker. Hadrons cross the ECAL with low energy loss and deposit most of their energy in the HCAL. The trajectory of their charged components is detected in the tracker. Neutrinos escape CMS undetected; their presence manifests nonetheless through the energy imbalance of the event in the transverse plane. The raw information from the subdetector systems is combined to reconstruct, identify and calibrate the relevant physics objects with the maximal efficiency; the algorithms used for this purpose are summarised in what follows.

The most general reconstruction algorithm in CMS is the Particle Flow (PF) [39]. The algorithm starts with the low-level information (hits in the silicon tracker and energy deposits in the calorimeters) to build basic elements (tracks and clusters). These are used as inputs to the reconstruction of the individual physics objects. The muon reconstruction is done first by associating the PF tracks to the tracks in the muon chambers. Then, electrons are built by linking tracks to clusters with a dedicated procedure which takes into account the effects of bremsstrahlung radiation. The remaining clusters with no assigned track are identified as photons or neutral hadrons; the former present electromagnetic deposits and the latter present hadronic deposits. The rest of the clusters associated to a track are identified as charged hadrons. The reconstructed objects reconstructed (electrons, muons, photons, charged hadrons and neutral hadrons) are combined to build more complex ones, such as jets and hadronically decaying τ_h leptons, as well as to estimate the missing transverse energy of the event, which reveals the presence of neutrinos.

3.4.1 Particle Flow Basic Elements

The PF algorithm presents two building blocks, the tracks and the clusters, that are used as input for the reconstruction of the physics objects.

- The **tracks** of charged particles are reconstructed from the hits in the silicon tracker layers making use of the *Combinatorial Track Finder* (CTF) algorithm [40], based on the Kalman filtering technique. The reconstruction follows an iterative approach to achieve a high efficiency and low fake rate. In the first iteration, only the reconstruction of isolated tracks which clearly originate from the primary vertex is attempted. If selected, the hits of the corresponding tracks are not considered at later stages. This reduction of combinatorial complexity allows the quality criteria applied in successive iterations to be progressively loosen, while targeting less evident tracks, such as radiating electrons or B-hadrons. A typical track reconstruction undergoes a total of about 12 iteration steps; it enables the algorithm to reconstruct tracks with p_T as low as 0.1 GeV and produced as far as 60 cm for the primary vertex.

- The **clusters** are constructed from the energy deposits in the calorimeters. The algorithm runs separately in the preshower, the ECAL and the HCAL subdetectors, in order to achieve a high efficiency even at low p_T and to be able to disentangle overlapping showers. It is based on an iterative clustering technique which follows the typical lateral shower profile. The local maxima of deposited energy are identified, to which neighbouring energy deposits are aggregated if their signals are larger than twice the standard deviation of the electronic noise. In practice, less than 5 iterations are needed for the algorithm to converge.

Tracks and clusters can be associated (or *linked*) to form PF *blocks*, based mostly on topological considerations. The properties of these blocks are evaluated to identify the final-state objects.

3.4.2 Muons

Muons are the first particle to be identified in CMS, as they benefit from clean signatures in the muon chambers and high identification efficiency. The muon tracks are reconstructed independently in the silicon tracker and in the muon chambers [41]. The former are reconstructed following the aforementioned PF procedure and are referred to as *tracker tracks*. The reconstruction of the latter, denoted as *standalone-muon tracks*, is PF-independent and relies solely on the information from the muon systems. In this case, the positions of the hits within the DT, CSC and RPC sensors are combined and fitted to form track segments, which give an initial estimation of the direction and the momentum of the muon. The track segments are then combined to reconstruct the muon track making use of a Kalman filter technique [40].

The global muon reconstruction in CMS proceeds by either matching tracker muons to standalone-muons (*inside-out*) or viceversa (*outside-in*). The *tracker muon tracks* are built inside-out by propagating tracker tracks with $p_T > 0.5$ GeV and total momentum $p > 2.5$ GeV to the muon system, where the presence of at least one muon segment at a compatible position is required. The *global muons*, in turn, are reconstructed outside-in by matching a standalone muon track to the tracker tracks after verifying the compatibility of their parameters; the global fit is performed on both tracks together. Owing to the large size of the muon chambers, the combined fit improves the momentum resolution with respect to the tracker-only fit for muons of $p_T > 200$ GeV. About 99% of the muons produced within the muon system acceptance are reconstructed with one of these two approaches, and very often as both, in which case they are merged into a single candidate.

The reconstructed muons undergo a set of selection criteria based on the quality of the tracks; these concern observables such as the track fit χ^2, the number of hits per track or the degree of matching between tracker and standalone tracks. Different

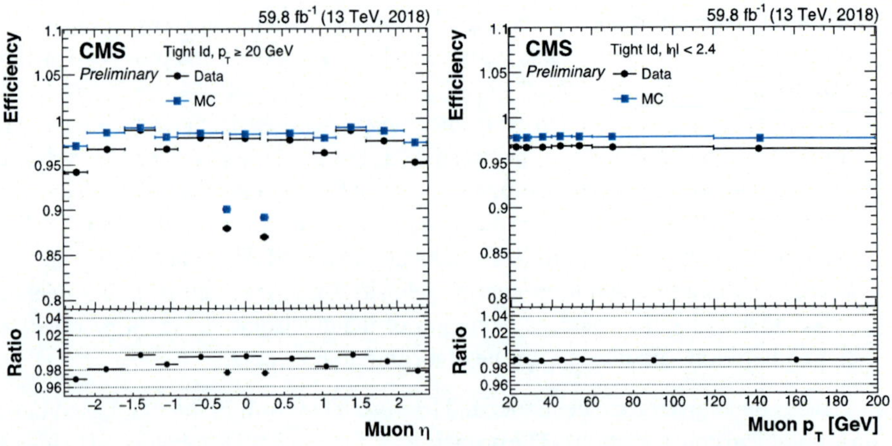

Fig. 3.24 Muon selection efficiency for the tight isolation and identification working points as a function of the muon η (left) and p_T (right), measured in 2018 data [42]. The difference in selection efficiency between the data and the simulation is applied as a scale factor in the physics analyses

levels of requirements result in different muon identification working points (loose, medium, tight) with increasing purity and decreasing efficiency. The information of the muon subdetectors is further analyzed to compute the isolation of the muon candidate, which serves as a handle to distinguish between prompt muons and those arising from weak decays within jets. The PF relative isolation is defined as the ratio between the sum of p_T of all PF candidates within a cone size $\Delta R < 0.4$ around the muon and the p_T of the muon itself. Decreasing cutoffs to this value define six increasingly tight muon isolation working points. In the analysis presented here, a variable p_T-dependent cone size is used to evaluate the isolation of the muons (see Sect. 6.1); it is beneficial for topologies in which the leptons are produced in the decay of boosted objects, emitted collinearly with the other decay products.

As a result of the upgrade of the muon system and the improvements of the algorithms (see Sect. 3.2.2), the performance of the muon reconstruction in Run 2 was comparable to Run 1 despite the increased instantaneous luminosity and pileup. The spatial resolution of the reconstructed hits lied between 50 and 300 μm, leading to a muon selection efficiency above 96% [42]. The efficiency as a function of p_T and η after applying the tightest PF identification and isolation criteria to the reconstructed muons can be found in Fig. 3.24. The strong CMS magnetic field together with the excellent resolution of the muon system makes the CMS detector ideally suited to reconstruct dimuon candidates in a mass interval ranging up to 150 GeV, instrumental in the Higgs boson discovery and other searches for heavy resonances.

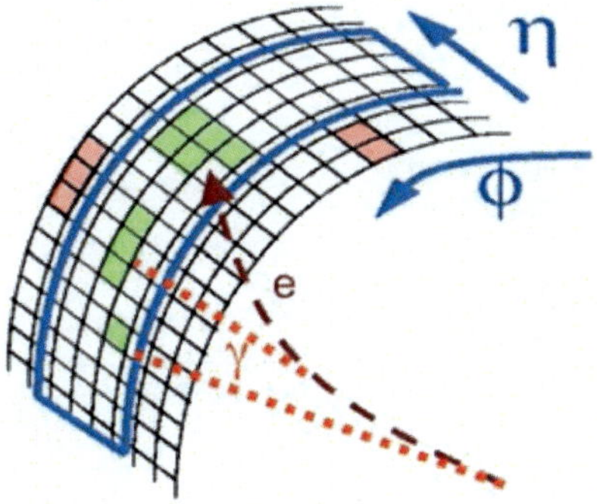

Fig. 3.25 Diagram of the electron trajectory in the CMS detector. The electron interacts with the material in the tracker and loses energy due to bremmstrahlung, emiting photons tangent to its trajectory along the ϕ direction due to the magnetic field. The energy deposits of the electron and the photons in the ECAL (green squares) are grouped together in superclusters, used in the GSF electron reconstruction algorithm

3.4.3 Electrons

After muons, the second particle to be reconstructed at CMS is the electron. Before reaching the ECAL, electrons lose a sizeable fraction of their energy in the tracker, mostly due to the non-Gaussian phenomenon of bremmstrahlung. At $\eta \approx 0$, an electron loses an average of 33% of its energy, and this fraction can reach ~86% at $\eta \approx 1.4$, where the material budget is the highest. This makes their reconstruction more complex: the tracker algorithm must take into account the energy loss and the clustering algorithm must collect the bremsstrahlung photon energy, which can be located far away from the electron interaction point in the ECAL. The electron reconstruction addresses these challenges with a dedicated tracking algorithm and an advanced energy clustering procedure.

The electron tracks in CMS are reconstructed with the *Gaussian-Sum Filter* (GSF) [43, 44] algorithm, which is independent from the PF iterative tracking procedure. The algorithm models the electron energy loss with a weighted sum of Gaussian PDFs to better capture the bremmstrahlung effects, instead of relying on a single Gaussian like the Kalman filter procedure of the PF algorithm. The electron energy deposits in the calorimeters are built by regrouping PF ECAL clusters in the so-called *superclusters*. The procedure consists in identifying a seed cluster, to which the energy deposits associated to the bremmstrahlung photons are aggregated. These typically spread along the ϕ direction due to the magnetic field, tangent to the electron trajectory, as schematically shown in Fig. 3.25.

The global electron is reconstructed by associating the GSF tracks to the superclusters following two approaches. The *ECAL-based* approach uses superclusters with $p_T > 4$ GeV as seeds for the electron GSF track finding. This procedure is better suited for high-p_T isolated electrons, as the low p_T superclusters are too small to contain all the bremmstrahlung radiation. In order to recover the soft electrons,

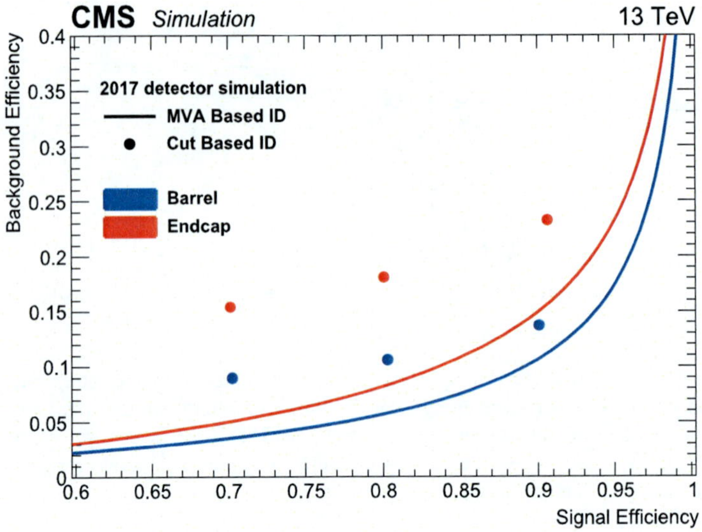

Fig. 3.26 Background efficiency as a function of the signal efficiency for the BDT-based electron identification algorithm in the ECAL barrel and endcap derived from the simulation with 2017 detector conditions. The cut-based selection working points are shown for comparison [45]

a *tracker-based* approach is used, which takes the GSF tracks with $p_T > 2$ GeV as seeds to the clustering. GSF tracks and PF superclusters are combined into an electron candidate if they fulfil a series of requirements on the matching quality. They are used to estimate the electron charge and momentum, the latter being derived from a the GSF track curvature and the supercluster energy. Owing to the high granularity of the tracker and the ECAL detectors, the energy resolution of electrons with p_T of 10-50 GeV is 0.1% (0.3%) in the barrel (endcaps) [45].

The electron candidates are subject to additional identification criteria to distinguish the genuine electrons from the misidentified jets, namely pions. The identification is based on Boosted Decision Trees (BDTs) with input variables related to the shower shape, the track quality, the track-cluster matching, the fraction of momentum lost due to bremsstrahlung and the isolation. The output of the BDT, representing the probability of the candidate to be a genuine electron, is used to define three levels of electron identification (loose, medium, tight) with decreasing selection efficiency and increasing purity. The performance of such discriminator is found in Fig. 3.26, displaying the identification efficiency as a function of the mistag rate for the jets faking electrons. The BDT-based identification visibly outperforms the previous cut-based identification in both the barrel and the endcaps [45]. Similarly to the muons, the isolation of the electrons is used to reject non-prompt electrons produced inside jets; it is typically evaluated within a cone size of $\Delta R = 0.3$, modified to a p_T-dependent variable size in this analysis (see Sect. 6.1).

3.4.4 Photons

In most cases, photons cross the tracker without interacting and deposit most of their energy (~97%) in the ECAL, where the shower can spread into multiple neighbouring crystals. Hence, the photon reconstruction [46] is seeded from ECAL superclusters with $E_{\mathrm{T}} > 10$ GeV which are not related to any GSF track. These superclusters must be isolated from other ECAL clusters or any track extrapolated from the tracker, and their energy deposit must present a distribution compatible with a photon shower. In some cases, photons interact with the tracker material and convert into electron-positron pairs before entering the ECAL. Converted photons are identified because the electron and positron leave bent trajectories in the tracking system, with momenta approximately parallel as a result of the photon being massless, and an energy deposit spread in the ϕ direction. The corresponding tracks are either seeded by displaced secondary vertices and extrapolated outward, or by ECAL superclusters and extrapolated inwards. The typical energy resolution for unconverted (converted) photons with $p_{\mathrm{T}} > 25$ GeV is around 1% (1.3-2.5%) in the barrel and 2.5% (3%) in the endcaps [46]. The high precision of the photon energy reconstruction enhances the selectivity of the $\mathrm{H} \rightarrow \gamma\gamma$ decay, one of the main drivers of the precise measurements of the Higgs boson properties.

Neutral mesons decaying to photons constitute the most substantial source of photon misidentification. Photons arising from mesons are typically collimated, commonly reconstructed as a single one. The contribution of this background is reduced relying heavily on the isolation of the photon candidate; it is defined either by using a cut-based approach on groups of discriminating variables or by employing multivariate techniques, in both cases making use of observables similar to the electron identification BDT.

3.4.5 Jets

Quarks and gluons arising from the hard interactions hadronize in the detector and produce several collimated particles, grouped together in cone-shaped jets. The momenta of the jets is estimated from the recollected hadronic products, which can be charged or neutral. These charged and neutral hadrons are reconstructed with the remaining PF elements after removing the electrons, muons and photons. As they generally deposit energy in the ECAL and in the HCAL, the clusters of both calorimeters are linked first. If the HCAL cluster is reconstructed within the tracker acceptance and is not linked to any other track, it is identified as a neutral hadron when no ECAL cluster is found. The remaining HCAL clusters are linked to one or several tracks, which may be linked to ECAL clusters too; they are reconstructed as charged hadrons.

Jets are reconstructed by clustering charged and neutral hadrons with the anti-k_{T} algorithm [47]. The clustering is done recursively by pairing the PF candidates

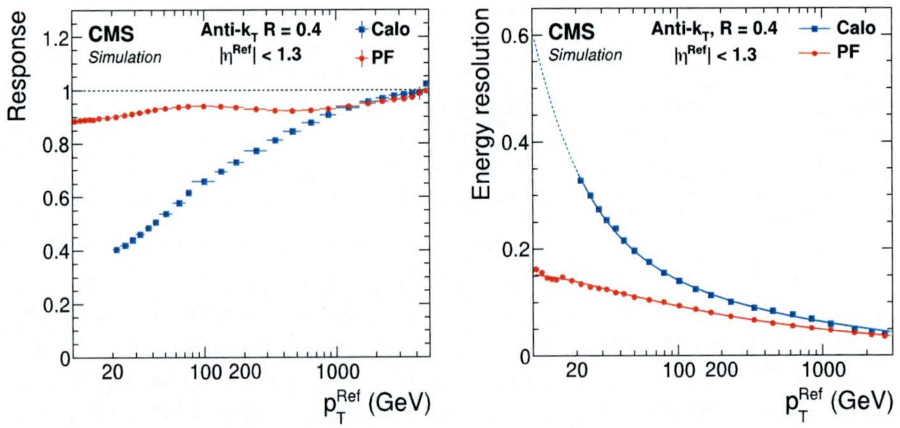

Fig. 3.27 Jet energy response (left), defined as the mean ratio of the reconstructed jet energy to the reference jet energy, and jet energy resolution (right), defined as the Gaussian width of the ratio between the corrected and reference jet energies, as a function of the reference jet p_T. Results are shown for jets in the barrel reconstructed with the anti-k_T algorithm with $R = 0.4$ following the PF and calorimeter-based approaches [39]

which are close to each other, according to a metric defined to produce jets of conic shape. The size of the cone is determined by the distance parameter R at which the algorithm operates, that can be set to $R = 0.4$ or $R = 0.8$. In both cases, the higher-p_T pairs are clustered first; this way the cone is built around the hardest particle of the event and the soft radiation or collinear parton splitting at the borders is suppressed. The jet four momentum is computed as the vector sum of the clustered PF candidates four momenta; it differs from that of the original parton due to theoretical uncertainties on the hadronization process and other experimental effects. A p_T- and η-dependent set of jet energy corrections is applied to calibrate the jet response taking into account the pileup and the detector noise, along with the response, non-linearity and inhomogeneity of the calorimeters. The typical reconstructed jet energy has a response of around 0.9 and a resolution lower than 15% for all p_T ranges, as shown in Fig. 3.27, where the improvement of the PF approach with respect to the calorimeter-based reconstruction is clearly visible. The jets are further subject to identification requirements aimed at suppressing those poorly reconstructed or due to instrumental noise in the calorimeters. Two increasingly stringent levels of selection are defined (loose, tight) based on the fraction of charged and neutral hadrons in the jet, the charged hadron multiplicity and the fraction of energy deposited in the ECAL.

Dedicated tagging algorithms are used to identify jets coming from b-quarks (b-*jets*), essential in analysis such as the one presented in this thesis, where the top quarks from the tH and $t\bar{t}$H processes decay almost exclusively to b-quarks. These algorithms exploit the distinct signature of B-mesons compared to lighter quarks. B-hadrons have a high mass and are long-lived; they decay at a secondary vertex with large particle multiplicity and present and electron or muon inside the jet in 40% of the cases. These features are exploited with the tracking and vertex variables used as

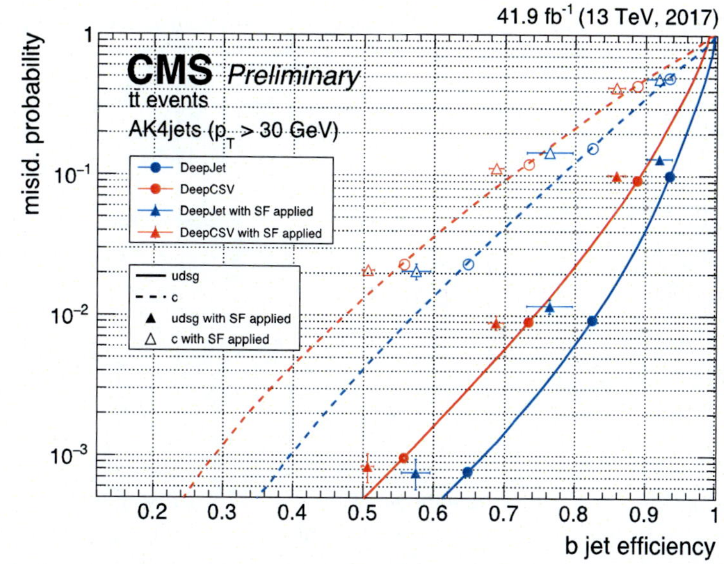

Fig. 3.28 Misidentification probability for c and light-flavour jets versus b-jet identification efficiency for various b-tagging algorithms measured in simulated $t\bar{t}$+jets events with 2017 detector conditions [48]

input to the Deep Neural Networks employed in the identification. Two algorithms are developed in CMS: *DeepCSV* and *DeepJet* [48, 49]. The latter presents an increasing complexity and results in the improved performance illustrated in Fig. 3.28. Three working points are defined (loose, medium, tight) corresponding to 10, 1 and 0.1% light jet misidentification probabilities, respectively.

Quark- and the gluon-induced jets can also be distinguished in CMS with the *Quark-Gluon Likelihood* [49] algorithm; it exploits the fact that gluon-induced jets are wider, with higher particle multiplicities and with a more uniform energy fragmentation than quark-induced jets.

3.4.6 Hadronic τ Leptons

Since the τ is the heaviest lepton ($m_\tau = 1777$ MeV [50]), it is the only one which can decay into hadrons, together with a single neutrino. It can also decay into an electron or a muon along with two neutrinos. The τ lepton branching ratios are listed Table 3.3. The τ lifetime is relatively short ($\tau_\tau = 2.9 \times 10^{-13}$ s [50]), meaning a 30 GeV τ lepton covers a distance of $\sim$1 mm before decaying, hence it is only possible to reconstruct its visible decays products in the detector. In around 1/3 of the cases, the τ lepton decays into an electron or a muon, reconstructed via the aforementioned respective algorithms. In the remaining cases, the τ lepton decays to charged hadrons (mostly $\pi^\pm$), commonly denoted as *prongs*, and neutral hadrons (π^0),

Table 3.3 Branching ratios (BR) of the τ lepton decays. The symbol $h^{\pm}$ indicates a charged pion. When the decay occurs via an intermediate resonance, the corresponding meson is shown [50]

Decay mode	Resonance	BR [%]
$\tau \to e\nu_e\nu_\tau$		17.8
$\tau \to \mu\nu_\mu\nu_\tau$		17.4
$\tau \to h^{\pm}\nu_\tau$		11.5
$\tau \to h^{\pm}\pi^0\nu_\tau$	$\rho(770)$	25.9
$\tau \to h^{\pm}\pi^0\pi^0\nu_\tau$	$a_1(1260)$	9.5
$\tau \to h^{\pm}h^{\mp}h^{\pm}\nu_\tau$	$a_1(1260)$	9.8
$\tau \to h^{\pm}h^{\mp}h^{\pm}\pi^0\nu_\tau$		4.8
Other hadronic decays		3.3

sometimes through an intermediate $\rho(770)$ or $a_1(1260)$ resonance. The neutrinos remaining undetected, it is not possible to evaluate exactly the momentum of the τ lepton before decaying; sophisticated algorithms are needed for this purpose.

Hadronically decaying τ leptons are reconstructed using the *hadrons-plus-strips* (HPS) algorithm [51]. The algorithm is seeded by the PF jet constituents reconstructed with the anti-k_T algorithm with distance parameter $R = 0.4$. The prongs from the τ decay are identified via the PF charged hadron constituents. The identification of π^0's is done via the $\pi^0 \to \gamma\gamma$ decay, which manifests as photon PF candidates or as electron candidates from photon conversions inside the jet. Their deposits in the ECAL are clustered dynamically into strips elongated in the ϕ direction, where the window size depends on the p_T of the cluster itself. The charged hadrons are combined with the strips to reconstruct the different decay modes, namely

- 1-prong from a single charged hadron with no strips,
- 1-prong+π^0 from a single charged hadron with a single strip,
- 1-prong+2π^0 from a single charged hadron with two strips,
- 3-prongs from three charged hadrons.

An schematic representation of the HPS reconstruction in the 1-prong+π^0 decay mode can be found in Fig. 3.29. Additional quality criteria are applied to the candidates to verify their compatibility with a τ decay: multiprong tracks have to originate from the same vertex, the constituents must lie within a p_T-dependent cone size, the invariant mass of the system has to be compatible with the intermediate resonances and the total electric charge has to be ± 1.

The τ_h reconstruction is complemented with identification methods to discriminate genuine τ_h from QCD-induced jets, muons and electrons. A DNN-based algorithm (*DeepTau* [52]) is used for this purpose; it combines the usage of high-level features of the reconstructed τ_h and low level information from the PF candidates within the τ_h cone. Different working points are defined based on the output score of the DNN: 8 in the discrimination against jets, 4 against muons and 8 against electrons. The sensitivity of the $H \to \o^+\o^-$ decay in the analysis presented in this thesis is significantly improved with the use of the DNN against jets, which outperforms the previous BDT-based identification techniques. The improvement is illustrated in Fig. 3.30, in which the different working points are shown.

3.4.7 *Missing Transverse Energy*

The colliding partons within the protons carry an unknown fraction of the proton momentum in the longitudinal direction; however, their momentum is negligible in the transverse plane. Hence, the conservation of momentum can be exploited in the transverse plane to infer the production of undetected particles, such as neutrinos or BSM weakly-interacting particles. An instrumental quantity is the *missing transverse momentum* $\vec{p}_{\mathrm{T}}^{\,\mathrm{miss}}$, defined as minus the vectorial sum of the transverse momenta of

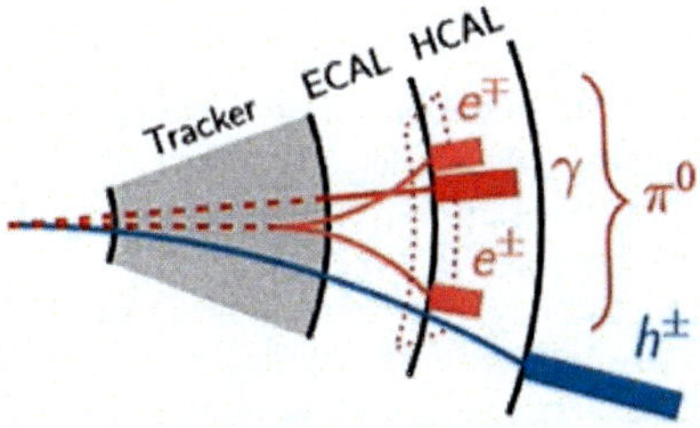

Fig. 3.29 Detector signature of the $\tau \rightarrow h^{\pm}\pi^{0}\nu_{\tau}$ decay reconstructed via the hadrons-plus-strips algorithm[51]. The reconstructed PF charged hadron ($h^{\pm}$) in the HCAL is combined with deposits of the neutral pion (π^0) decays in the ECAL clustered into a strip with p_{T}-dependent size

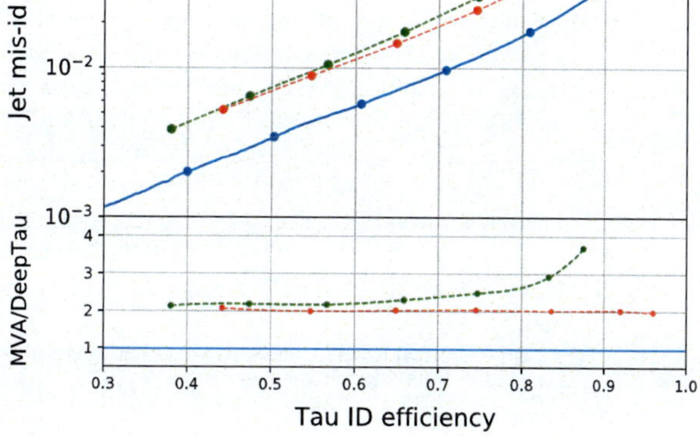

Fig. 3.30 Jet misidentification probability versus τ_h identification efficiency for various τ_h identification algorithms measured in $t\bar{t}$+jets simulated events with 2017 detector conditions. The DNN-based DeepTau algorithm (in blue) outperforms the previous BDT-based algorithms (in green and red) [52]

all the reconstructed PF objects,

$$\vec{p}_{\mathrm{T}}^{\,\mathrm{miss}} = -\sum_{i=1}^{N} \vec{p}_{\mathrm{T}}^{\,i} \,. \tag{3.9}$$

A precise measurement of this quantity requires the CMS detector to cover almost the full solid angle, and to ensure that as many particles as possible are accounted for. The inefficiencies arising from the tracking or clustering algorithms and the non-linearities in the calorimeter response might introduce biases in the determination of $\vec{p}_{\mathrm{T}}^{\,\mathrm{miss}}$. Thus, the energy corrections applied to jets are also propagated to the $\vec{p}_{\mathrm{T}}^{\,\mathrm{miss}}$ computation. After corrections, the relative energy and ϕ angular resolution of the missing transverse energy lies under 20% for all p_{T} ranges [39].

References

1. ATLAS Collaboration (2012) Observation of a new particle in the search for the Standard Model Higgs boson with the ATLAS detector at the LHC. Phys Lett B 716:1–29
2. CMS Collaboration (2012) Observation of a new boson at a mass of 125 GeV with the CMS experiment at the LHC. Phys Lett B 716:30–61
3. Mobs E (2018) The CERN accelerator complex. Aug 2018. https://cds.cern.ch/record/2636343. Accessed 29 Apr 2020
4. ATLAS Collaboration (2008) The ATLAS experiment at the CERN Large Hadron Collider. JINST 3(08):S08003–S08003
5. CMS Collaboration (2008) The CMS experiment at the CERN LHC. JINST 3(08):S08004–S08004
6. LHCb Collaboration (2008) The LHCb detector at the LHC. JINST 3(08):S08005–S08005 (2008)
7. ALICE Collaboration (2008) The ALICE experiment at the CERN LHC. JINST 3(08):S08002–S08002
8. Bruning OS, Collier P, Lebrun P, Myers S, Ostojic R, Poole J, Proudlock P (2004) LHC design report. CERN Yellow Reports: Monographs. CERN, Geneva
9. CMS Collaboration (2018) Measurement of the inelastic proton-proton cross section at $\sqrt{s} =$ 13 tev. J High Energy Phys 2018(7):161
10. HiLumi Collaboration (2020) The HL-LHC project. https://hilumilhc.web.cern.ch/content/hl-lhc-project/. Accessed 30 Apr 2020
11. CMS Collaboration (2020) Public luminosity results. https://twiki.cern.ch/twiki/bin/view/CMSPublic/LumiPublicResults. Accessed 30 Apr 2020
12. Neutelings I (2020) CMS Wiki pages, how to draw diagrams in LaTeX with TikZ. https://wiki.physik.uzh.ch/cms/latex:tikz. Accessed 30 Apr 2020
13. Sakuma T (2020) Cutaway diagrams of CMS detector, May 2019. https://cds.cern.ch/record/2665537. Accessed 30 Apr 2020
14. CMS Collaboration (1997) The CMS magnet project: technical design report. Technical report CERN-LHCC-97-010. CMS-TDR-1, CERN, Geneva
15. Klyukhin VI et al (2010) The CMS magnetic field map performance. IEEE Trans Appl Supercond 20(3):152–155
16. CMS Collaboration (1997) The CMS tracker system project : technical design report. Technical report CERN-LHCC-98-006. CMS-TDR-5, CERN, Geneva

17. CMS Collaboration (2000) The CMS tracker: addendum to the technical design report. Technical report CERN-LHCC-2000-016. CMS-TDR-5-add-1, CERN, Geneva
18. CMS Collaboration (2014) Description and performance of track and primary-vertex reconstruction with the CMS tracker. JINST 9(10):P10009–P10009
19. CMS Collaboration (2012) CMS technical design report for the pixel detector upgrade. Technical report CERN-LHCC-2012-016. CMS-TDR-11, CERN, Geneva
20. CMS Collaboration (2017) The Phase-1 upgrade of the CMS pixel detector. Technical report CMS-CR-2017-135. 06, CERN, Geneva
21. CMS Collaboration (2017) Test beam performance measurements for the Phase I upgrade of the CMS pixel detector. JINST 12(05):P05022–P05022
22. CMS Collaboration (1997) The CMS electromagnetic calorimeter project: technical design report. Technical report CERN-LHCC-97-033. CMS-TDR-4, CERN, Geneva
23. CMS Collaboration (2013) The CMS ECAL performance with examples. Technical report CMS-CR-2013-430, CERN, Geneva
24. CMS Collaboration (2007) Energy resolution of the barrel of the CMS electromagnetic calorimeter. JINST 2(04):P04004–P04004
25. CMS Collaboration (2019) CMS ECAL response to laser light. Technical report CMS-DP-2019-005, CERN, Geneva
26. CMS Collaboration (2008) Laser monitoring system for the CMS lead tungstate crystal calorimeter. Nucl Instrum Meth A 594:292–320
27. CMS Collaboration (2017) The Phase-2 upgrade of the CMS Endcap Calorimeter. Technical report CERN-LHCC-2017-023. CMS-TDR-019, CERN, Geneva
28. CMS Collaboration (1997) The CMS hadron calorimeter project: technical design report. Technical report CERN-LHCC-97-031. CMS-TDR-2, CERN, Geneva
29. CMS Collaboration (2009) The CMS barrel calorimeter response to particle beams from 2 to 350 GeV/c. J Phys: Conf Ser 160:012056
30. CMS Collaboration (2012) CMS technical design report for the Phase 1 Upgrade of the Hadron Calorimeter. Technical report CERN-LHCC-2012-015. CMS-TDR-10, CERN, Geneva
31. CMS Collaboration (1997) The CMS muon project: technical design report. Technical report CERN-LHCC-97-032. CMS-TDR-3, CERN, Geneva
32. CMS Collaboration (2018) Performance of the CMS muon detector and muon reconstruction with proton-proton collisions at $\sqrt{s} = 13$ TeV. JINST 13(06):P06015–P06015
33. CMS Collaboration (2015) CMS technical design report for the Muon Endcap GEM upgrade. Technical report CERN-LHCC-2015-012. CMS-TDR-013, CERN, Geneva
34. Ahmed W et al (2017) The triple GEM detector control system for CMS forward muon spectrometer upgrade. JINST 12(02):P02003, 15 p
35. CMS Collaboration (2020) Summaries of CMS cross section measurements. https://twiki.cern.ch/twiki/bin/view/CMSPublic/PhysicsResultsCombined. Accessed 4 May 2020
36. CMS Collaboration (2013) CMS technical design report for the Level-1 trigger upgrade. Technical report CERN-LHCC-2013-011. CMS-TDR-12, CERN, Geneva
37. Zabi A et al (2017) The CMS Level-1 Calorimeter trigger for the LHC Run II. JINST 12(01):C01065–C01065
38. CMS Collaboration (2020) Performance of the CMS level-1 trigger in proton-proton collisions at $\sqrt{s} = 13$TeV. J Instrum 15(10):P10017–P10017
39. CMS Collaboration (2017) Particle-flow reconstruction and global event description with the CMS detector. JINST 12(10):P10003–P10003
40. Frühwirth R (1987) Application of Kalman filtering to track and vertex fitting. Nucl Instrum Meth A 262(2):444–450
41. CMS Collaboration (2018) Performance of the CMS muon detector and muon reconstruction with proton-proton collisions at $\sqrt{s} = 13$ TeV. JINST 13(06):P06015–P06015
42. CMS Collaboration (2019) Muon reconstruction performance during Run II. Technical report CMS-DP-2019-022, CERN, Geneva
43. Adam W, Frühwirth R, Strandlie A, Todor T (2005) Reconstruction of electrons with the gaussian-sum filter in the CMS tracker at the LHC. Technical report CMS-NOTE-2005-001, CERN, Geneva

44. CMS Collaboration (2015) Performance of electron reconstruction and selection with the CMS detector in proton-proton collisions at $\sqrt{s} = 8$ TeV. JINST 10(06):P06005–P06005
45. CMS Collaboration (2018) Electron and Photon performance in CMS with the full 2017 data sample and additional 2016 highlights for the CALOR 2018 conference. Technical report CMS-DP-2018-017, CERN, Geneva
46. CMS Collaboration (2015) Performance of photon reconstruction and identification with the CMS detector in proton-proton collisions at $\sqrt{s} = 8$ TeV. JINST 10(08):P08010–P08010
47. Cacciari M, Salam GP, Soyez G (2008) The anti-k_t jet clustering algorithm. J High Energy Phys 2008(04):063
48. CMS Collaboration (2018) Performance of the DeepJet b tagging algorithm using 41.9 /fb of data from proton-proton collisions at 13 TeV with Phase 1 CMS detector. Technical report CMS-DP-2018-058, CERN, Geneva
49. CMS Collaboration (2018) Identification of heavy-flavour jets with the CMS detector in pp collisions at 13 TeV. JINST 13(05):P05011–P05011
50. Tanabashi M et al (2018) Review of particle physics. Phys Rev Lett D 98:030001
51. CMS Collaboration (2018) Performance of reconstruction and identification of τ leptons decaying to hadrons and ν_τ in pp collisions at $\sqrt{s} = 13$ tev. JINST 13(10):P10005–P10005
52. CMS Collaboration (2019) Performance of the DeepTau algorithm for the discrimination of taus against jets, electron, and muons. Technical report CMS-DP-2019-033, CERN, Geneva

Chapter 4
Performance and Optimization of the Level-1 τ_h Trigger

As described in Sect. 3.3, the CMS detector presents a sophisticated two-stage trigger system that reduces the event rate to an acceptable level, while retaining the potentially interesting physics events with the highest efficiency possible. To guarantee the success of the ambitious CMS physics program in the harsher experimental conditions of the LHC Run 2 with respect to Run 1, the L1 trigger hardware and architecture were upgraded during LS1 [1]. The new calorimeter trigger features an increased granularity and a time-multiplexed design to guarantee a global view of the inputs and to remove the regional boundaries in the object reconstruction. The algorithms are implemented in powerful FPGAs; their increased sophistication is complemented with the improved correlation capabilities of the global trigger, building the grounds for analysis-targeted seeds to enhance the selectivity of the physics signals.

The upgraded trigger includes a dedicated L1 algorithm to reconstruct hadronically decaying τ leptons for the first time. The τ_h decay products involve charged and neutral hadrons that deposit energy in both the electromagnetic and hadronic calorimeters. Hence, the τ_h reconstruction inherits features from the e/γ and jet algorithms at L1, but is adapted to capture the unique topology of the τ_h. The relevant energy deposits are clustered dynamically to minimize the influence of pileup; secondary clusters, expected from the multiple objects in the final state, can be aggregated under proximity conditions. L1 τ_h candidates suffer from a significant QCD-induced jet background contamination which, together with the increasingly harsh pileup environment, can boost the output rate of the trigger. This background is efficiently removed upon evaluation of the isolation of the candidate, as QCD-induced jets typically result in a wider energy deposit.

The physics case for a L1 τ_h trigger is mostly centred on the Higgs boson, which decays into a τ pair with a BR of around 6%. In turn, these τ leptons undergo a hadronic decay in 2/3 of the cases. Thus, the sensitivity of the analyses involving τ leptons, such as the $\mathrm{t\bar{t}H}$ production in multileptonic final states, crucially relies on the capability to identify and efficiently select the τ_h candidates at trigger level. As the luminosity and pileup levels have been gradually increased during Run 2 operations

C. Martin Perez, *Search for the Higgs Boson Produced in Association with Top Quarks with the CMS Detector at the LHC*, Springer Theses,
https://doi.org/10.1007/978-3-030-90206-3_4

(see Figs. 3.3 and 3.4), the L1 τ_h trigger algorithm has evolved to maintain high selection efficiency for the benefit of the physics analyses and high background rejection for an acceptable trigger rate. As part of my thesis work, I was a leading contributor to the measurement of the performance of the L1 τ_h trigger with the data collected in 2017 and 2018. I was also deeply involved in the optimization of the algorithm to face the extreme experimental conditions expected towards the end of Run 2. The trigger developed herein granted the collection of an unprecedented dataset of 137 fb^{-1} in the LHC Run 2 era; it enabled several physics milestones to be achieved, in particular the observation of the $H \rightarrow \tau^+\tau^-$ decay in 2017 [2] and, more indirectly, the observation of the $t\bar{t}H$ process in 2018 [3].

In view of the upcoming Run 3, the physics case of the L1 τ_h algorithm is getting increasingly focused on the collection of rarer processes such as the double Higgs boson production; an upper limit of around 7.7 times on the SM cross section of this process is expected at the end of Run 3, assuming the foreseen integrated luminosity of 300 fb^{-1} is collected [4]. The precise measurements of the properties of the $H \rightarrow \tau^+\tau^-$ decay, leading to interesting probes of the SM such as the CP structure of the Yukawa coupling between the Higgs boson and the τ lepton [5], is also a physics goal during Run 3. To enhance the selectivity of these signals, I developed tailored L1 τ_h trigger algorithms that better exploit their detector signature and allow for the trigger thresholds to be loosened. These triggers complement the classic double-τ_h seed in the menu: they cover an enlarged phase space and they improve the sensitivity to rarer processes whose selection is limited by the current trigger configuration.

This chapter is structured as follows. First, an overview of the Level-1 τ_h trigger algorithm is given in Sect. 4.1. The measurement of the performance of the algorithm in 2017 data is presented in Sect. 4.2. The changes to the algorithm at the end of 2017 and during 2018 are explained in Sect. 4.3; the resulting performance is depicted at the end of the section. The development of novel L1 τ_h seeds for Run 3 is presented in Sect. 4.4. An outlook on the future improvements to the L1 τ_h trigger is given in Sect. 4.5. Looking further into the future, I designed an innovative machine-learning-based L1 τ_h algorithm for the endcaps of the CMS detector during the HL-LHC era, described in Chap. 5. The work presented in this section has been shown in two international conferences: the 14th Pisa Meeting on Advanced Detectors (Italy) in 2018 [6] and the Lake Louise Winter Institute (Canada) in 2019 [7]. It was included in the CMS publication featuring the performance of the L1 trigger in Run 2 [8], and in two CMS detector notes describing the 2017 [9] and 2018 [10] L1 τ_h trigger performance.

4.1 The Level-1 τ_h Trigger Algorithm

The Level-1 τ_h trigger is the first level of selection of τ leptons decaying hadronically in the CMS detector. As such, is must comply with the two-fold requirement of high signal efficiency to maximize the acceptance to the physics analysis, and large

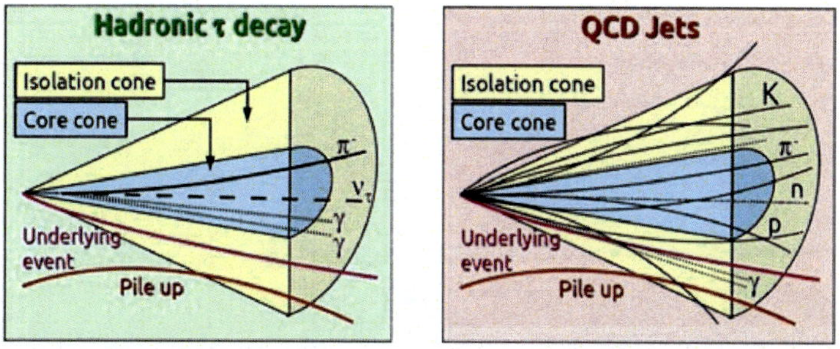

Fig. 4.1 Sketch of the experimental signature of hadronically decaying τ leptons (left) compared to QCD-induced jets (right). The former presents narrow and collimated energy deposits, while the latter is generally broader and with higher particle multiplicity

background rejection to keep the rate under control; this must be done within the $\sim$1 μs latency allocated to the algorithm. The challenge of triggering on hadronically decaying τ leptons is the presence of multiple energy deposits arising from the one, two or three charged or neutral pions present in the decay. The τ_h signature typically results in a narrow and collimated jet, contrary to the quark and gluon jets, which are generally broader and with a higher particle multiplicity, as illustrated in Fig. 4.1. Given that the L1 trigger has no access to the tracking information, the τ_h reconstruction algorithm is built relying solely on the inputs from the ECAL and the HCAL. With this limited information, the τ_h candidate must be identified and distinguished from the overwhelming jet background by means of its isolation. The identification is particularly demanding under high pileup and instantaneous luminosity conditions, as these yield additional hadronic activity in the detector that has to be disentangled from the signal. The calorimeter inputs and algorithmic steps of the L1 τ_h trigger are summarized in the following.

4.1.1 Calorimeter Inputs

The L1 τ_h are reconstructed in the calorimeter trigger, which is also responsible for the reconstruction of e/γ, jets and energy sums, all built from the same calorimeter inputs. The information from the ECAL, HCAL and HF subdetectors is encoded in the so-called *trigger primitives*, digital quantities synchronous with the 40 MHz bunch crossing of the LHC. These are organized in *trigger towers* (TTs), each carrying the sum of the ECAL and HCAL energies in a localized detector region. In the barrel, each TT is associated to 5 $\times$ 5 ECAL crystals and one HCAL readout unit, with an extension of 0.087 $\times$ 0.087 in the $\eta \times \phi$ directions. This constitutes 17 TTs in the η direction and 72 TTs in the ϕ direction for each half-barrel, totalling 1224 TTs, each tagged by a pair of indices ($i\eta$, $i\phi$) following the convention shown in Fig. 4.2. The endcap presents a more complex TT configuration, as the ECAL crystals in this region present different geometries, depicted in Fig. 4.3. The size and number of

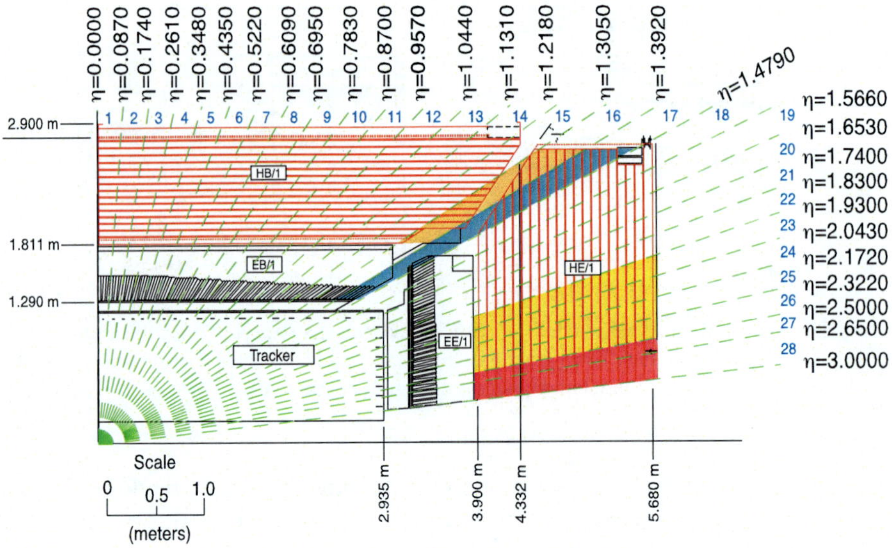

Fig. 4.2 Sketch of the trigger tower (TT) arrangement in the barrel and the endcaps, represented in the $r - z$ plane. Each TT groups inputs from the ECAL and HCAL subdetectors. The regions are tagged by their corresponding $i\eta$ coordinate [11]

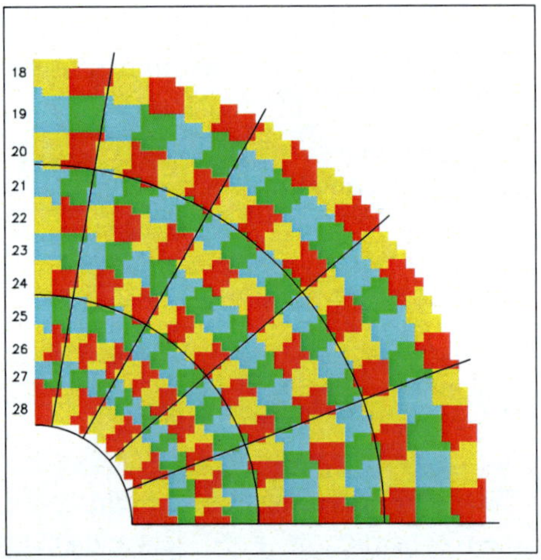

Fig. 4.3 Sketch of the trigger tower (TT) arrangement in the endcaps, represented in the $x - y$ plane. The regions of same colour represent one TT, tagged by their $i\eta$ coordinate. Each small square within the TT corresponds to an ECAL crystal [11]

ECAL crystals contained in each TT increase with η, matching the corresponding readout unit of the HCAL, summing a total of 11×72 in the $\eta \times \phi$ directions; this yields to 792 TTs in the endcap, again identified by $(i\eta, i\phi)$ coordinates. The inputs from the HF have a coarser granularity, with 4 TTs in the η direction and 18 TTs in the ϕ direction, amounting to a total of 72 TTs.

For each TT, the projection onto the transverse plane of the momentum vector originating in the detector centre and pointing to the calorimeter cells is computed. The trigger primitives are transmitted to the Layer-1 of the calorimeter trigger (see

Fig. 3.17), that performs a series of preprocessing operations before sending the data to Layer-2. In particular, a calibration procedure is applied to the TTs, separately for the ECAL and for the HCAL, depending on the η position and on E_T itself. The ECAL and HCAL energies are summed and the trigger primitives are sorted. The whole detector information is sent to one of the 9 processing nodes of the Layer-2 (see Fig. 3.17), where the reconstruction algorithms of the different calorimetric objects (L1 e/γ, L1 τ_h, L1 jet, L1 E_T^{miss} and other sums) are implemented.

4.1.2 Algorithm Steps

The L1 τ_h trigger at Layer-2 receives the calorimeter inputs from Layer-1 and reconstructs the τ_h candidates in 4 steps:

1. collection of the τ_h energy deposits in the calorimeters by **clustering** the pertinent TTs, thus defining the "clusters",
2. potential **merging** of two clusters into a single L1 τ_h candidate,
3. **calibration** of the clusters to improve the τ_h energy scale and resolution,
4. application of **isolation** criteria to reject the QCD-induced jet background.

The description of the algorithm reported here corresponds to the version deployed online during 2017 and 2018 data-taking. It is implemented in the MP7 cards of the Layer-2 of the trigger system, embedded within FPGAs, whose internal connections are configured with a set of instructions encoded in VHDL (Very high speed integrated circuits Hardware Description Language). In order to evaluate and optimize the algorithm, an *emulator* is used, i.e. a C++ code that replicates the VHDL implementation of the firmware. The emulator is built to reproduce the online algorithms at 100% accuracy; it is also used to monitor the correct functioning of the trigger during data-taking and to simulate the trigger reconstruction.

Clustering

The first step of the algorithm consists in identifying the energy deposits of the τ_h decay products in the detector. The charged pions expected in the τ_h decays interact with the detector material and can deposit energy both in the ECAL and in the HCAL subdetectors. The neutral pions decay into γ pairs; they can either arrive directly to the ECAL surface or convert into electrons in their passage through the tracker material. Hence, the τ_h energy deposits are found both in the ECAL and HCAL, and within elongated areas in the ϕ direction due to the bremsstrahlung radiation and the separation of the charged hadrons due to the magnetic field. The identification of the τ_h footprint is very challenging in the presence of pileup, which typically produces low-energy and diffuse deposits in the calorimeter that are entangled with the genuine τ_h signature. Being the τ_h a collimated object, the τ_h reconstruction follows a dynamic clustering procedure where only small groups of TTs are selected to minimize the impact of pileup.

The dynamic clustering procedure is sketched in Fig. 4.4. The L1 τ_h candidate is initiated by the so-called *seed*, a local maximum of energy in a region extending over

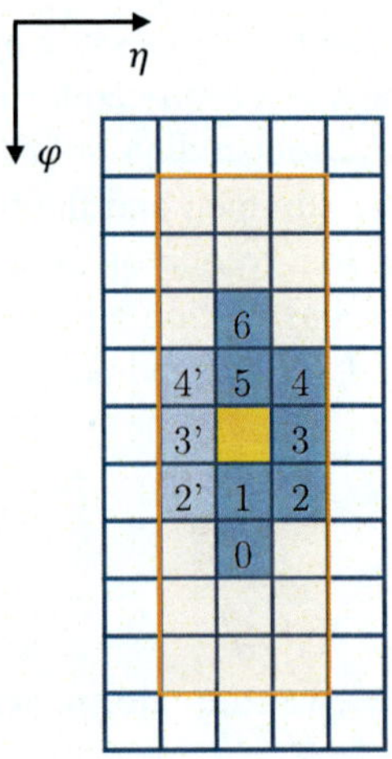

Fig. 4.4 Schematic representation of the clustering step of the L1 τ_h algorithm, where each square denotes a TT. The algorithm is initiated by a seed (in yellow) containing the maximum energy deposit in the orange area. Any of the 10 TTs indicated in light and dark blue can be added to the seed to form a protocluster. Either the primed or the non-primed TTs (2, 3, 4) are removed from the protocluster during the lateral trimming [12]

3 TTs along the η direction and 9 TTs along the ϕ direction. Valid seed towers have $E_\mathrm{T} \geq 2$ GeV and are either in the barrel or in the endcaps, but not in the forward calorimeter. The 8 TTs within one unit in $i\eta$ and $i\phi$ of the seed are grouped to the seed if they satisfy $E_\mathrm{T} \geq 1$ GeV, forming the so-called *protocluster*. The two TT with the same $i\eta$ position as the seed but at 2 units distance in $i\phi$ can also be added to the protocluster provided the TT in between is part of the protocluster. As the τ_h typically presents a highly asymmetric deposit in η, lateral trimming is performed: the outermost side of the protocluster presenting the lowest E_T sum is removed, sacrificing only a small fraction of the total energy deposited by the candidate. In the most general case, the position assigned to the cluster corresponds to the centre of the seed TT; an offset of 1/4 of the TT size in the η and/or ϕ direction can be applied depending on the cluster shape to improve the precision. The shift in ϕ is done towards the highest energetic part of the cluster, while the shift in η is done towards the side of the active TTs remaining after the trimming.

Merging

The clustering procedure previously described is very efficient at collecting the typical energy deposits of a single hadron issued from the τ decay, but has an intrinsic limitation in decays containing multiple hadrons, in particular the 3-prongs decays. These decays are more widely spread in the detector due to the opening angle between the particles at the vertex, sometimes amplified by the magnetic field. The effect is illustrated in Fig. 4.5, where the typical deposits for the 1-prong, 1-prong+π^0 and 3-prongs decay modes are compared. Since the clustering procedure is shared with the L1 e/γ algorithm, the enlargement of the area considered for the clustering cannot be envisioned, as it would make the algorithms more sensitive to pileup. To

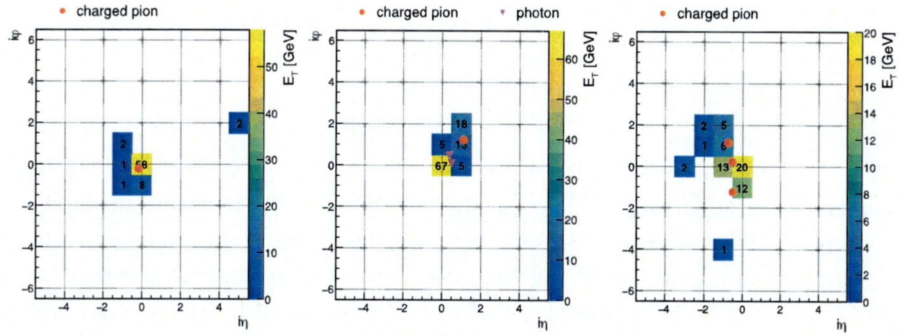

Fig. 4.5 Examples of energy deposits in the calorimeters originating from the 1-prong (left), 1-prong+π^0 (middle) and 3-prongs (right) decays of the τ lepton. The generator-level charged pion and photon are overlaid. Decays with multiple objects present wider energy depositions collected during the merging step of the L1 τ_h algorithm [12]

recover the fraction of energy loss during clustering, the L1 τ_h algorithm identifies and reconstructs a *secondary cluster* and merges it with the main one into a single τ_h candidate.

Secondary clusters are constructed in the same way as the main ones, but considering seeds within a smaller window of 3×3 TTs. A secondary cluster will be associated to the main one if its seed is found in the upper or lower ϕ region with respect to the main seed, in one of the eight positions sketched in Fig. 4.6. If multiple secondary clusters are found, the one with the highest energy is merged to the main cluster. The average fraction of merged L1 τ_h candidates is 15%; this value is largely dependent on the decay mode and the p_{T} of the τ_h, since the decay products are more

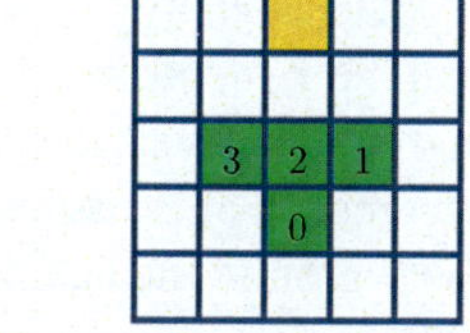

Fig. 4.6 Schematic representation of the merging step of the L1 τ_h algorithm, where each square denotes a TT. Secondary clusters from multiple prong τ_h decays are merged to the main one, whose se ed is indicated in yellow, if their seed is found in one of the 8 positions indicated in green, located in the upper or lower ϕ direction as expected from the bending of the prong trajectory by the magnetic field [12]

collimated at high energies. The low p_T τ_h (< 40 GeV) 3-prongs decays from are merged in 35% of the cases, while high-p_T τ_h (> 80 GeV) 1-prong and 1-prong+π^0 decays are merged less than 10% of the times [12].

Energy calibration
The sum of the energies of the TTs contained in the L1 τ_h candidate is denoted as the *raw energy* (E_T^{raw}). Despite the energy of the single TTs being already calibrated at Layer-1, the raw energy does not reflect precisely the real energy carried by the τ_h for several reasons. First, the clustering procedure induces losses in the energy recollection, especially in the forward region, where the TTs have a more complex geometry. Second, the merging procedure can induce systematic differences between merged and unmerged candidates that must be accounted for and corrected. Third, several detector effects can deteriorate the resolution of the L1 τ_h candidate, such as the non-linearities in the response of the calorimeters across the energy and η regions and/or the differences in the ECAL and HCAL energy response. To correct for these effects, a second energy calibration specific to the τ_h candidate is applied.

The energy calibration consists in correcting the raw energy by a multiplicative factor c, which depends on the value of the raw energy itself and the $i\eta$ coordinate. The calibration takes into account whether the candidate is a merged cluster (represented by the flag i^{merged}) and whether the seed presents deposits in the ECAL detector (represented by the flag i^{EM}). The calibrated energy is expressed as

$$E_T = c\,(E_T^{\text{raw}}, i\eta, i^{\text{merged}}, i^{\text{EM}}) \cdot E_T^{\text{raw}} \,. \tag{4.1}$$

The calibration constants c are derived with a BDT in exclusive intervals of the input variables. The ratio of the uncalibrated energy to the offline reconstructed momentum $E_T^{\text{raw}}/p_T^{\text{offline}}$ is used as the target in the training; the inverse of the predicted value is applied as a correction. The effect of the calibration is illustrated in Fig. 4.7, which shows the ratios of the L1 raw and calibrated energies to the offline reconstructed p_T. The calibrated candidates present a response centred in 1, with a relative resolution of ~20%. These calibration factors have been re-derived for 2018 data-taking; the method followed is explained in Sect. 4.3.1.

Isolation
Jets originating from quarks and gluons constitute the main background to the τ_h reconstruction and selection. As the jet energy deposits tend to spread out and have a higher particle multiplicity than the τ_h deposits (see Fig. 4.1), an adequate handle to identify and reject the jet background consists in evaluating the isolation of the L1 τ_h candidate. This is done at L1 by means of the so-called *isolation energy* E_T^{iso}, which evaluates the calorimeter activity around the candidate, as sketched in Fig. 4.8. It is defined as the difference between the energy sum in a region of TTs extending 6×9 in $i\eta \times i\phi$ and the uncalibrated energy E_T^{raw} of the τ_h candidate, i.e.

$$E_T^{\text{iso}} = E_T^{6\times 9} - E_T^{\text{raw}} \,. \tag{4.2}$$

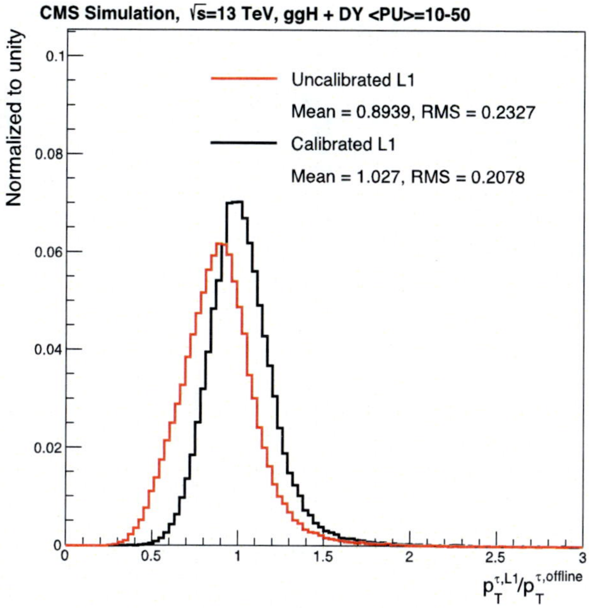

Fig. 4.7 Comparison of the energy response of the L1 τ_h candidate, defined as the ratio between the L1 E_T and the offline reconstructed p_T, for the uncalibrated and calibrated candidates, obtained from simulated ggH and DY events. The values of the mean and the RMS of the distributions are indicated

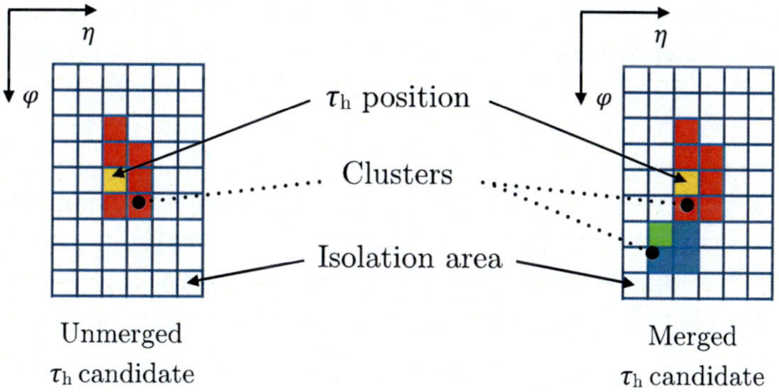

Fig. 4.8 Schematic representation of the region considered in the evaluation of the isolation energy (E_T^{iso}) of the L1 τ_h candidate, defined as the difference between the energy deposit in a 6×9 TT window and the energy deposit of the unmerged (left) or merged (right) L1 τ_h candidate [12]

The uncalibrated energy is used in the subtraction to ensure that homogeneous quantities are compared and that the isolation energy is not biased by the application of calibration constants.

A comparison of the isolation energy for the τ_h signal and for the jet background is presented in Fig. 4.9. As expected, larger isolation energies are observed for the jet background, meaning the associated hadronic activity has a wider extension.

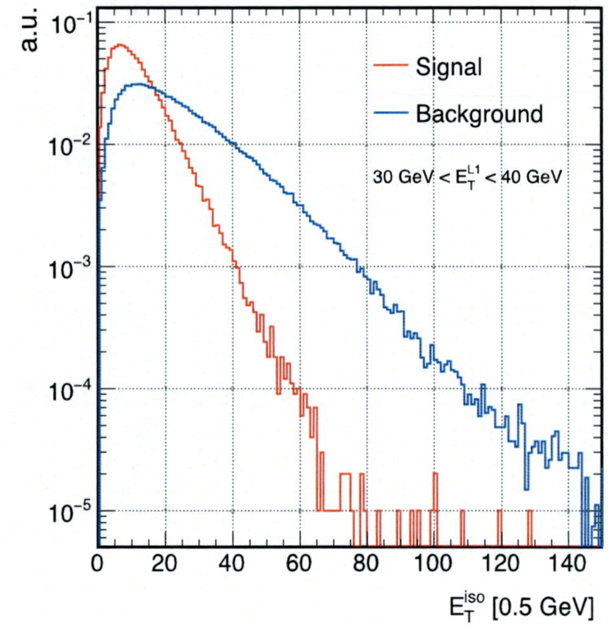

Fig. 4.9 Distributions of the isolation energy ($E_\mathrm{T}^\mathrm{iso}$) of τ_h candidates (in red), obtained from VBF H $\rightarrow \tau\tau$ simulated events, and jet candidates (in blue), obtained from zero bias events recorded in 2016. The genuine τ_h presents a softer $E_\mathrm{T}^\mathrm{iso}$ distribution than the QCD-induced jets [12]

Therefore, by requiring the L1 τ_h candidate to be isolated, the jet background can be rejected. A L1 τ_h candidate is considered isolated when its isolation energy does not exceed a certain threshold ξ, i.e. $E_\mathrm{T}^\mathrm{iso} < \xi$. The choice of the threshold value is a trade-off between the signal efficiency and the background rejection. Given the imperfect energy collection of the candidate during clustering, the value of the threshold depends on the energy of the candidate. It is also η-dependent due to the different TT geometries in the barrel and the endcaps. The dependence on the pileup is estimated by means of a L1 pileup estimator, defined as the number of active TTs (n_TT) in the central region of the detector ($|i\eta| \leq 4$), whose value is on average proportional to the number of primary vertices reconstructed in the event [12]. In conclusion, the isolation threshold ξ is expressed as

$$\xi \equiv \xi(E_\mathrm{T}^\mathrm{raw}, i\eta, n_\mathrm{TT}) \ . \tag{4.3}$$

The values of the threshold are derived in simulated events in separate intervals of the input variables. In each of these intervals, the value of ξ is chosen according to a targeted signal efficiency, which is flat across $i\eta$ and n_TT but evolves as a function of $E_\mathrm{T}^\mathrm{raw}$. As the background events have a steeply falling E_T spectrum, larger background rejection is achieved by applying a tighter isolation threshold at low energies, progressively relaxed until reaching a 100% efficiency. This is the so-called *relaxation* of the isolation, defined as

Table 4.1 Parameters used in the relaxation of the isolation of the L1 τ_h algorithm in different eras of the Run 2 operations. The targeted efficiencies for each isolation scheme follows Eq. 4.4 (see text)

Year	Era	Isolation option	ε_0 [%]	A [GeV]	B [GeV]
2016	B–G	21	80	20	50
	H	5	70	25	50
2017	B–F	22	70	25	70
2018	A–B	22	70	25	70
	C–D	32	84	30	75

$$\varepsilon = \begin{cases} \varepsilon_0 & \text{if } E_T^{raw} < A \\ \varepsilon_0 + \dfrac{1-\varepsilon_0}{B-A}(E_T^{raw} - A) & \text{if } E_T^{raw} \geq A \text{ and } E_T^{raw} < B \\ 1 & \text{if } E_T^{raw} \geq B \end{cases}, \tag{4.4}$$

A constant efficiency ε_0 is targeted below an energy threshold A, linearly increased until the maximum efficiency is reached at an energy threshold B, after which the efficiency is maintained. The values of the parameters ε_0, A and B constitute the so-called *isolation option*; they are optimized according to the expected pileup and instantaneous luminosity profiles. The parameters used in the Run 2 operations are shown in Table 4.1: increasingly tight isolation schemes were deployed as the experimental conditions turned harsher.

4.1.3 Main Level-1 τ_h Seeds

From all the reconstructed L1 τ_h objects, the 12 candidates with the highest E_T (regardless if they are isolated or not) are transmitted from the Layer-2 of the calorimeter trigger to the μGT. Likewise, the μGT receives the L1 e/γ, L1 jet, L1 sums and L1 muon candidates. The decision of accepting or rejecting an event is conducted based on requirements on the multiplicity and kinematic properties of these L1 objects, encoded into the L1 seeds. As far as the L1 τ_h are concerned, the seeds used during Run 2 are classified in three groups:

- **Single-τ_h triggers** require the presence of one L1 τ_h candidate satisfying certain energy and position criteria, mostly related to E_T and η. They are denoted as `L1_SingleTauXerY`, with X the E_T threshold and Y the maximum $|\eta|$ value allowed (generally $|\eta| < 2.1$). If the candidate is required to be isolated, it is indicated as `Iso` in the seed name. These triggers profit from a wide physics acceptance and can be used in the search of different physics signals, such as the boosted Higgs boson predicted by supersymmetric theories. The requirement of a single object results however in a higher trigger rate, meaning tighter thresholds of ~120 GeV have to be applied on the E_T of the candidate.

- **Double-τ_h triggers** require the simultaneous presence of two L1 τ_h candidates, both satisfying the same energy and position criteria. To be able to loosen the E_T threshold on these seeds, it is usually required that both L1 τ_h candidates are isolated. These seeds are denoted as `L1_DoubleIsoTauXerY`, following the same convention as the single-τ_h seeds. This type of triggers target more specific signal topologies, like the $H \rightarrow \tau_h\tau_h$ decays, thus they are most extensively used in analyses. The simultaneous presence of two τ_h allows for the trigger thresholds to be loosened down to ~32 GeV.
- **Cross τ_h+X triggers** require the simultaneous presence of one L1 τ_h (typically isolated) together with other L1 object(s) of different kind. These seeds include the tags `EG`, `Mu`, `Jet` or `ETM` if they are associated to L1 e/γ, L1 muon, L1 jet or L1 E_T^{miss}, respectively; they follow the previous conventions on the E_T and η requirements. In some cases, the seeds encode requirements on the angular separations between the objects, such as ΔR (`dR`). Similarly to the double-τ_h triggers, they target specific physics signals with high efficiency and a low trigger rate. The $H \rightarrow \tau_h e$ ($H \rightarrow \tau_h\mu$) decays are selected with e/γ+τ_h (μ+τ_h) cross triggers, while charged Higgs boson searches make use of E_T^{miss}+τ_h triggers. The thresholds on the e/γ+τ_h (μ+τ_h) triggers are low, typically ~22 (~18) GeV for the electron/γ (muon) leg and ~26 GeV for the τ_h leg.

The performance of some benchmark seeds used in 2017 is shown in Sect. 4.2.3. For trigger objects with high background contamination, like in the case of the L1 τ_h, the double-τ_h and lepton+τ_h cross triggers are the most efficient seeds. In the case of the $t\bar{t}H$ multilepton analysis presented in this thesis, these triggers were used to collect the data analyzed in the final states containing two τ_h and containing one τ_h plus one lepton, respectively, as described in Chap. 6. Novel L1 τ_h seeds that enhance the acceptance to the $t\bar{t}H$, VBF, ggH and double Higgs boson productions have been studied as part of this thesis work; they are presented in Sect. 4.4.

4.2 Measurement of the Performance in 2017 Data

In 2017, the LHC achieved a record instantaneous luminosity of 2.1×10^{34} cm^{-2}s^{-1}, doubling the peak value achieved during 2016 operations. Maximum pileup values of more than 60 were delivered at times, while in 2016 they rarely exceeded 50. Additionally, the LHC suffered frequent beam dumps in 2017, caused by electron clouds generated by tightly packed bunches interacting with frozen gas in the beam pipe. To mitigate the effect, the LHC moved to a special "8b4e" filling scheme in September 2017, where the standard 48 bunch trains were replaced by mini-trains of 8 filled bunches followed by 4 empty slots to suppress the formation of electron clouds. As this filling scheme allows a maximum of 1916 filled bunches in the LHC, the instantaneous luminosity was levelled to around 1.55×10^{34} cm^{-2}s^{-1} to keep the pileup under 60. Besides the evolving LHC running conditions, the trigger had to operate on an ageing detector whose response degrades over time, in particular in the forward region, where the radiation damage is significant. The robustness of the algorithm in these complicated conditions was measured with the data collected in 2017, an essential operation to verify that the expected performance had been achieved and to monitor the behaviour of the algorithm during CMS operations. The results presented here are derived using the fill dataset collected in 2017 at $\sqrt{s} = 13$ TeV, corresponding to an integrated luminosity of 40.9 fb^{-1}. The resolution in position and energy is presented first, followed by the trigger efficiency and rates.

4.2.1 Resolution in Position and Energy

Upon evaluation of the position resolution of the L1 τ_h candidates, the η and ϕ coordinates of the L1 object, computed from the TT seed explained in Sect. 4.1.2, are compared to the associated offline reconstructed τ_h. Given the effect of the magnetic field on the trajectory of the charged hadrons, the point of emission of the offline τ_h at its production vertex does not match the point where the object impacts the ECAL surface. To take this into account, the offline τ_h position is defined as the energy-weighted average of the position of its signal components at the entrance of the ECAL subdetector. Offline τ_h are considered only if they are geometrically matched to at least one L1 τ_h candidate within $\Delta R < 0.5$. In the case where multiple L1 τ_h candidates are found, the one closest to the offline reconstructed τ_h is selected, so that at most one L1 τ_h candidate is associated to each offline τ_h. The difference in η and ϕ direction between the inclusive L1 τ_h seeds (with no isolation requirement) and their offline reconstructed counterparts is shown in Fig. 4.10, separately for the barrel and for the endcaps. A full width at half maximum resolution of about 0.08 rad in η and 0.1 rad in ϕ is observed. The worse resolution for the latter is explained by the bending of the charged particles in the magnetic field; it causes a spread of the energy deposits which distorts the accuracy of the estimation of the energy barycentre as the seed TT position.

Likewise, the calibrated energy E_T of the L1 τ_h candidate is compared to the p_T of its offline reconstructed counterpart. The energy response, namely the ratio between the two, is presented in Fig. 4.11, separately for the barrel and the endcaps. In both detector regions, a response centred in 1 is found, with an energy resolution of $\sim$20%. The evolution of the energy resolution, defined as the RMS over the mean of the response distribution, is shown in Fig. 4.12a as a function of the offline reconstructed p_T. Owing to the improved resolution of the ECAL and HCAL subdetectors at high energies, the resolution of the τ_h is likewise benefited. Similarly, the resolution as a function of the offline η position is presented in Fig. 4.12b. Stability in the energy resolution is observed across the whole detector, slightly degraded in the transition region between the barrel and the endcaps. A worse resolution is observed in the endcaps (19–27%) than in the barrel (14–25%) for offline p_T values from 30 to

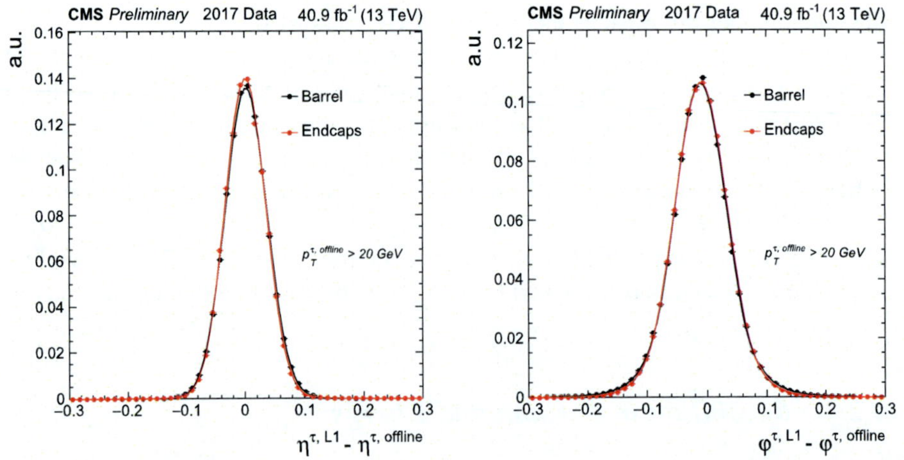

(a) η position (b) ϕ position

Fig. 4.10 Difference in **a** η position and **b** ϕ position between the inclusive L1 τ_h candidates and the offline reconstructed τ_h, for the barrel ($|\eta| < 1.305$) and for the endcaps ($|\eta| > 1.479$). The offline τ_h position is computed at the entrance of the ECAL subdetector. The fits to the data are performed with a symmetric two-sided Crystal Ball function

Fig. 4.11 Energy response, defined as the ratio between the L1 τ_h E_T and the offline reconstructed τ_h p_T, of the inclusive L1 τ_h candidates in the barrel ($|\eta| < 1.305$) and in the endcaps ($|\eta| > 1.479$)

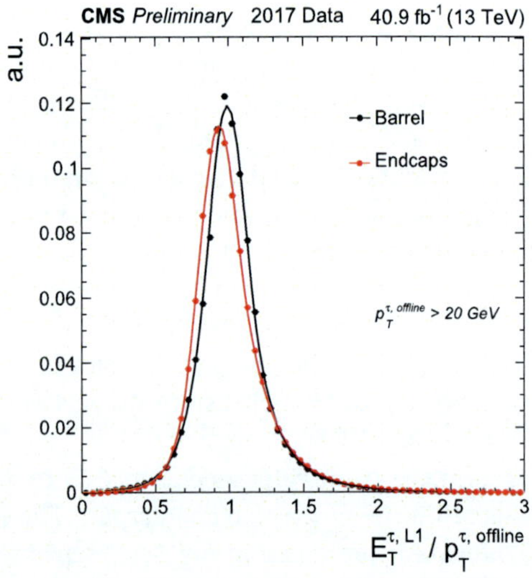

110 GeV due to the inhomogeneous size of the TTs, the higher pileup density and the radiation damage in the forward region of the detector.

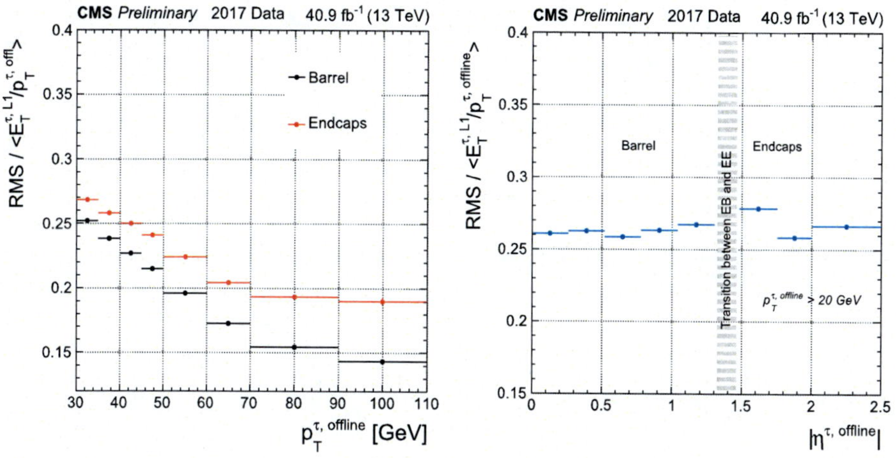

(a) (b)

Fig. 4.12 Energy resolution, defined as the RMS over the mean of the response distribution, of the inclusive L1 τ_h candidates as a function of the **a** p_{T} and **b** $|\eta|$ position of the offline reconstructed τ_h, separately for the barrel ($|\eta| < 1.305$) and for the endcaps ($|\eta| > 1.479$)

4.2.2 *Selection Efficiency*

The efficiency of the L1 τ_h trigger is measured with τ_h candidates from a sample of $\mathrm{Z} \rightarrow \tau\tau \rightarrow \mu\nu_\mu\nu_\tau\ \tau_h\nu_\tau$ selected with the *tag-and-probe* method. The method is particularly suited for performance measurements, as the clean signature provided by the muon is exploited in the trigger and offline selections and the kinematics of the $\mathrm{Z} \rightarrow \tau\tau$ decays are well known. The method consists in requiring the presence of a muon (the *tag*), which satisfies the basic identification and isolation criteria of the $\mathrm{H} \rightarrow \tau\tau$ analysis [2]. It must fire the single-muon HLT path requiring an isolated muon of $p_{\mathrm{T}} > 24$ GeV and $|\eta| < 2.1$, as well as an offline selection of $p_{\mathrm{T}} > 27$ GeV and $|\eta| < 2.1$. The muon must be found in association with an offline reconstructed τ_h candidate (the *probe*) of opposite charge, which fulfils the conditions $p_{\mathrm{T}} > 20$ GeV and $|\eta| < 2.1$. No trigger requirement is applied on the τ_h candidate; the data sample obtained this way is unbiased and can be used to measure the L1 τ_h selection efficiency. The kinematics of the selected τ_h and μ system must be compatible with the Z boson decay: the invariant mass is required to be in the interval $40 < m(\mu, \tau_h) < 80$ GeV, corresponding to the position of the peak of the distribution in this variable for $\mathrm{Z} \rightarrow \tau\tau$ events.

The contribution of the background events from $\mathrm{Z} \rightarrow \mu\mu$ is suppressed by vetoing events with more than one muon, while the contribution from $\mathrm{t\bar{t}}$ processes is suppressed by rejecting events with at least one jet passing the medium working point of the *b*-tagging discriminator. The W+jets production is reduced by applying the requirement on the transverse mass of the muon of $m_{\mathrm{T}}^{\mu} < 30$ GeV, defined as

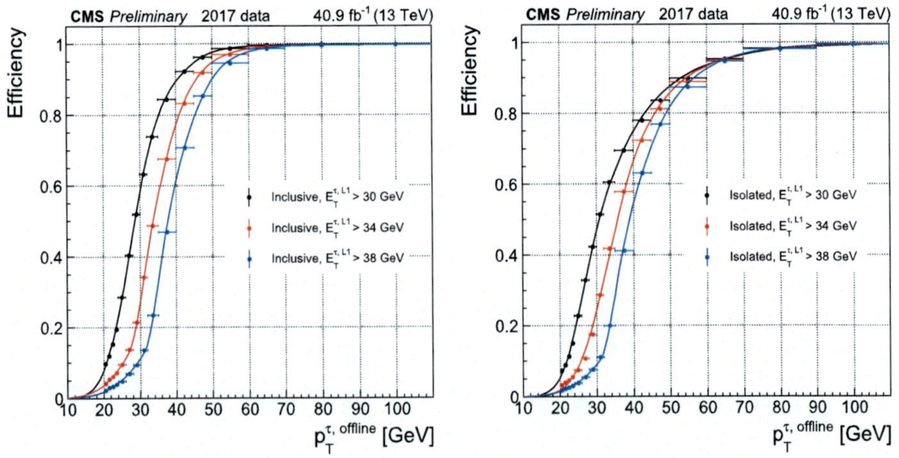

(a) Inclusive (b) Isolated

Fig. 4.13 Efficiency as a function of the offline reconstructed τ_h p_T for **a** inclusive and **b** isolated L1 τ_h candidates, for different L1 E_T benchmark thresholds. The fits to the data are performed with a cumulative distribution of a Crystal Ball function, convolved with the linear relaxation function in the case of isolated L1 τ_h candidates

$$m_T^\mu = \sqrt{(p_T^\mu + p_T^{\text{miss}})^2 - (\vec{p}_T^\mu + \vec{p}_T^{\text{miss}})^2}\,. \tag{4.5}$$

After these requirements, a residual fraction of ~20% events from background processes is present in the sample; they mainly originate from QCD-induced jets misidentified as a τ_h, which are mostly relevant in the low p_T region [12].

The measured selection efficiency of a single inclusive or isolated τ_h candidate as a function of the offline reconstructed p_T is presented in Fig. 4.13. Offline τ_h are matched to one and only one L1 τ_h with a $\Delta R < 0.5$ requirement. The efficiency is shown for the different L1 E_T thresholds, typically used in the isolated double-τ_h seeds during 2017 data-taking. The excellent response resolution previously shown ensures a sharp rise of the efficiency. It reaches a flat plateau of 100% efficiency also in presence of the isolation criterion, as a result of the relaxation of the isolation at high E_T. A 90% efficiency is achieved for inclusive (isolated) L1 τ_h at an offline p_T of 46 (55) GeV for a L1 threshold of 34 GeV, well adapted to physics analyses. The efficiencies measured in data are compared to the ones measured with simulated events in Fig. 4.14 for a L1 τ_h threshold of 30 GeV applied both to inclusive and isolated L1 τ_h candidates. A VBF H $\rightarrow \tau\tau$ Monte Carlo sample with a flat pileup distribution between 28 and 62 is used for this purpose. Good agreement is observed between the data and the simulation, with a slightly higher efficiency observed for isolated L1 τ_h in data; it is ascribed to small differences in the simulation of the collision conditions and particle reconstructions, typically corrected for with scale factors at analysis level.

The measured selection efficiency of a single inclusive τ_h candidate as a function of the η position in the detector is shown in Fig. 4.15a for different benchmark L1

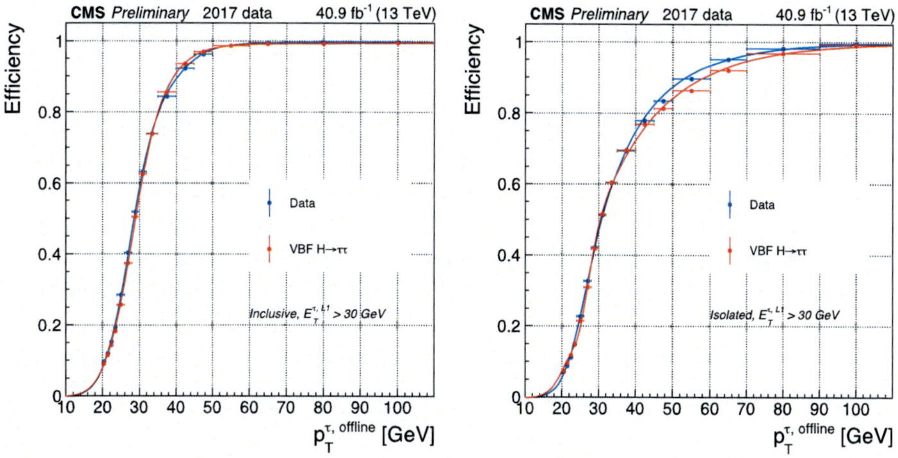

(a) Inclusive (b) Isolated

Fig. 4.14 Comparison of the efficiency measured in data and in VBF $H \rightarrow \tau\tau$ simulated events as a function of the offline reconstructed τ_h p_T for **a** inclusive and **b** isolated L1 τ_h candidates, for a L1 E_T threshold of 30 GeV. The fits to the data and the simulation are performed with a cumulative distribution of a Crystal Ball function, convolved with the linear relaxation function in the case of isolated L1 τ_h candidates

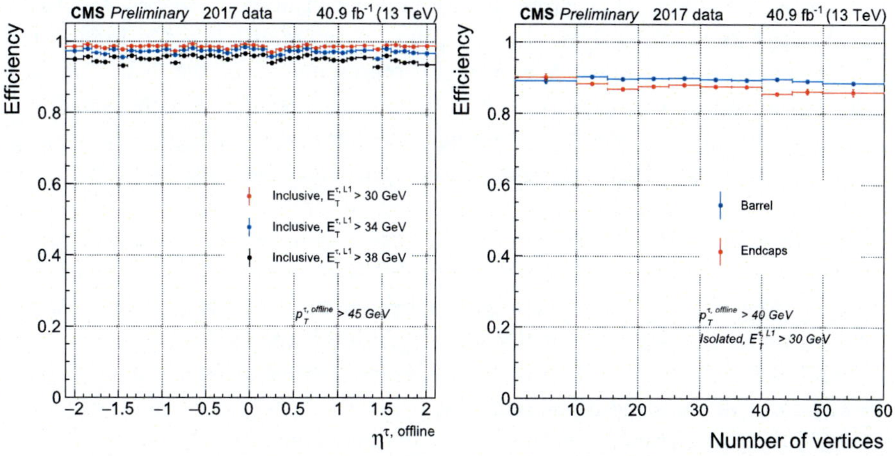

(a) η position (b) Number of vertices

Fig. 4.15 **a** Efficiency as a function of the η position of the offline reconstructed τ_h for different L1 E_T thresholds applied on inclusive L1 τ_h candidates. **b** Efficiency as a function of the number of reconstructed vertices in the event for a L1 E_T threshold of 30 GeV applied on isolated L1 τ_h candidates thresholds. Stability over the whole detector is observed, with selection efficiencies surpassing 95% for offline τ_h with $p_T > 45$ GeV, only slightly worsened in the barrel-endcap transition regions. The integrated L1 selection efficiency for isolated L1 τ_h seeds as a function of the number of vertices in the event is illustrated in Fig. 4.15b for a L1 threshold of 30 GeV, separately for the barrel and for the endcaps. The number of vertices is used as an estimator of the number of pileup collisions in the event. A stable efficiency larger than 85% is maintained across the range of pileup collisions delivered during 2017 operations considering a L1 threshold of 30 GeV and offline τ_h with p_T larger than 40 GeV. The excellent performance demonstrates that the dynamic clustering, the pileup estimator and the isolation scheme result in an algorithm capable of identifying and removing the softer contributions due to pileup while maintaining a high signal efficiency.

4.2.3 *Trigger Rates*

The main L1 seeds involving τ_h candidates during 2017 data-taking are shown in Table 4.2, together with their associated trigger rates. These correspond to the unprescaled seeds with the lowest thresholds involving a τ_h candidate present in the menu. The rates reported are measured for run number 305064 (14.10.2017 at 15:33:37 to 15.10.2017 at 04:53:01), with a recorded integrated luminosity of 383.0 pb^{-1} and an average pileup of 40, representative of the 2017 conditions. The quoted rates correspond to the maximum values: they are measured at the beginning of the run, when the instantaneous luminosity is higher, amounting to 1.4×10^{34} $cm^{-2}s^{-1}$ in this case. The rate of the double-τ_h seed (`L1_DoubleIsoTau32er2p1`), extensively used in physics analysis involving Higgs boson searches, is shown in Fig. 4.16 as a function of the average pileup. The rate is shown for several consecutive runs (305059 to 305064), with pileup values ranging from 20 to 60. As expected, a linear behaviour of the rate is observed, proving once again the resilience of the algorithm against pileup.

Table 4.2 shows how all the L1 τ_h triggers encode a restriction of $|\eta| < 2.1$ (`er2p1`) to the τ_h candidates. The restriction is applied to keep the rates under control, since higher levels of noise and background are typically present in the forward regions of the detector. Thus, a high efficiency is achieved in the more central part of the detector, where most of the physics signals are found. Besides, requiring the L1 τ_h candidates to be isolated (`Iso`) helps reduce the rate but also improves the purity of the τ_h objects used in the physics analyses. The rate reduction achieved in the double-τ_h trigger when applying the isolation requirement is illustrated in Fig. 4.17 for a run of 2017 with an average pileup value of 55. The dependence of the rate on the L1 threshold is evaluated on the so-called *zero bias* events. Such data collection has been selected using a set of triggers that are synchronized on a specific proton bunch crossing: it constitutes an unbiased sample of the calorimeter activity in *pp* collisions. For a typical trigger threshold of 30 GeV, a rate reduction of a

Table 4.2 List of the most used unprescaled triggers involving τ_h during 2017 data-taking. The table shows the name of the seed, the requirements and the maximum rate observed in a reference run of 2017, with a maximum instantaneous luminosity of 1.4×10^{34} cm^{-2}s^{-1} and an average pileup of 40

Level-1 seed	Maximum rate
`L1_SingleTau120er2p1`	1.8 kHz
One L1 τ_h with $E_T > 120$ GeV and $\lvert\eta\rvert < 2.1$	
`L1_DoubleIsoTau32er2p1`	7.9 kHz
Two isolated L1 τ_h with $E_T > 32$ GeV and $\lvert\eta\rvert < 2.1$	
`LooseIsoEG22er2p1_IsoTau26er2p1_dR_Min0p3`	2.1 kHz
One isolated L1 τ_h with $E_T > 26$ GeV and $\lvert\eta\rvert < 2.1$	
One isolated L1 e/γ with $E_T > 22$ GeV and $\lvert\eta\rvert < 2.1$	
$\Delta R(\tau_h, e/\gamma) > 0.3$	
`L1_Mu18er2p1_IsoTau26er2p1`	0.8 kHz
One isolated L1 τ_h with $E_T > 26$ GeV and $\lvert\eta\rvert < 2.1$	
One L1 muon with $E_T > 18$ GeV and $\lvert\eta\rvert < 2.1$	
`L1_IsoTau40er_ETM100`	1.7 kHz
One isolated L1 τ_h with $E_T > 40$ GeV and $\lvert\eta\rvert < 2.1$	
L1 $E_T^{miss} > 100$ GeV	

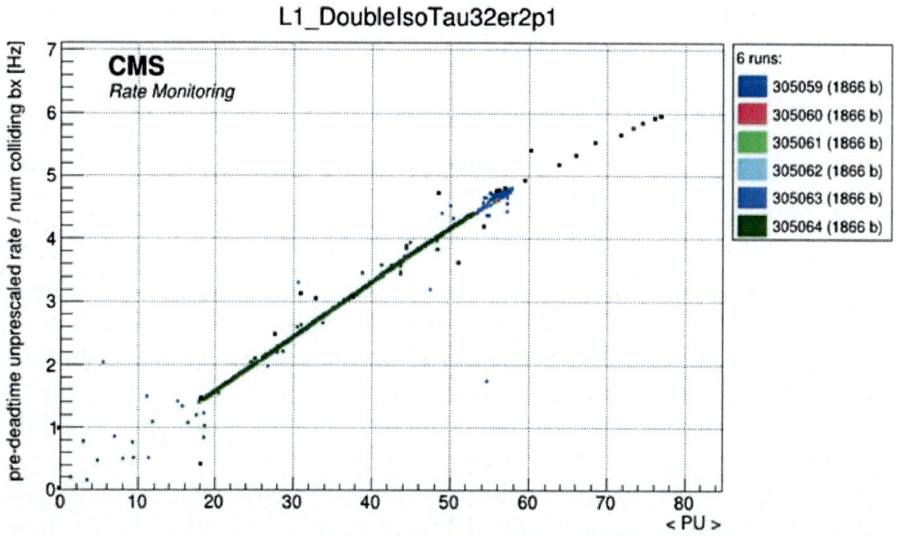

Fig. 4.16 Rate of the benchmark double-τ_h seed (`L1_DoubleIsoTau32er2p1`, see Table 4.2) as a function of the average pileup measured in the several runs during 2017 data-taking. The rate of the trigger has a linear dependence with pileup

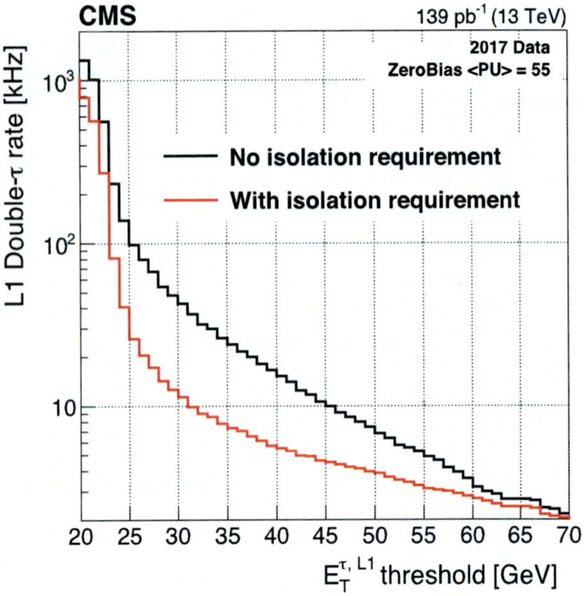

Fig. 4.17 Rate of the L1 double-τ_h trigger as a function of the E_T threshold on the L1 τ_h, for inclusive and isolated L1 τ_h candidates, measured with zero bias events recorded in a reference run of 2017. The L1 τ_h candidates have a restriction of $|\eta| < 2.1$

factor ~4 can be achieved when applying the isolation requirement to the double-τ_h candidates; this factor is crucial to fit the rate of the τ_h triggers in the values allowed by the CMS data acquisition system.

4.3 Optimization Towards 2018 Data-Taking and Performance

The results shown in Sect. 4.2 summarize the excellent performance of the L1 τ_h trigger in 2017. High efficiency and stability were observed in data, showing that the algorithm worked in the evolving experimental conditions of 2017 and was well reproduced by the simulation. The beam instabilities encountered during 2017 had been largely mitigated in 2018; the return to the nominal 48b filling scheme was possible. This scheme has the advantage that larger number of colliding bunches are used, hence providing higher instantaneous luminosity without increasing the pileup level. The instantaneous luminosity delivered in 2018 was higher than in most of 2017, with values of $\sim 2.0 \times 10^{34}$ cm^{-2}s^{-1}. An average pileup of ~40 was delivered in both years; however slightly higher peak values were reached in 2018, surpassing 70 at times. Owing to the changes in the running conditions, the L1 τ_h algorithm was optimized at the end of 2017 and during the first half of 2018. A highly efficient algorithm was crucial at this point, as the largest integrated luminosity was foreseen for 2018. The necessary changes in the algorithm were aimed at maintaining the adequate thresholds and rates achieved in 2017; they were mostly devoted to the derivation of new calibration constants and the optimization of isolation schemes adapted to the slightly higher pileup levels. The improvements of the algorithm are shown in this section; the resulting performance is summarized at the end.

4.3.1 Derivation of Calibration Factors

As mentioned in Sect. 4.1.1, the calibration of the L1 τ_h is a two-stage process: Layer-1 applies an inclusive calibration to all calorimeter objects, while Layer-2 after it performs a τ_h-specific calibration. Changes in the first stage of the calibration were proposed at the end of Run 2; they lead to consistent changes in the second stage, described herein. Likewise, the calibration process is typically revisited before each data-taking year, as the detector properties, readout methods and reconstruction algorithms evolve. In particular, the new readout electronics of the HCAL upgrade in 2017 had to be tested at this point. The changes and optimizations of the L1 τ_h calibration explained in this section were deployed online at the end of 2017 and were used throughout 2018 data-taking.

The Layer-1 of the calorimeter trigger applies calibration factors to the TT energies sent by the ECAL and the HCAL. These are derived a function of the TT E_{T} and $|i\eta|$, separately for the two subdetectors. The corrections are obtained from simulated events by comparing the p_{T} of the generator-level object (a photon for the ECAL and a pion for the HCAL) to the sum of the TT E_{T} in a square centred around the object. Within each interval in E_{T} and $|i\eta|$, the ratio of these two quantities is computed, and the value of the calibration scale factor is traditionally obtained from the average of the resulting distribution. The calibrations factors obtained this way are denoted as *mean* parameters. An alternative approach to the derivation of the calibration constants in Layer-1 was proposed at the end of 2017. Given the large tails observed in these distributions, illustrated in Fig. 4.18 for a specific bin, the derivation of the calibration factors obtained from the mode -and not the mean- of the distribution was proposed. The corrections obtained this way are denoted as *mode* parameters. As the L1 e/γ, jets and τ_h algorithms take as inputs the calibrated TTs from Layer-1, the impact of the new calibration methods on the resolution of these objects had to be tested to make an educated choice.

To compensate for the changes in the Layer-1 inputs, a new set of calibration factors for the L1 τ_h were derived on top. They were obtained and tested on simulated samples of VBF H $\rightarrow \tau\tau$ events, assuming a flat pileup distribution from 28 to 62. For consistency, the new sets of mean and mode parameters were compared to the one used during 2017 data-taking, which was mean-based. The performance of the three calibration methods is compared in Fig. 4.19. The figure shows the response for the calibrated L1 τ_h objects (after Layer-1 and Layer-2 calibrations) and the uncalibrated L1 τ_h (after Layer-1 calibrations alone). A relative resolution of 22% is achieved with the mean-based approach, similar to the one observed for 2017, slightly better than the 23% of the new mode-based approach. The fact that both parameters show a similar performance means that the Layer-2 calibrations can compensate for the differences in the Layer-1 calibrations. The slightly degraded performance

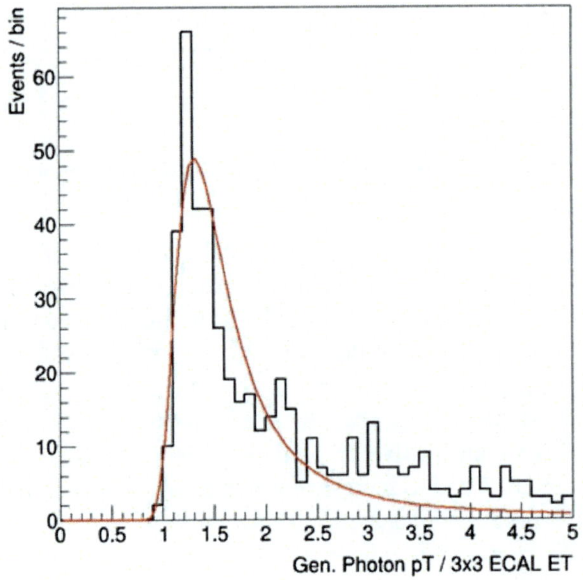

Fig. 4.18 Ratio between the generated photon p_T and the E_T deposited in a 3 × 3 TT square centred around the photon, evaluated in the ECAL for $|i\eta| = 18$ from single-photon simulated events. The mean (mode) based Layer-1 parameters are obtained from the arithmetic mean (mode) of the red curve, fitted with a Landau distribution convolved with a Gaussian

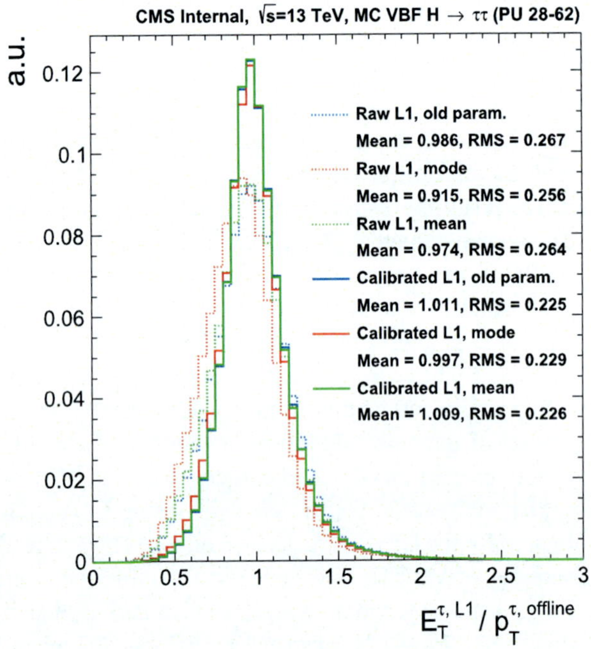

Fig. 4.19 Response of the L1 τ_h, defined as the ratio between the L1 E_T and the offline reconstructed τ_h p_T, obtained from simulated VBF H → $\tau\tau$ events, separately for the Layer-1 parameters online during 2017 ("old parameters") and the proposed parameters for 2018 data-taking, obtained from the mean-("mean") and mode-based ("mode") approaches (see text). The mean and the RMS for the uncalibrated and calibrated L1 τ_h candidates are indicated

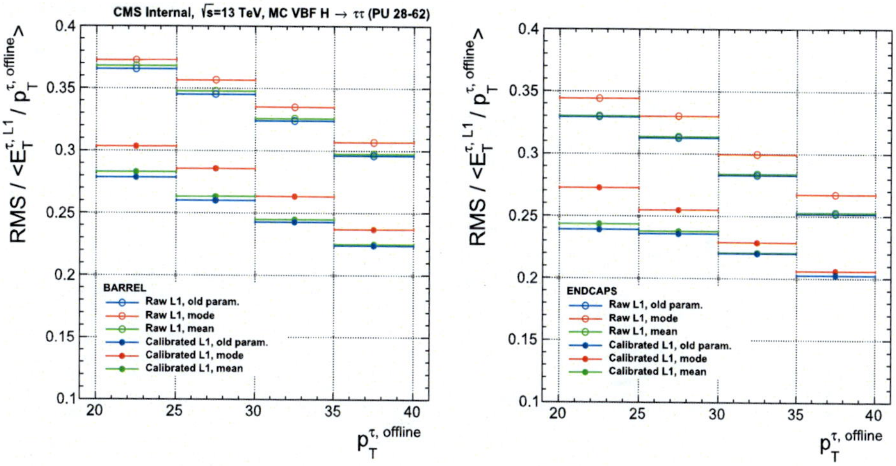

Fig. 4.20 Resolution of the L1 τ_h candidates, defined as the RMS over the mean of the response, as a function of the offline reconstructed τ_h p_T, obtained from simulated VBF H $\rightarrow \tau\tau$ events, separately for the Layer-1 parameters online during 2017 ("old parameters") and the proposed parameters for 2018 data-taking, obtained from the mean- ("mean") and mode-based ("mode") approaches (see text)

of the mode parameters becomes more evident when the resolution is evaluated in different bins of offline p_T, shown in Fig. 4.20 for offline p_T values ranging from 20 to 40 GeV. As the typical di-τ_h trigger thresholds range between 30 and 40 GeV, a precise calibration is instrumental in this region to enhance the overall trigger efficiency. It is noteworthy that the resolution for all parameter sets is better in the endcaps than in the barrel, contrary to what was observed with the data collected in 2017. The effect is a result of the improved resolution of the hadronic objects in the endcap after the upgrade of the readout electronics of the HCAL at the end of 2017 [13].

The di-τ_h rates associated to the new sets of corrections can be found in Fig. 4.21a as a function of the L1 threshold. They are obtained from zero bias events recorded in a high pileup run at the end of 2017, representative of the expected conditions for 2018. For a typical L1 seed with a 30 GeV threshold on two isolated τ_h, the mean parameters result in a ~10% higher rate than the mode parameters. As the allocated L1 τ_h bandwidth is constrained regardless of the calibration factors, the selection efficiency for the two parameter sets is evaluated at a fixed rate of 14 kHz, a value typically allocated to di-τ_h seeds. At this fixed rate, the mode-based approach requires a threshold of 27 GeV, namely 1 GeV looser than the mean-based approach. The L1 τ_h selection efficiencies for the corresponding thresholds are shown in Fig. 4.21b. They are computed in simulated VBF H $\rightarrow \tau\tau$ events; the L1 and offline τ_h are matched within a cone size of $\Delta R = 0.3$. A slightly higher efficiency is observed with the mean-based calibrations for an offline p_T larger than 30 GeV, as expected from the fact that it provides a better resolution. Regardless of the slight difference between the

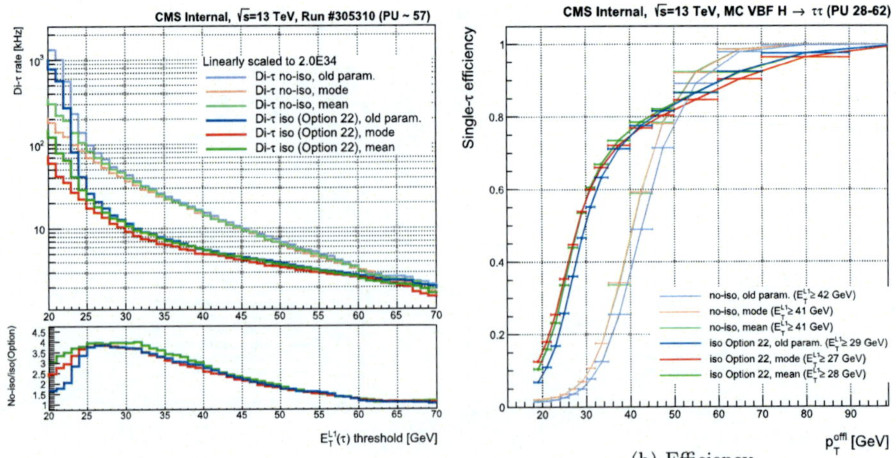

(a) Rate (b) Efficiency

Fig. 4.21 **a** Isolated and inclusive double-τ trigger rates as a function of the L1 τ_h threshold, obtained from zero bias events recorded in a high pileup run of 2017. **b** Single L1 τ_h efficiency as a function of the offline reconstructed τ_h p_{T}, obtained from simulated VBF H $\rightarrow$ $\tau\tau$ events for a fixed rate. Results are presented separately for the Layer-1 parameters online during 2017 ("old parameters") and the proposed parameters for 2018 data-taking, obtained from the mean- ("mean") and mode-based ("mode") approaches (see text)

two new approaches, both sets of corrections visibly enhance the selection efficiency with respect to the calibration factors online during 2017.

The results presented in this section show that the traditionally used mean-based corrections are slightly favoured by the L1 τ_h compared to the newly-proposed mode approach. They result in an improved energy resolution and efficiency, at the cost of a moderate increase of rate. Similar studies evaluating the performance of the L1 jets and L1 $E_{\mathrm{T}}^{\mathrm{miss}}$ were conducted. In this case, both methods showed a comparable performance; however the mode parameters were preferred, as the associated rates were lower, which is crucial for highly pileup-dependent objects. In the case of the L1 e/γ trigger, the mode-based approach showed an improved energy resolution, essential for multiple analyses at CMS where the electrons and photons are drivers of the physics sensitivity. Thus, the best compromise for all calorimeter objects was found with the mode-based Layer-1 corrections, which were deployed online during 2018 data-taking. The resulting performance with 2018 data is presented in Sect. 4.3.3.

4.3.2 Pileup Resilient Isolation

The evaluation of the τ_h isolation towards an optimal jet background rejection is highly dependent on the pileup and luminosity profiles delivered by the LHC during operation. As the number of simultaneous interactions at the core of CMS grows, the presence of calorimeter activity is enhanced: tighter isolation schemes must be applied to keep the trigger rate at an acceptable level and make sure the signal purity is maintained. The relaxation of the isolation (see Eq. 4.4) is revisited before each data-taking year according to the foreseen LHC conditions; the evolution of the isolation parameters throughout Run 2 is shown in Table 4.1. The table shows how an increase in luminosity pushes the start of the relaxation to higher E_{T} values to maintain the nominal rates. For most of 2016 data-taking, when an instantaneous luminosity of $\sim 1.2 \times 10^{34}$ cm^{-2}s^{-1} was delivered, values of $\varepsilon_0 = 80\%$, $A = 20$ GeV and $B = 50$ GeV were used. For 2017, when the instantaneous luminosity was increased to $\sim 1.5 \times 10^{34}$ cm^{-2}s^{-1}, the isolation criterion was tightened to $\varepsilon_0 = 70\%$, $A =$ 25 GeV and $B = 80$ GeV. With instantaneous luminosities of $\sim 2 \times 10^{34}$ cm^{-2}s^{-1} foreseen for 2018, a systematic study of possible relaxation options was performed. Complementary developments in the isolation algorithm were carried out to mitigate the effect of the upwards pileup distributions on the output rate of the algorithms. The studies and changes deployed for 2018 data-taking are summarized in the following.

Study of New Relaxation Schemes
Starting from the mode-based calibrated L1 τ_h candidates, different parameters for the relaxation of the isolation were studied in view of 2018 operations. The isolated τ_h efficiency resulting from the set of the most suitable options is presented in Fig. 4.22b; the corresponding relaxation parameters are outlined in Table 4.3. The L1 thresholds applied to derive these efficiencies are obtained for a fixed double-τ_h rate of 13 kHz, measured with zero bias events in a high pileup run of 2017 and shown in Fig. 4.22a. The effect of the different relaxation parameters is visible in the curves. Relaxation schemes that start from a lower initial efficiency, such as option 1, rise more sharply and reach the 100% plateau at a earlier stage, whereas isolation schemes like option 22, which start at a 70% efficiency, rise slower and reach the plateau at a later stage. Evidently, the phase space coverage is very different in both cases and, for our purposes, an enhanced acceptance at low p_{T} values is favoured. Thus, relaxation option 22, the one used for 2017 operations, was still kept online at the beginning of 2018.

New isolation schemes were likewise tested for the cross triggers involving a τ_h (in particular for the L1 $e/\gamma + \tau_h$ trigger) in view of developing a dedicated second isolation working point. The study is motivated by the lower thresholds applied on the τ_h leg ($\sim$26 GeV) in these triggers with respect to the thresholds of the di-τ_h trigger ($\sim$32 GeV). The cross triggers could benefit from looser isolation schemes than option 22; dedicated studies of different options were conducted to assess the potential improvement. The benchmark seed used for this purpose is `LooseIsoEG22er2p1_IsoTau26er2p1_dR_Min0p3`, with a typical rate of around 2 kHz. The threshold on the e/γ leg was kept at a value of 22 GeV, while the threshold on the τ_h leg remained under 30 GeV but was allowed to evolve with the different relaxation options, maintaining the rate at the expected value. The efficiency was evaluated in VBF H $\rightarrow \tau\tau$ simulated events.; the optimal performance was provided by option 4 ($\varepsilon_0 = 60\%$, $A = 25$ GeV, $B = 50$ GeV). However, the observed improvement with respect to option 22 was marginal ($\sim$2%). Besides, it

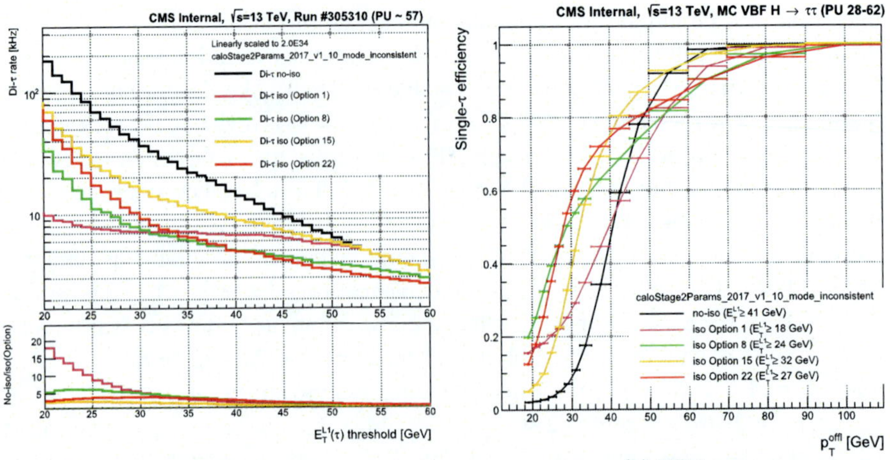

(a) Rate (b) Efficiency

Fig. 4.22 **a** L1 double-τ trigger rates as a function of the L1 τ_h threshold, obtained from zero bias events recorded in a high pileup run of 2017. **b** Single L1 τ_h efficiency as a function of the offline reconstructed τ_h p_T, obtained from simulated VBF H $\rightarrow \tau\tau$ events at a fixed rate. The results are shown for different sets of relaxation parameters in the isolation (see Table 4.3) and for the case where no isolation requirement is applied

Table 4.3 Parameters corresponding to different options considered for the relaxation of the isolation. ε_0 is the minimal efficiency, A is the start of the linear relaxation and B is the point where the 100% efficiency is reached (see Eq. 4.4). The associated L1 double-τ_h rates and single-τ_h efficiencies are be found in Fig. 4.22

Isolation option	ε_0 [%]	A [GeV]	B [GeV]
1	10	25	50
4	60	25	50
8	50	25	60
15	70	20	50
22	70	25	70

required a threshold on the τ_h leg of 29 GeV, namely 3 GeV higher than the one encoded in the seed of reference. Hence, isolation option 22 remained the generic scheme for all L1 τ_h triggers at the beginning of 2018.

Improvement of the Pileup Resilience

One of the most extensively used seeds in the CMS physics analysis is `L1_DoubleIsoTau32er2p1`; it requires two isolated L1 τ_h with $E_T > 32$ GeV within $|\eta| < 2.1$, mostly targeted at the H $\rightarrow \tau\tau$ decays. The rate of this seed is expected to be proportional to the average pileup, as represented in the linear fit of Fig. 4.23. Such behaviour was observed in 2016 and most of 2017 operations, when the pileup levels were rarely exceeding the value of 60. In preparation for 2018 running conditions, the linearity of the trigger rate in extreme experimental condi-

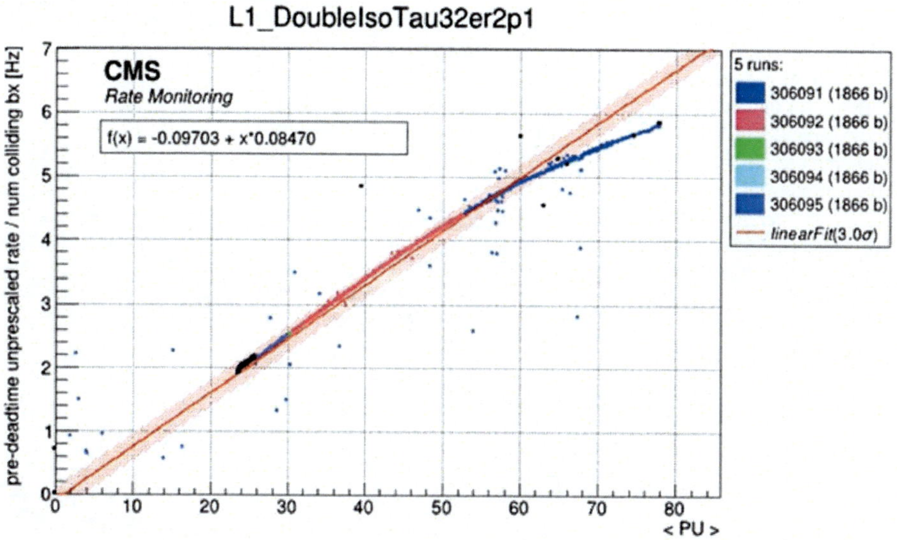

Fig. 4.23 Rate of the `L1_DoubleIsoTau32er2p1` trigger as a function of the average pileup for different high pileup runs during 2017 operations. The expected linear behaviour is overlaid in red. At the end of 2017, deviations from the linear behaviour at pileup levels larger than 60 were observed

tions was probed with high pileup runs ($\langle \text{PU} \rangle = 53 - 80$) that pushed the algorithms way beyond their nominal design. Among others, the rate of the L1 double-τ_h seed showed a significant deviation from the expected linear behaviour, with ~13% less rate than expected at $\langle \text{PU} \rangle \approx 78$, as illustrated in Fig. 4.23. This suboptimal behaviour would translate into a similar loss of double-τ_h events for physics analysis during 2018 operations; it was corrected with improvements in the pileup resilience of the algorithm, described in what follows.

The first change to the algorithm is related to the dependence of the isolation on the pileup profile. As described in Sect. 4.1.2, the upper cut on the isolation energy of the τ_h candidate is derived as a function of E_T^{raw}, $i\eta$ and the pileup estimator n_{TT}, in exclusive intervals of these input variables. The range and the boundaries of these intervals are partially driven by the statistical power of the simulated sample used to derive the isolation cuts. The saturated behaviour of the L1 double-τ_h seed was found to be related to the limited statistics at high values of n_{TT} in the MC sample, generated according to the nominal 2017 operations. Besides, a suboptimal binning of the n_{TT} variable was leading inaccurate estimations of the isolation cuts. As MC samples with higher pileup values became available, the number of n_{TT} bins in which the isolation cut was derived was doubled; finer binning was allocated to the uppermost bins, critical to protect the algorithm from upward fluctuation in pileup. The effect on the rate of the double-τ_h seed as a function of pileup is found in Fig. 4.24. The new n_{TT} scheme was able to restore ~15% of the rate for $\langle \text{PU} \rangle$ ~75, granting the desired physics acceptance in 2018 operations.

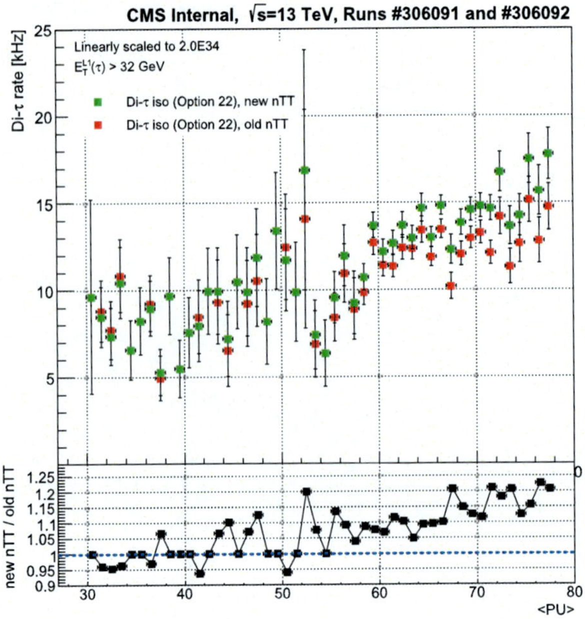

Fig. 4.24 Isolated L1 double-τ_h trigger rate as a function of the average pileup, for the old (red) and new (green) n_{TT} binning scheme used in the derivation of the isolation (see text), obtained from zero bias events recorded in high pileup runs in 2017. A ~15% rate is recovered at pileup levels larger than 60 with the new scheme (see Fig. 4.23)

During spring 2018, the isolation scheme was further revisited to withstand pileup levels of up to 90 simultaneous interactions. The isolation cut was relaxed as a function of n_{TT} to recover the lost selection efficiency at high pileup, similarly to what was already done in the algorithm as a function of E_T. The relaxation was applied in the high n_{TT} region, separately for the barrel and for the endcaps, based on the desired efficiency gain in each bin. The increase in rate was compensated with a tighter relaxation scheme as a function of E_T, tagged as option 31 ($\varepsilon_0 = 84\%$, $A = 30$ GeV and $B = 75$ GeV). The changes in the isolation algorithm resulted a few percent gain in efficiency at pileup values larger than 60, illustrated in Fig. 4.25a. It came at the cost of a slightly higher rate (~1 kHz), depicted in Fig. 4.25b, but with a more linear dependence on pileup. The isolation option 32, encoding the relaxation of the isolation at high n_{TT}, was deployed online in June 2018 and was maintained until the end of the year. It granted a stable response of the algorithm against pileup values peaking at 70 during the second half of 2018 data-taking; the excellent performance of the trigger throughout the year is demonstrated in the next section.

4.3.3 Performance with 2018 Data

The performance of the L1 τ_h trigger was measured with the complete dataset recorded in 2018, corresponding to an integrated luminosity of 57 fb^{-1}. The methods used are the same as the ones explained in Sect. 4.2 for 2017 data. A selected set of results is presented in this section; the complete performance measured as part of this

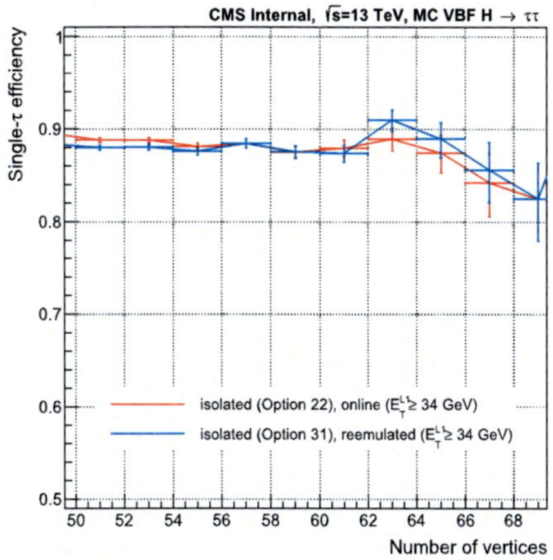

(a) Efficiency

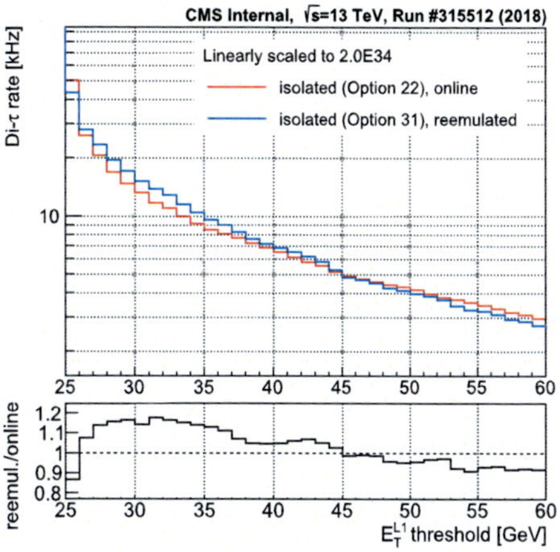

(b) Rate

Fig. 4.25 **a** Single L1 τ_h efficiency as a function of the number of reconstructed vertices, obtained from simulated VBF H $\rightarrow \tau\tau$ events. **b** L1 double-τ trigger rates as a function of the L1 τ_h threshold, obtained from zero bias events recorded in 2018. The results are shown for the isolation scheme used at the beginning of 2018 (option 22) and for the scheme adapted for high pileup deployed in spring 2018 (option 31), which includes the linear relaxation of the isolation as a function of n_{TT} (see text)

thesis work can be found in Ref. [10]. The energy response of the L1 τ_h is shown in Fig. 4.26a. The evolution of the energy resolution as a function of the offline reconstructed p_{T} is presented in Fig. 4.26b. Contrary to the behaviour observed in 2017, the L1 τ_h energy resolution in 2018 was found to be better in the endcaps (16-24%) than in the barrel (19-26%) for offline p_{T} values ranging 30 to 110 GeV. A similar trend was observed for other hadronic objects: the effect was ascribed to the upgrade of the readout electronics of the endcap hadronic calorimeter at the end of 2017 [13] and/or to the change in the HCAL trigger primitives compression from a non-linear to a linear scheme for 2018 data-taking. The sharp selection efficiencies for inclusive and isolated L1 τ_h remained unaffected by the change, as seen in Fig. 4.27. A 90% selection efficiency is reached for inclusive (isolated) L1 τ_h at 49 (56) GeV for a L1 threshold of 34 GeV. The changes in the algorithm resulted in a resilient trigger against pileup, demonstrated in Fig. 4.28. An efficiency higher than 90% is achieved for events with up to 60 reconstructed vertices considering a L1 threshold of 30 GeV and an offline p_{T} larger than 50 GeV.

The necessary changes implemented in the L1 τ_h algorithm in view of 2018 data-taking presented in this section lead to an outstanding performance of the trigger in 2018. This year of data-taking was pivotal to enhance the sensitivity of physics analyses in CMS, as almost half of the full Run 2 dataset was recorded during this period. Regardless of the increase of instantaneous luminosity delivered by the LHC and the ageing of the detectors, the level of physics acceptance and reliability achieved in 2018 was comparable to 2017. The nominal thresholds of the L1 τ_h

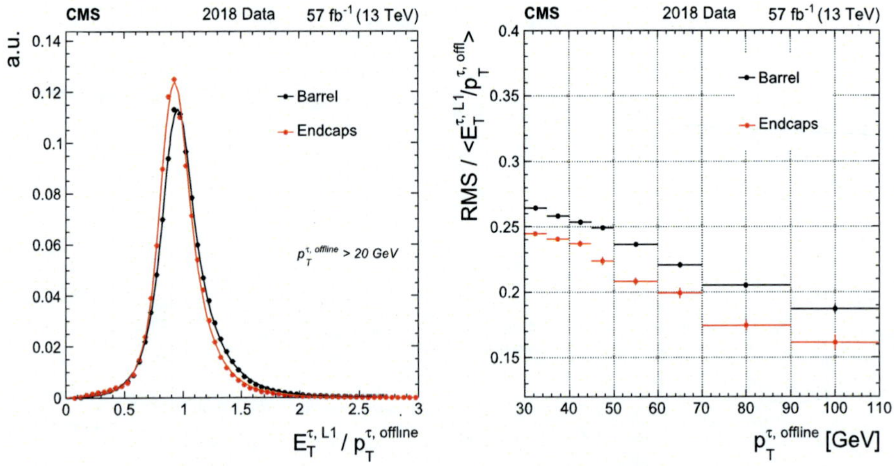

(a) Response (b) Resolution

Fig. 4.26 **a** Energy response, defined as the ratio between the L1 τ_h E_T and the offline reconstructed τ_h p_T, of the inclusive L1 τ_h candidates. **b** Energy resolution, defined as the RMS over the mean of the response distribution, of the inclusive L1 τ_h candidates as a function of the p_T of the offline reconstructed τ_h, separately for the barrel ($|\eta| < 1.305$) and for the endcaps ($|\eta| > 1.479$)

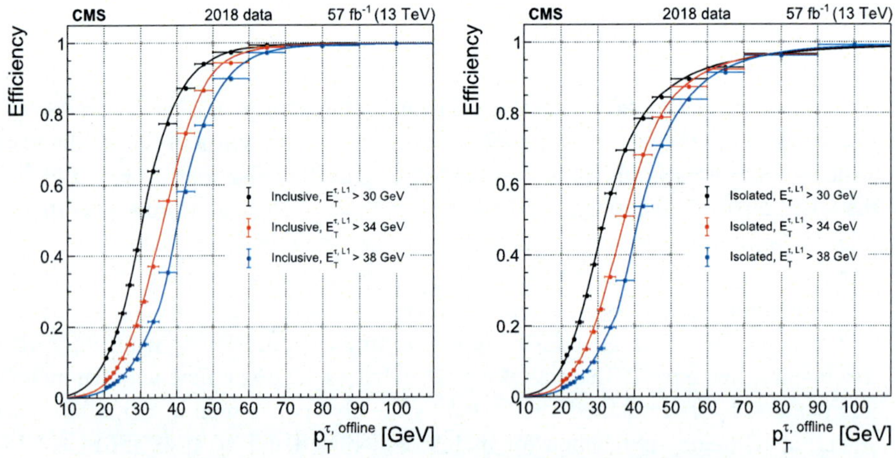

(a) Inclusive (b) Isolated

Fig. 4.27 Efficiency as a function of the offline reconstructed τ_h p_T for **a** inclusive and **b** isolated L1 τ_h candidates, for different L1 E_T benchmark thresholds. The fits to the data are performed with a cumulative distribution of a Crystal Ball function, convolved with the linear relaxation function in the case of isolated L1 τ_h candidates

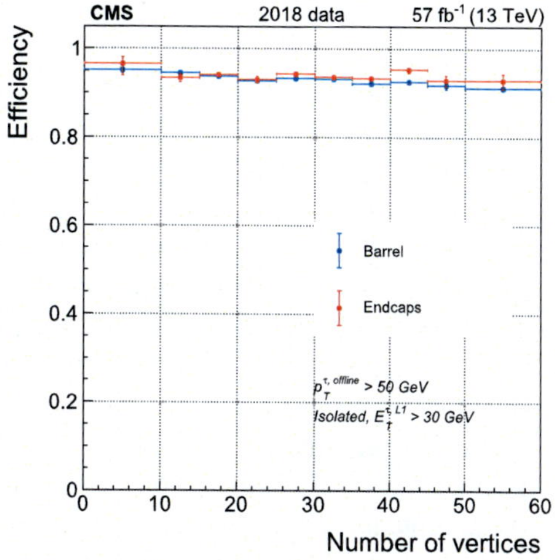

Fig. 4.28 Efficiency as a function of the number of reconstructed vertices in the event for isolated L1 τ_h candidates for a L1 E_{T} threshold of 30 GeV, separately for the barrel ($|\eta| < 1.305$) and for the endcaps ($|\eta| > 1.479$)

seeds (32 GeV for the di-τ_h trigger) were maintained during 2018, well adapted to the variety of physics searches including τ_h in the final state. Stable selection efficiencies were granted by a flexible algorithm which was able to respond to upwards pileup fluctuations at the end of Run 2, peaking at values of 70 at times. Alongside the excellent performance observed in 2017 (see Sect. 4.2), an unprecedented and high-quality τ_h dataset was collected by CMS during Run 2, certainly favouring the high sensitivity of the tH and $\mathrm{t\bar{t}H}$ analysis presented in this manuscript.

In view of Run 3, additional changes to the algorithm are being implemented as this thesis is being written; they are briefly described in Sect. 4.5. Run 3 foresees a total integrated luminosity of 300 fb^{-1} and hence presents a great opportunity to improve the searches for signals that are suboptimally captured by the current trigger configuration. Two dedicated seeds have been studied to improve the sensitivity to the $\mathrm{H} \rightarrow \tau^+\tau^-$ process; their optimization and performance is presented in the next section.

4.4 Studies of New Level-1 τ_h Triggers for Run 3

After the L1 upgrade, sophisticated correlation algorithms could be run at the µGT. These include the evaluation of topological conditions ($\Delta\eta$, ΔR), charge correlations and invariant masses amongst the L1 objects. Analysis-targeted seeds were designed by encoding requirements on these correlation variables exploiting the properties of the physics processes of interest, in a similar way to what is done at the analysis level. Upon the inclusion of these new conditions to the seed, the E_{T} thresholds on the objects can be loosened while maintaining the same output rate as the original seed. An example of such seeds is the VBF trigger [8], which for the first time targets the VBF Higgs boson production mode. This signal is characterized by the presence of two jets with very high invariant mass; the feature is exploited at trigger level with an $m_{\rm jj}$ requirement included in the seed. The VBF trigger can additionally include requirements on other L1 objects ($E_{\rm T}^{\rm miss}$, τ_h) originating from the Higgs boson decay [8]. In 2017, it provided ~40% additional VBF $\rm H \rightarrow \tau\tau$ events with respect to the double-τ_h trigger alone [14].

In view of the upcoming Run 3, significant effort is being put on the development of seeds that enhance the efficiency to the rarer physics processes, such as $\rm HH \rightarrow 4b/bb\tau\tau$, $\rm W \rightarrow 3\pi$, $\tau \rightarrow 3\mu$ and long lived particles decaying to muons, for instance. The L1 seeds designed in this thesis are intended to enlarge the acceptance to the $\rm H \rightarrow \tau\tau$ process in both the single and double Higgs boson production modes; the latter is expected to reach the observation during the HL-LHC era. The first trigger developed requires the presence of two τ_h but applies different thresholds on the leading and subleading τ_h. This "asymmetric" double-τ_h seed is aimed at enhancing the acceptance to the $\rm H \rightarrow \tau^+\tau^-$ process when the leading τ_h carries a significantly higher energy than the subleading τ_h. The second one requires the simultaneous presence of two τ_h along with an additional jet, expected in many di-τ_h processes; the handle on the jet allows the thresholds on the two τ_h to be loosened.

4.4.1 Asymmetric Double-τ_h Triggers

The L1 double-τ_h triggers used throughout Run 2 apply the same energy thresholds to both τ_h legs. A symmetric cut is suboptimal, as by definition the two products of the $\rm H \rightarrow \tau\tau$ decay will have distinct $p_{\rm T}$ spectra: applying a tighter threshold on the leading object and a looser threshold on the subleading one in principle can impact the performance only positively. The energy spectra of the L1 τ_h candidates in different Higgs boson production modes can be found in Fig. 4.29. The leading τ_h resulting from the ggH process presents a steeply falling spectra compared to the others; in turn, the subleading τ_h in the VBF and $\rm t\bar{t}H$ processes present softer energy distributions. The former could potentially be disfavoured by the inclusion of an aggressive asymmetric seed, while the latter could see its signal acceptance significantly enhanced. The most optimal threshold configuration should provide acceptance gain across all Higgs boson processes, possibly enhanced in the most sensitive categories. The exploration of several configurations is presented in what follows.

Different sets of thresholds are identified assuming a maximum rate of 9 kHz allocated to the asymmetric di-τ_h seed. It corresponds to the rate of the reference symmetric seed (`L1_DoubleIsoTau32er2p1`) measured in the zero bias during a run in 2017 with a high average pileup of 57; the same run is used to evaluate the rates of the new seeds. The resulting sets of thresholds are given in Fig. 4.30, where the curves for other values of the rate are likewise illustrated for comparison. The measured rates correspond to isolated L1 τ objects following the isolation option 22, the one used for most of the data-taking in Run 2. A restriction of $|\eta| < 2.1$ is applied to the τ_h, as is done in the reference seed. The seeds studied are denoted as

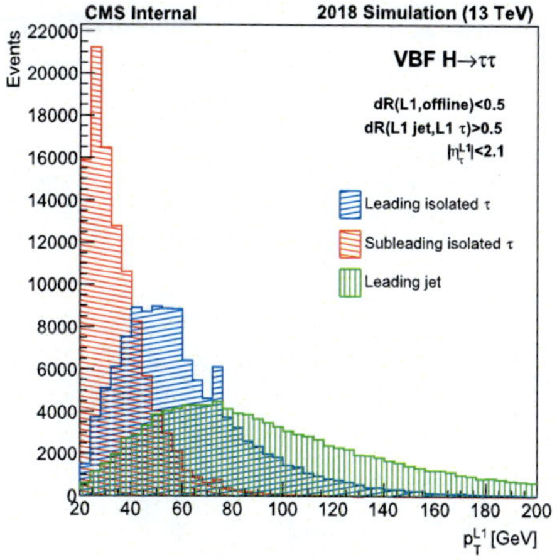

(a) VBF with H $\to \tau\tau$

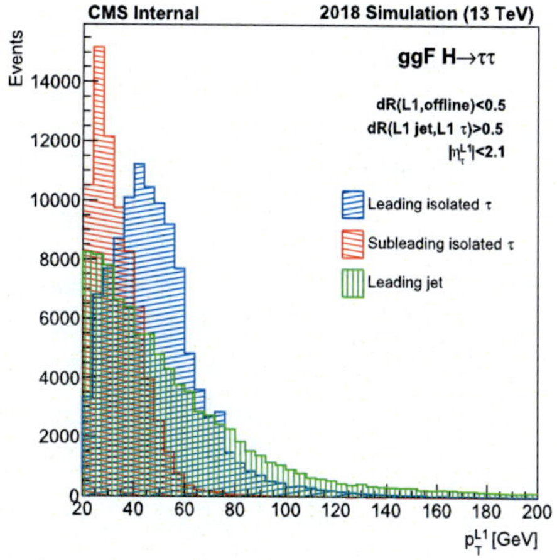

(b) ggH with H $\to \tau\tau$

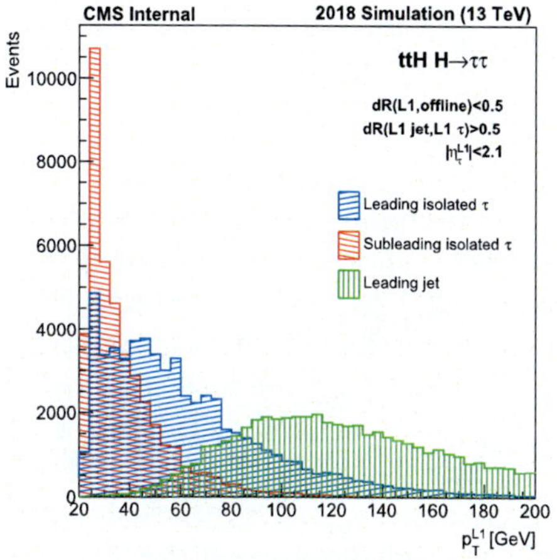

(c) $\mathrm{t\bar{t}H}$ with H $\to \tau\tau$

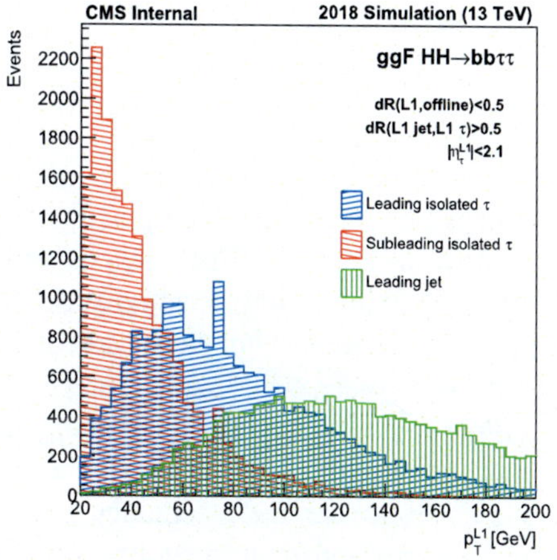

(d) ggHH with HH $\to \mathrm{bb}\tau\tau$

Fig. 4.29 Distributions of p_T of the leading L1 τ_h, the subleading L1 τ_h and the leading L1 jet, obtained in simulated events of different Higgs boson production processes. An isolation requirement is applied to the L1 τ_h candidates. A $\Delta R > 0.5$ requirement is applied between L1 jets and L1 τ_h candidates

`L1_IsoTauXXIsoTauYYer2p1`, encoding a threshold X on the leading L1 τ_h and a threshold Y on the subleading L1 τ_h, grouped under the notation (X, Y) in what follows. While restricting the threshold on the leading L1 τ_h to a maximum value of 50 GeV, five asymmetric seed candidates are identified, corresponding to the threshold sets (40, 30), (42, 28), (44, 27), (46, 26) and (48, 24) GeV.

As the ultimate goal is to maximize the selection efficiency of the H $\to \tau\tau$ decay, the gain with the new asymmetric seed (`L1_IsoTauXXIsoTauYYer2p1`) is computed with respect to the seed of reference (`L1_DoubleIsoTau32er2p1`). It is evaluated in simulated events of the VBF, ggH and $\mathrm{t\bar{t}H}$ processes with H $\to \tau\tau$. The

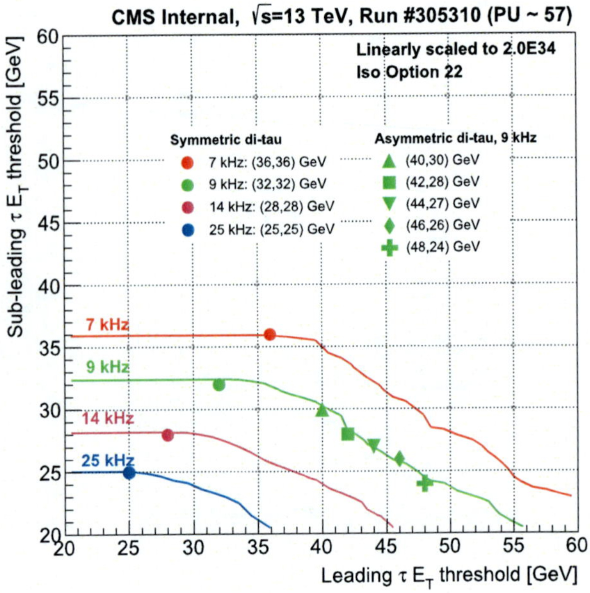

Fig. 4.30 Curves of equal rate for different L1 thresholds applied on leading and subleading τ_h, recorded with zero bias events in 2017. The L1 double-τ_h symmetric thresholds for different rates are indicated with circles. Five potential asymmetric double-τ_h triggers are studied for a fixed rate of 9 kHz (in green), with the corresponding thresholds being indicated with different symbols

acceptances are computed upon application of the offline selections of the $H \rightarrow \tau\tau$ analysis at CMS [2]. The analysis presents three subcategories that require 0, 1 or 2 jets, with increasing sensitivity, denoted as 0-jet, 1-jet and 2-jet, respectively. The offline selection cuts in these categories are shown in Table 4.4; they are represented with the boolean `OfflTau32` in what follows. With a L1 τ_h threshold of 32 GeV, the offline selection requires the leading τ_h to have $p_T > 50$ GeV (18 GeV higher than L1) and the subleading τ_h to have a $p_T > 40$ GeV (8 GeV higher than L1). Following this rationale, the acceptance gain of the asymmetric seed is evaluated assuming offline selections that to evolve accordingly, i.e. with thresholds of X+18 GeV and Y+8 GeV on the leading and subleading τ_h, respectively. The asymmetric offline selections are shown in Table 4.4 under the boolean `OfflTauXY`.

Assuming the asymmetric seed to replace the symmetric one, the acceptance gain is defined separately for the 0-, 1- and 2-jet categories as

$$g^{XY}_{0j,1j,2j} = \frac{N\,[\text{ L1TauXY \& OfflTauXY}_{0j,1j,2j}\,]}{N\,[\text{ L1Tau32 \& OfflTau32}_{0j,1j,2j}\,]}\,. \tag{4.6}$$

The function compares the number of events that fire the L1 asymmetric seed (boolean `L1TauXY`) and its corresponding offline selections (boolean `OfflTauXY`) in the numerator to the number of events that pass the current L1 symmetric seed (boolean `L1Tau32`) and its corresponding offline selections (boolean `OfflTau32`) in the denominator. In both cases, the L1 τ_h are required to match geometrically the corresponding offline τ_h with $\Delta R < 0.5$. An overlap removal between the offline reconstructed jets and τ_h of $\Delta R > 0.5$ is applied in categories with at least one jet.

Table 4.4 Offline p_T selections associated to the L1 selections of the `L1_DoubleIsoTau32er2p1`, `L1_IsoTauXXIsoTauYYer2p1`, `L1_DoubleIsoTauXXer2p1` and `L1_DoubleIsoTauYYer2p1JetZZdR0p5` triggers, in that order (see text). The selections are shown for the 0-jet, 1-jet and 2-jet categories of the $H \to \tau\tau$ analysis [2] and for the $bb\tau\tau$ category of the HH analysis [15]

	Offline p_T selections per category [GeV]			
Boolean	0-jet	1-jet	2-jet	HH
`OfflTau32`	$p_T^{\tau 1} > 50$	$p_T^{\tau 1} > 50$	$p_T^{\tau 1} > 50$	$p_T^{\tau 1} > 40$
	$p_T^{\tau 2} > 40$	$p_T^{\tau 2} > 40$	$p_T^{\tau 2} > 40$	$p_T^{\tau 2} > 40$
		$p_T^{jet1} > 30$	$p_T^{jet1} > 30$	$p_T^{jet1} > 20$
			$p_T^{jet2} > 30$	$p_T^{jet2} > 20$
`OfflTauXY`	$p_T^{\tau 1} > X + 18$	$p_T^{\tau 1} > X + 18$	$p_T^{\tau 1} > X + 18$	
	$p_T^{\tau 2} > Y + 8$	$p_T^{\tau 2} > Y + 8$	$p_T^{\tau 2} > Y + 8$	
		$p_T^{jet1} > 30$	$p_T^{jet1} > 30$	
			$p_T^{jet2} > 30$	
`OfflTauX`	$p_T^{\tau 1} > X + 18$	$p_T^{\tau 1} > X + 18$	$p_T^{\tau 1} > X + 18$	$p_T^{\tau 1} > X + 10$
	$p_T^{\tau 2} > X + 8$	$p_T^{\tau 2} > X + 8$	$p_T^{\tau 2} > X + 8$	$p_T^{\tau 2} > X + 10$
		$p_T^{jet1} > 30$	$p_T^{jet1} > 30$	$p_T^{jet1} > 20$
			$p_T^{jet2} > 30$	$p_T^{jet2} > 20$
`OfflTauYJetZ`	$p_T^{\tau 1} > Y + 18$	$p_T^{\tau 1} > Y + 18$	$p_T^{\tau 1} > Y + 18$	$p_T^{\tau 1} > Y + 10$
	$p_T^{\tau 2} > Y + 8$	$p_T^{\tau 2} > Y + 8$	$p_T^{\tau 2} > Y + 8$	$p_T^{\tau 2} > Y + 10$
		$p_T^{jet1} > Z + 10$	$p_T^{jet1} > 30$	$p_T^{jet1} > 20$
			$p_T^{jet2} > 30$	$p_T^{jet2} > 20$
			$p_T^{jet} > Z + 10$	$p_T^{jet} > Z + 10$

The acceptance gain for the proposed asymmetric seeds defined this way is shown in Table 4.5, separately for each Higgs boson signal and for each analysis category. A general loss of acceptance is observed in VBF and ggH events for all possible L1 asymmetric thresholds; it is especially noticeable for the ggH process in the 0-jet category, as expected from the fact that the leading and subleading τ_h carry comparable energies. A moderate acceptance gain of up to ~7% is observed for $t\bar{t}H$ events with very aggressive thresholds of X = 48 GeV and Y = 24 GeV. For the sake of comparison, the gain was also evaluated keeping the same offline selections as the current ones (`OfflTau32`) in the numerator of Eq. 4.6. The largest gain of the acceptance in this case does not surpass ~2% for any signal or threshold set considered. With such marginal acceptance gain, the double-τ_h asymmetric seeds were not studied any further. The gain of these seeds could potentially be enhanced by designing a second isolation working point for the subleading τ_h leg, with a faster relaxation adapted to the lower threshold it is subject to.

Table 4.5 Acceptance gain of the L1_IsoTauXXIsoTauYYer2p1 trigger with respect to the L1_DoubleIsoTau32er2p1 trigger for different sets of (X, Y) thresholds. The gain of the VBF, ggH and $t\bar{t}$H processes is evaluated in the 0-jet, 1-jet and 2-jet categories of the H $\rightarrow \tau\tau$ analysis [2]

(X, Y)	Category	Acceptance gain		
		VBF H $\rightarrow \tau\tau$ (%)	ggH H $\rightarrow \tau\tau$ (%)	$t\bar{t}$H H $\rightarrow \tau\tau$ (%)
(40, 30)	0-jet	−8	−30	−2
	1-jet	−7	−15	−2
	2-jet	−6	−7	−2
(42, 28)	0-jet	−3	−32	+3
	1-jet	−3	−11	+3
	2-jet	−2	−2	+3
(44, 27)	0-jet	−4	−37	+3
	1-jet	−3	−14	+3
	2-jet	−2	−4	+3
(46, 26)	0-jet	−5	−43	+3
	1-jet	−5	−17	+3
	2-jet	−3	−6	+3
(48, 24)	0-jet	−2	−43	+7
	1-jet	−1	−15	+7
	2-jet	+1	−2	+7

4.4.2 Double-τ_h + Single-Jet Triggers

In most of the H $\rightarrow \tau_h\tau_h$ candidate events used in the analyses, the di-τ_h pair is accompanied with jets, either arising from the hard-scattering process or from initial- or final-state QCD radiation. Thus, the acceptance to these analyses can be enlarged by including the requirement of the presence of an additional jet to the double-τ_h trigger; the thresholds on the τ_h candidates can be loosened this way. Other physics signals such as the HH boson production in the bb$\tau\tau$ final state can benefit from a di-τ_h+jet seed, as two *b*-jets are expected in the final state. However, a careful treatment has to be given to seeds requiring the simultaneous presence jets and τ_h. The object identification is ambiguous at the μGT, as both are similar purely calorimetric objects: all the genuine τ_h enter the L1 jet collection, while a fraction of jets generated from quarks and gluons enter the L1 τ_h collection. To mitigate this effect, an overlap removal is applied between the objects [8], meaning only the L1 jet candidates which are more than $\Delta R = 0.5$ apart from the L1 τ_h candidates are retained. The resulting seed is denoted as L1_DoubleIsoTauYYer2p1JetZZdR0p5. It requires two τ_h with $E_T >$ Y and one non-overlapping jet with $E_T >$ Z. The L1 p_T distributions for these objects in different Higgs boson processes are shown in Fig. 4.29. In general, the leading L1 jet has a harder spectrum than the leading L1 τ_h, hence a tighter

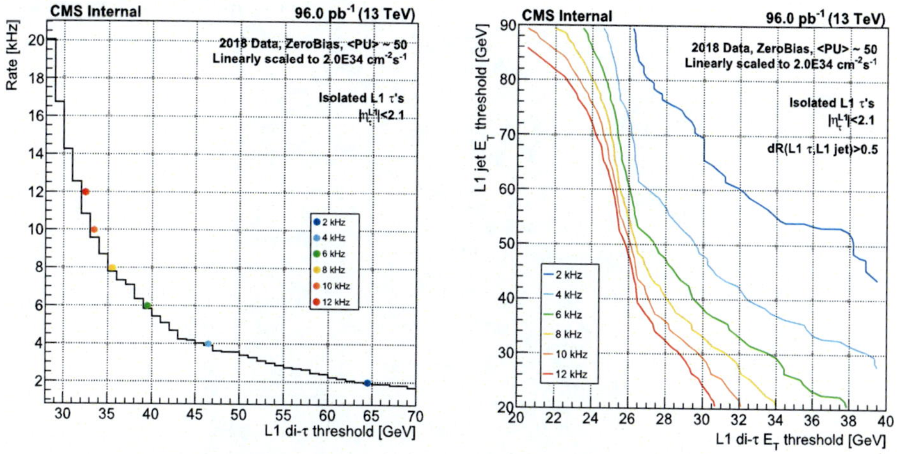

(a) L1_DoubleIsoTauXXer2p1 (b) L1_DoubleIsoTauYYer2p1JetZZdR0p5

Fig. 4.31 **a** Rate of the L1_DoubleIsoTauXXer2p1 trigger as a function of the X threshold. **b** Curves of equal rate as a function of the Y and Z thresholds of the L1_DoubleIsoTauYYer2p1JetZZdR0p5 trigger. The rates are obtained from zero bias events recorded in 2018

threshold on the former compared to the later (Z > Y) is expected to improve the acceptance to these signals.

The di-τ_h+jet seed studied here is not foreseen to replace the current double-τ_h seed (L1_DoubleIsoTau32er2p1) but to serve as a complement to it. However, the associated thresholds of the di-τ_h seed are tightened to be able to shift part of its allocated bandwidth to the new di-τ_h+jet seed. The resulting double-τ_h seed is denoted as L1_DoubleIsoTauXXer2p1; it encodes a threshold $E_T > X$ on both τ_h legs, with X > 32 GeV. The rates corresponding to different values of X are shown in Fig. 4.31a, measured in zero bias events in a run of 2018 with an average pileup of 50. Likewise, the rates corresponding to different values of Y and Z in the di-τ_h+jet seed are shown in Fig. 4.31b for the same run. Assuming a rate A allocated to the double-τ_h seed and a rate B allocated to the di-τ_h+jet seed, their simultaneous deployment would yield a total rate $C = A + B - (A \cup B)$, where $(A \cup B)$ is their shared rate. Figure 4.32 shows the different sets of (X, Y, Z) thresholds allowed for a total maximum rate of $C = 12$ kHz, corresponding to the rate of the di-τ_h reference seed (X = 32 GeV) for that run. For example, thresholds of Y = 26 GeV and Z = 58 GeV yield to a di-τ_h+jet rate of ~5 kHz. It would require a threshold of X = 35 GeV for the di-τ_h seed in order to keep the total rate under 12 kHz, 3 GeV tighter than the current configuration.

The acceptance gain with the new triggers (L1_DoubleIsoTauYYer2p1JetZZdR0p5 || L1_DoubleIsoTauXXer2p1) with respect to the reference double-τ_h seed alone (L1_DoubleIsoTau32er2p1) was evaluated for Higgs boson signals in the single and double production modes. In the case of the single Higgs boson processes (VBF, ggH, $t\bar{t}$H), the acceptance is

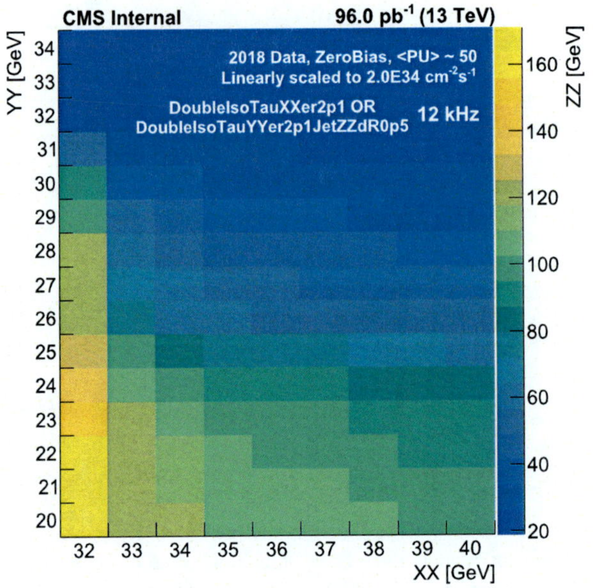

Fig. 4.32 L1 threshold sets (X, Y, Z) of the `L1_DoubleIsoTauXXer2p1` and `L1_DoubleIsoTauYYer2p1JetZZdR0p5` seeds for a fixed shared rate of 12 kHz. The thresholds are obtained from zero bias events recorded in 2018

computed separately for the three categories of the $H \rightarrow \tau\tau$ analysis [2], requiring 0, 1 or 2 jets. In the 0-jet channel, the di-τ_h+jet seed is not considered, as no jet is expected. Since the X threshold on the di-τ_h seed is tighter than 32 GeV by definition, the usage of this seed leads to an acceptance loss in this category; it has to be quantified nonetheless to make an educated choice of thresholds. In this category, the acceptance gain is

$$g_{0j}^{XYZ} = \frac{N\ [\ L1TauX\ \&\ OfflTauX_{0j}\ ]}{N\ [\ L1Tau32\ \&\ OfflTau32_{0j}\ ]}\ . \tag{4.7}$$

In the 1-jet and 2-jet categories, the di-τ_h+jet seed is included together with the double-τ_h seed, and the acceptance gain is

$$g_{1j,2j}^{XYZ} = \frac{N\ [\ (L1TauX\ \&\ OfflTauX_{1j,2j})\ ||\ (L1TauYJetZ\ \&\ OfflTauYJetZ_{1j,2j})\ ]}{N\ [\ L1DiTau32\ \&\ OfflTau32_{1j,2j}\ ]}\ . \tag{4.8}$$

Upon evaluation of the acceptance in the gluon fusion double Higgs boson production, the offline selections associated to the bb$\tau\tau$ category are inspired from the corresponding analysis of 2016 data [15]. The acceptance gain in this case is defined as

$$g_{\text{HH}}^{\text{XYZ}} = \frac{\text{N [(L1TauX \& OfflTauX}_{\text{HH}}\text{) || (L1TauYJetZ \& OfflTauYJetZ}_{\text{HH}}\text{)]}}{\text{N [L1DiTau32 \& OfflTau32}_{\text{HH}}\text{]}} . \tag{4.9}$$

The `L1DiTau32`, `L1TauX` and `L1TauYJetZ` booleans in Eqs. 4.7–4.9 are set to true when the event fires the `L1_DoubleIsoTau32er2p1`, `L1_DoubleIsoTauXXer2p1` and `L1_DoubleIsoTauYYer2p1JetZZdR0p5` triggers, respectively. Similarly, the booleans `OfflTau32`, `OfflTauX` and `OfflTauYJetZ` fire when the event passes the respective offline selections. The acceptance gain is thus computed as the ratio between the number of events that pass either of the new triggers and their hypothetical offline selections over the number of events that pass the current trigger and its current associated offline selection. Similarly to the asymmetric double-τ_h seed studies, the offline selections associated to the new triggers are set in accordance with the L1 thresholds following the rationale of the H $\rightarrow \tau\tau$ [2] analysis. The dependence of the offline τ_h cuts on the L1 τ_h cuts remain the same; the corresponding offline jet p_T selections are incorporated, set at values 10 GeV higher than the L1 thresholds. The evaluation of the acceptance gain in HH $\rightarrow$ bbøø events uses the offline selections from the corresponding 2016 analysis [15]. The offline leading τ_h, subleading τ_h and leading jet must pass thresholds which are 10 GeV higher than the L1 selection, as shown in Table 4.4.

The acceptance gain for the VBF, ggH and t$\bar{\text{t}}$H signals with H $\rightarrow \tau\tau$ in the 0-, 1- and 2-jet categories is shown in Fig. 4.33. The gain is presented as a function of the X and Y thresholds: the corresponding Z threshold can be inferred from Fig. 4.32. In the 0-jet category, the biggest acceptance loss is found for the ggH process, as expected from the softer p_T spectra of the τ_h. In the 1- and 2-jet categories, an adequate simultaneous gain is found for the three signals with thresholds around X $\sim$ 36 GeV and Y $\sim$ 25 GeV applied on the τ_h legs of the double-τ_h and di-τ_h+jet triggers, respectively. The most optimal set of thresholds for the three processes across all categories are (34, 26, 64), (32, 24, 134) and (35, 20, 110) for VBF, ggH and t$\bar{\text{t}}$H production, respectively. Similarly, the acceptance gain of the HH $\rightarrow$ bbøø signal in its dedicated category can be found in Fig. 4.34. The most optimal threshold set is found to be (45, 25, 64) in this case. The acceptance gains to the different physics signals in the most optimal configurations for each of them are shown in Table 4.6.

The largest global acceptance gain with respect to the current double-τ_h trigger considering all physics signals is provided by the threshold set (34, 26, 64), namely the combination of the modified double-τ_h seed `L1_DoubleIsoTau34er2p1` and new di-τ+jet seed `L1_DoubleIsoTau26er2p1Jet64dR0p5` seeds. The new trigger configuration translates into a tighter threshold on the current double-τ_h seed by 2 GeV to reallocate some of its bandwidth to the new double-τ_h+jet seed. The latter presents a relaxed threshold on the τ_h legs down to 26 GeV, along with a reasonable jet threshold of 64 GeV. The new seeds would significantly enhance the acceptance to the VBF and t$\bar{\text{t}}$H processes, with gains of ~35% in the H $\rightarrow \tau^+\tau^-$ categories requiring at least one jet, while sacrificing only ~12% acceptance in the 0-jet category. The acceptance gain to the ggH and HH processes amounts to ~10% in categories with at least one jet; however, it represents a ~25% loss for the ggH process in the 0-jet category. Being the 1-jet and 2-jet categories the most sensitive

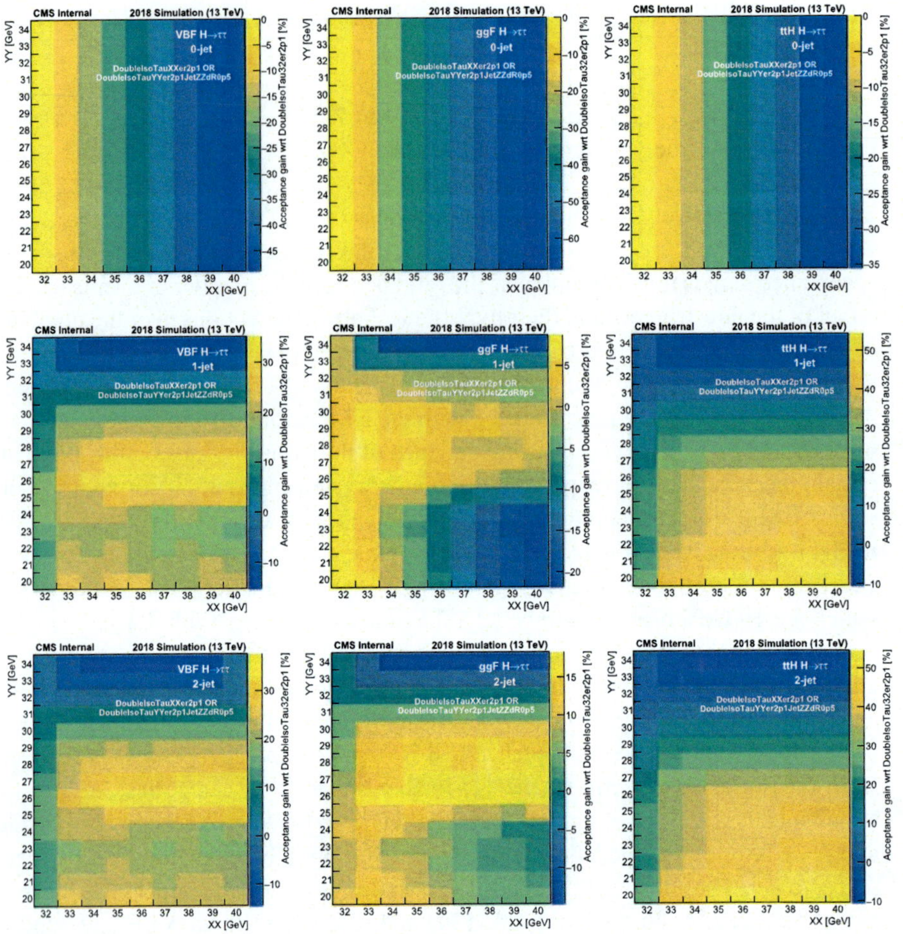

Fig. 4.33 Acceptance gain of the `L1_DoubleIsoTauXXer2p1` and `L1_DoubleIsoTauYYer2p1JetZZdR0p5` triggers with respect to the `L1_DoubleIsoTau32er2p1` trigger evaluated in VBF (left column), ggH (middle column) and $t\bar{t}$H (right column) $H \to \tau\tau$ events in the 0-jet (top row), 1-jet (middle row) and 2-jet (bottom row) categories of the $H \to \tau\tau$ analysis [2]. The acceptance is shown as a function of the X and Y thresholds; the corresponding Z threshold can be inferred from Fig. 4.32

of the $H \to \tau^+\tau^-$ analysis [2], the implementation of these triggers can result in a significant improvement in the measurement of the Higgs boson coupling to the τ lepton during Run 3, leading to stringent constraints on the CP structure of this sector. It would also enhance the double Higgs boson selection in the bb$\tau\tau$ final state, one of the most sensitive, certainly favouring the upper limit of ~7 times the SM cross section expected for Run 3. The potential deployment of the new di-τ_h+jet seed is being coordinated with the HLT as this manuscript is being written.

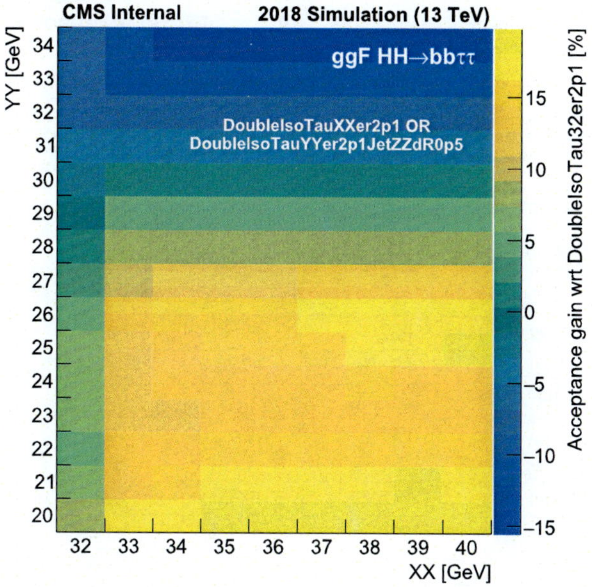

Fig. 4.34 Acceptance gain of the `L1_DoubleIsoTauXXer2p1` and `L1_DoubleIsoTauYYer2p1JetZZdR0p5` triggers with respect to the `L1_DoubleIsoTau32er2p1` trigger evaluated in gluon fusion HH events in the bbøø final state in its dedicated analysis category [15]. The acceptance is shown as a function of the X and Y thresholds; the corresponding Z threshold can be inferred from Fig. 4.32

4.5 Conclusions and Outlook

The instantaneous luminosity delivered by the LHC doubled the nominal value during 2017 and 2018 operations. Record pileup levels of up to 70 were reached at times and changing filling schemes were significantly affecting the detector operations. In this context, making a fast selection of the genuine τ_h deposits with high efficiency while relying solely on the calorimeter inputs is not an easy task: an overwhelming QCD-induced jet background has to be rejected and the calorimetric activity associated to pileup and detector noise in the forward region has to be disentangled. The work developed in this thesis was crucial to tackle these challenges and monitor the correct functioning of the trigger during data-taking. The necessary changes introduced in the algorithm resulted in an increased reliability and physics reach of the trigger. The L1 double-τ_h thresholds were maintained at their nominal 32-34 GeV range regardless of the extreme conditions, and energy resolutions below 27% were achieved across all detector regions and energy ranges. The excellent resolution granted sharp selection efficiencies that reached 90% for an offline p_T selection as low as ~55 GeV, well adapted to the $H \to \tau\tau$ searches at CMS. Stable selection efficiencies of ~90% were ensured for pileup levels as high as 60 with the optimizations introduced in the isolation scheme. The excellent L1 τ_h trigger performance granted an unprecedented

Table 4.6 Acceptance gain of the `L1_DoubleIsoTauXXer2p1` and `L1_DoubleIsoTauYYer2p1JetZZdR0p5` triggers with respect to the `L1_DoubleIsoTau32er2p1` trigger. The gain of the VBF, ggH and $t\bar{t}H$ processes is evaluated in the 0-jet, 1-jet and 2-jet categories of the $H \to \tau\tau$ analysis [2]. The acceptance to the HH signal in the $bb\tau\tau$ final state is evaluated in its dedicated analysis category [15]

		Acceptance gain			
(XX, YY, ZZ)	Category	VBF $H \to \tau\tau$	ggH $H \to \tau\tau$	$t\bar{t}H$ $H \to \tau\tau$	$HH \to bb\tau\tau$ (%)
Best set VBF (34, 26, 64)	0-jet	−15%	−25%	−10	
	1-jet	+32%	+7%	+38	
	2-jet	+33%	+17%	+38	
	HH				+15
Best set ggH (32, 24, 134)	0-jet	0%	0%	0	
	1-jet	+15%	+8%	+24	
	2-jet	+16%	+11%	+24	
	HH				+7
Best set $t\bar{t}H$ (35, 20, 110)	0-jet	−22%	−35%	−15	
	1-jet	+24%	−4%	+49	
	2-jet	+26%	+11%	+49	
	HH				+19

τ_h dataset to enrich the Higgs boson analysis during Run 2 and, in particular, the tH and $t\bar{t}H$ productions studied in this manuscript.

In view of future data-taking operations, I developed two new trigger seeds to enhance the selectivity of the $H \to \tau^+\tau^-$ process in both the single and double Higgs boson production modes. In particular, the inclusion of a di-τ_h+jet trigger with a threshold of 26 GeV on the τ_h legs and a threshold of 64 GeV on the jet leg was found to enhance significantly the selectivity of these processes. An acceptance gain with respect to the current di-τ_h seed alone of up to ~33% and ~38% was found for the VBF and $t\bar{t}H$ signals, respectively, while the $HH \to bb\tau\tau$ would benefit from a ~10% gain. The enhanced statistics would help constrain further the coupling of the Higgs boson to the τ leptons as well as the Higgs boson self-coupling. It will likewise improve the sensitivity to the tH and $t\bar{t}H$ processes during Run 3, with the goal of studying the CP structure of the top Yukawa sector and possibly ruling out BSM scenarios contemplating modified top-Higgs couplings.

In order to maintain the excellent performance in Run 3, several areas of improvement have been identified for the algorithm. The L1 τ_h calibration factors will clearly be revisited in the new experimental conditions, possibly envisioning data-driven methods for their derivation. Likewise, the necessary re-exploration of the different isolation options is underway to adapt the algorithm to the expected higher pileup

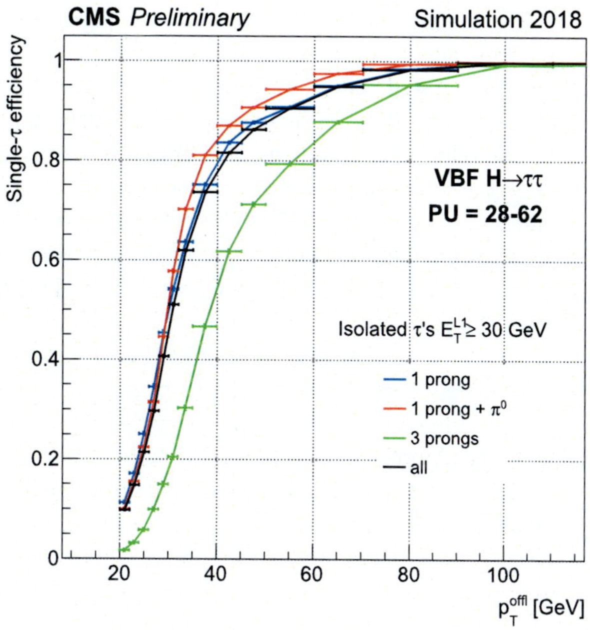

Fig. 4.35 Single-τ_h efficiency as a function of the offline τ_h p_T for different decay modes. The efficiency is evaluated for isolated candidates for a L1 threshold of 30 GeV in simulated VBF $H \to \tau^+\tau^-$ events with 2018 detector conditions. A suboptimal selection efficiency of the 3-prongs decay is observed

levels. Tighter isolation schemes will be studied for the merged L1 τ_h candidates, which present a higher isolation energy; it would result in an overall rate reduction while affecting a small fraction of the signal candidates. The change could be complemented with the use of alternative functions (non-linear) in the relaxation of the isolation as an additional handle to improve the signal efficiency. However, the biggest efforts are devoted to improving the suboptimal selection efficiency of the 3-prongs candidates, depicted in Fig. 4.35. To do so, the τ_h decay modes could be identified at L1 with quantities such as the merging flag, the shape of the cluster, the isolation energy, the ratio of HCAL-to-ECAL energies or the energy spread. By adding these observables to the calibration and isolation inputs, the selection efficiency of the multi-prong decays could be enhanced.

Additional L1 τ_h seeds are being studied to enlarge the acceptance to physics phenomena that suffer from lower signal efficiencies, such as $HH \to bb\tau\tau$ or $W \to \pi^+\pi^+\pi^-$. For what concerns the former, besides the di-τ_h+jet seed presented herein, semileptonic decay modes of the $\tau\tau$ pair could be captured with seeds requiring one τ_h plus one e/γ or muon, together with one or two jets expected from the b-quarks. For what concerns the latter, charged pions can be triggered and reconstructed as 1-prong τ_h, but the acceptance to the $W \to \pi^+\pi^+\pi^-$ decay is limited with the current L1 τ_h trigger, as the pions are typically very soft. Seeds requiring two isolated τ_h with an upper cut on their invariant mass can encode looser thresholds on the τ_h ($\sim$28 GeV). The selection efficiency of this signal can be improved by developing a modified version of the standard L1 τ_h reconstruction, specifically targeted at the 1-prong decay, identified making use of the discriminating observables previously mentioned.

The L1 trigger foresees the ultimate upgrade for the HL-LHC era [16], where an instantaneous luminosity of 5×10^{34} cm^{-2}s^{-1} is envisioned. After 10 years of operation, it will result in an unparalleled dataset of 3000 fb^{-1}, opening the door to a rich physics program including high precision measurements and BSM searches. Intense pileup environments of 140 are foreseen; to achieve the physics goals, the latency of the trigger system is extended from 3.8 to 12.5 μs and the maximum bandwidth is enlarged from 100 to 750 kHz. The new L1 trigger will incorporate tracking and high-granularity calorimetry for the first time, exploited by modern processors carrying machine-learning and Particle Flow algorithms to select very specific final states. Several L1 τ_h algorithms are being studied for the upgrade [16]; the underlying strategies vary from calorimeter-based to full Particle Flow reconstruction. As part of this thesis work, I developed a L1 τ_h algorithm restricted to the endcap, where the aged calorimeters will be replaced by the High-Granularity Calorimeter (HGCAL) [17]. The algorithm is built on calorimeter inputs and does not make use of tracking information; however, the transverse and longitudinal segmentation of HGCAL is enough to reconstruct efficiently the τ_h candidate, reject the pileup and even identify the individual decay modes. The development and performance of this innovative algorithm is explained in Chap. 5.

References

1. CMS Collaboration (2013) CMS technical design report for the level-1 trigger upgrade. Technical Report CERN-LHCC-2013-011. CMS-TDR-12, CERN, Geneva, Jun 2013
2. CMS Collaboration (2018) Observation of the Higgs boson decay to a pair of tau leptons with the CMS detector. Phys Lett B 779:283–316
3. CMS Collaboration (2018) Observation of $t\bar{t}$H production. Phys Rev Lett 120:231801
4. CMS Collaboration (2018) Prospects for HH measurements at the HL-LHC. Technical Report CMS-PAS-FTR-18-019, CERN, Geneva
5. CMS Collaboration (2020) Analysis of the CP structure of the Yukawa coupling between the Higgs boson and τ leptons in proton-proton collisions at $\sqrt{s} = 13$ TeV. Technical Report CMS-PAS-HIG-20-006, CERN, Geneva
6. Martin Perez C (2019) The CMS level-1 τ lepton and Vector Boson Fusion triggers for the LHC Run II. Nucl Instrum Meth A 936:370–371
7. International conference (2019) Lake Louise Winter Institute, Feb 2019. https://www.ualberta.ca/physics/research/lake-louise-winter-institute.html
8. CMS Collaboration (Oct 2020) Performance of the CMS level-1 trigger in proton-proton collisions at $\sqrt{s} = 13$TeV. J Instrum 15(10):P10017–P10017
9. CMS Collaboration (2018) Level-1 τ trigger performance in 2017 data. Technical Report CMS-DP-2018-006, CERN, Geneva, Feb 2018
10. CMS Collaboration (2018) Level-1 calorimeter trigger performance. Technical Report CMS-DP-2018-040, CERN, Geneva, Jul 2018
11. CMS Collaboration. CMS TriDAS project: technical design report, volume 1: the trigger systems. Technical Report CERN-LHCC-2000-038. CMS-TDR-6-1, CERN, Geneva
12. Cadamuro L (2017) Search for Higgs boson pair production in the $b\bar{b}\tau^+\tau^-$ decay channel with the CMS detector at the LHC. PhD thesis at Université Paris Saclay, Sep 2017
13. CMS Collaboration (2012) CMS technical design report for the phase 1 upgrade of the hadron calorimeter. Technical Report CERN-LHCC-2012-015. CMS-TDR-10, CERN, Geneva, Sep 2012

14. Amendola C (2019) Vector Boson Fusion trigger and study of events featuring di-tau pairs and b jets in the CMS experiment. PhD thesis at Université Paris Saclay, Sep 2019
15. CMS Collaboration (2018) Search for Higgs boson pair production in events with two bottom quarks and two tau leptons in proton-proton collisions at $\sqrt{s} = 13\text{TeV}$. Phys Lett B 778:101–127
16. CMS Collaboration (2020) The phase-2 upgrade of the CMS level-1 trigger. Technical Report CERN-LHCC-2020-004. CMS-TDR-021, CERN, Geneva, Apr 2020
17. CMS Collaboration (2017) The phase-2 upgrade of the CMS endcap calorimeter. Technical Report CERN-LHCC-2017-023. CMS-TDR-019, CERN, Geneva, Nov 2017

Chapter 5
Development of a Level-1 τ_h Trigger for the HGCAL Detector in the HL-LHC

The HL-LHC constitutes the Phase 2 of the LHC operation, following the ongoing Phase 1, and is scheduled to start in 2027. In its nominal configuration, it will run at a centre-of-mass energy of 14 TeV and it will reach an instantaneous luminosity of $5 \times 10^{34}\ \text{cm}^{-2}\text{s}^{-1}$, corresponding to an average pileup of 140. After 10 years of operations, it is expected to deliver an integrated luminosity of 3000 fb^{-1}. The machine will be possibly pushed up to an instantaneous luminosity of $7.5 \times 10^{34}\ \text{cm}^{-2}\text{s}^{-1}$ in its ultimate configuration, resulting in pileup values as high as 200 and a total integrated luminosity of 4000 fb^{-1}. The unprecedented dataset will be fully exploited with an ambitious physics programme including high-precision measurements and searches for BSM physics. A significantly improved characterization of the Higgs boson sector is foreseen; it contemplates a per-cent level precision on the couplings to fundamental fields, the observation of rare decays ($\text{H} \rightarrow \mu\mu$, $\text{H} \rightarrow \text{Z}\gamma$) and production modes (HH), and the search for forbidden decays ($\text{H} \rightarrow e\gamma$). For the $\text{t}\bar{\text{t}}\text{H}$ production, projections estimate a ~10% precision on the measurement of the cross section in the multilepton final state [1]. The inclusive study will be complemented with differential cross section measurements as a function of the Higgs boson p_T in the search of anomalous Higgs boson trilinear couplings [2]. Upon combinations of the tH process in several final states, it is foreseen that negative values of κ_t will be ruled out with a significance over 5σ [3]. Details on the physics reach of the $\text{t}\bar{\text{t}}\text{H}$ and tH production modes during the Phase 2 are given in Section 7.6.

Integrating ten times more luminosity than the LHC, the Phase 2 era will pose significant challenges in terms of radiation tolerance and in-time event pileup on the detectors. Major consolidations and upgrades of the detector are planned for LS3 to maintain the high physics selectivity, incorporating higher granularity detectors along with robust readout electronics. The strip and pixel tracking detectors will be replaced [4] alongside the readout electronics for the barrel calorimeter [5] and the muon system [6]. Additional RPC and GEM detectors will be installed in the forward region of the muon chambers to extend the coverage and provide overall

C. Martin Perez, *Search for the Higgs Boson Produced in Association with Top Quarks with the CMS Detector at the LHC*, Springer Theses,
https://doi.org/10.1007/978-3-030-90206-3_5

redundancy of the muon spectrometer [6]. A precision timing detector will be placed in front of the barrel and endcap calorimeters [7] towards an enhanced discrimination of the interaction vertices. The calorimetry in the forward region, already degraded with the current operations, will be significantly improved with the installation of the HGCAL [8]. It will be mounted with radiation-hard components based on silicon sensors and scintillator technology; they will grant unprecedented transverse and longitudinal segmentation that will facilitate innovative calorimetry à la Particle Flow. With a coverage up to $|\eta| = 3$, the HGCAL enhances the acceptance to very forward processes; examples of the physics goals in the forward region are the double parton scattering for the study of transverse and longitudinal parton correlations, and the $\tau^\pm \rightarrow \mu^\pm\mu^\mp\mu^\pm$ decay in the search for charged lepton-flavour violation. An overview of the HGCAL design is given in Sect. 5.1.

Along with the subdetector upgrades, a complete replacement of the CMS trigger and data acquisition system is planned. In particular, the upgrade of the L1 system [9], explained in Sect. 5.3, is designed not only to maintain the signal efficiencies pursued during Phase 1, but also to enhance the selection of new physics. The algorithms at L1 will resemble the offline reconstruction methods; they will profit from increased detector granularity, the inclusion of tracking information for the first time at L1 and modern FPGAs to run Particle Flow reconstruction and machine-learning identification techniques. A flexible and modular architecture is envisioned to adapt to the evolving HL-LHC running conditions and the physics needs. On this basis, the reconstruction of hadronically decaying τ_h leptons is explored with a variety of algorithms using different detector inputs. As part of this thesis work, I designed the first L1 τ_h trigger algorithm based on the highly granular trigger primitives from the HGCAL, described in Sect. 5.2. The fine segmentation of the forward detector is exploited with novel machine-learning τ_h reconstruction methods; it provides an energy resolution and selection efficiencies comparable to the ones achieved during Run 2 at pileup levels as high as 200. The algorithm and performance is presented in Sect. 5.4. The work presented in this chapter was included in the Technical Design Report of the Phase 2 Upgrade of the CMS Level-1 Trigger [9].

5.1 The HGCAL Design

The CMS endcap calorimeters will be replaced by the end of Phase 1, as both the ECAL and the HCAL in the forward region will have suffered from irreparable radiation damage after integrating 500 fb^{-1}. The major challenge is to preserve, and possibly improve, the high sensitivity in the busy forward region. The new devices must withstand unprecedented radiation levels and be granular enough to disentangle the pileup-induced particles from the particles originating from the hard scatter. For this purpose, the HGCAL [8] will become the new Calorimeter Endcap (CE), with a coverage of $1.5 < |\eta| < 3.0$. The upgraded design was first presented in the Phase 2 Technical Proposal in 2015 [10]; further developments and optimizations performed since are included in the HGCAL Technical Design Report published in 2018 [8].

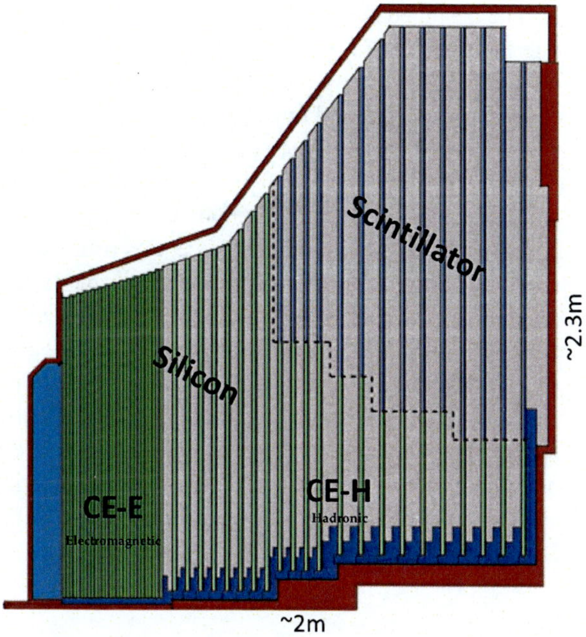

Fig. 5.1 Schematical view of the longitudinal cross section of the HGCAL detector. The electromagnetic compartment (CE-E) has 28 layers of silicon active material (in green) alternated with CuW, Cu and Pb absorbers (in grey). It is followed by the hadronic compartment (CE-H), with 22 layers of silicon or scintillator (in blue) active material alternated with steel absorber (in grey) [8]

A sketch of the HGCAL geometry is shown in Fig. 5.1. The HGCAL is a sampling calorimeter using silicon and scintillators as active material, capable of withstanding very high luminosities. These are mounted in a common detector system, subdivided into the electromagnetic (CE-E) and hadronic (CE-H) components. Silicon is the main active material; it will be used in the CE-E and the most forward part of the CE-H, where the radiation is expected to be higher (up to 10^{16} neutrons/cm^2 or 2 MGy radiation dose[1]). It is transversely segmented into hexagon cells of 0.5 or 1.0 cm^2 surface, with an active thickness of 120 to 300 μm depending on the detector region. Highly-segmented plastic scintillators are used in the lower occupancy sector of the CE-H, coupled to silicon photomultipliers (SiPM) for readout, with varying sizes of 4 to 30 cm^2 depending on the η position.

The longitudinal sampling of the 52 layers of the detector follows the physics principles of shower development. In the CE-E, a total of 28 layers of silicon are alternated with layers of CuW, Cu and Pb absorbers, covering a radiation length of $28X_0$ and a nuclear interaction length of $1.3\lambda_i$. The CE-H is composed of 24 layers of active material, out of which 14 present both silicon and scintillator technology. They are alternated with steel absorbers, totalling a nuclear interaction length $8.5\lambda_i$. Areas of 600 m^2 and 500 m^2 are covered by the silicon sensors and scintillator tiles, respectively, totalling 6 000 000 and 400 000 readout channels, respectively. The full system is cooled down to $-30\,^\circ$C to mitigate the radiation damage.

[1] A Gray is a unit of ionizing dose defined as the absorption of 1 joule of radiation energy per kg of matter.

The designed described above is tagged as version 9 (v9) and corresponds to the one included in the HGCAL TDR [8]. Certain elements have changed since then: the main differences are a slightly lower number of layers in the CE-H (22 instead of 24) and a simplified envelope of the detector close to the beam axis. The results presented in this chapter are derived with the latest version of the HGCAL geometry, tagged as version 10 (v10).

The achieved lateral granularity and the confinement ensured by the dense absorbers result in individual shower discrimination, while the fine longitudinal segmentation improves the pileup rejection, the particle identification and the energy resolution. The high-precision timing capabilities of the silicon sensors are used as an extra dimension in the event reconstruction, granting a full 5-dimensional view (3D-position, energy, time) of the calorimeter objects. The detector provides with around 990 000 channels exploited for trigger purposes; it constitutes more than a factor 500 compared to Phase 1. The raw data are processed in the HGCAL readout electronics to build highly-granular trigger primitives, as explained in the next section. These trigger primitives are sent to the central Level-1 trigger, described in Sect. 5.3, where the L1 τ_h algorithm I developed would potentially be implemented.

5.2 The HGCAL Trigger Primitive Generator

The raw input data stream from the HGCAL correspond to ~300 TB/s, while the allocated Level-1 rate for the upgraded system is 750 kHz. A significant data reduction is required in the generation of the trigger primitives; it is done within a maximum latency of 5μs. To maintain the fine segmentation of the data readout while meeting the bandwidth limitation, only half (alternate) electromagnetic layers of the HGCAL present a trigger readout. The creation of the trigger primitives is conducted in two stages in the *Trigger Primitive Generator* (TPG), sketched in Fig. 5.2. The first stage is implemented in the on-detector electronics, or *front-end* (FE), where the raw data are transformed into trigger cells which undergo energy summation, compression and selection. The process runs in two custom ASICs: the HGCROC (High-Granularity Calorimeter ReadOut Chip) and the ECON-T (Endcap CONcentrator TPG). The outgoing data are sent to the off-detector electronics, or *back-end* (BE), located in the service cavern, where the second processing operation takes place. The trigger cells are seeded and clustered in the longitudinal and transverse directions to form the so-called *3D-clusters*, that are passed to the central L1 trigger system. The operation runs on general-purpose Serenity ATCA platforms [11], a generic motherboard common to other subsystems.

The front-end electronics receive the raw data from the silicon sensors and the SiPMs and digitize it. In the case of the silicon sensors, the collected charges are summed into *trigger cells* (TC) by grouping 3×3 of the $0.5\,\text{cm}^2$ silicon cells or 2×2 of the $1.0\,\text{cm}^2$ silicon cells, covering an area of $\sim 4\,\text{cm}^2$ in both cases. Modules are formed by grouping 48 TCs together; the ones mounted with the smaller (larger) silicon cells are denoted as high (low) density modules, or HDM (LDM), sketched in

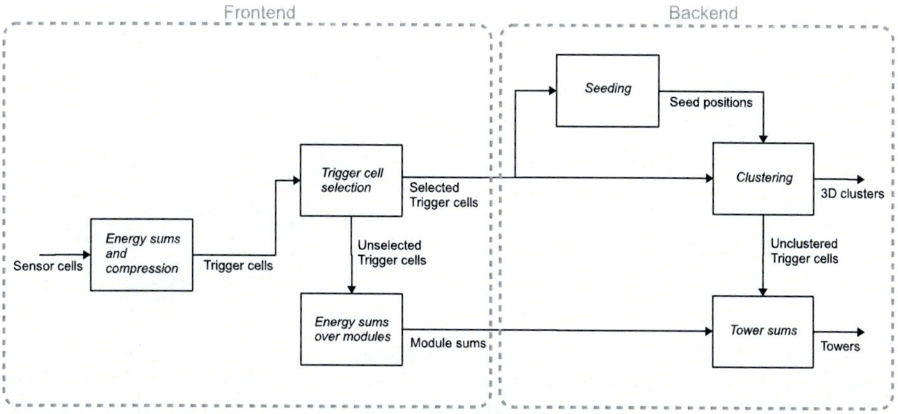

Fig. 5.2 Steps of the HGCAL trigger primitive generator in the front-end and back-end electronics. The energy deposits in the silicon sensors and the scintillator tiles are grouped to form trigger cells (TCs), used as input to the seeding and clustering algorithms that build the 3D-clusters sent to the central L1 trigger system. The unselected or unclustered TC energies are collected in tower sums to ensure no energy is lost [8]

Fig. 5.3. In the case of the scintillator, the TCs are formed by grouping tiles covering an azimuthal angle of 2.5°, corresponding to dimensions from 4 to 10 cm, and a similar radial extension. The charge deposited in the TCs is compressed to 7 bits; no timing information is kept, as the precise timing data would require an excessively large bandwidth. Only TCs above a certain energy threshold are selected to be sent out of the detector: a value of 1 to 2 MIP/sin θ (or 2 $\mathrm{MIP_T}$)[2] is set for this threshold, depending on the trigger occupancy. The energy loss resulting from the selection is collected by summing the unselected channels in each module to form the *module sums*. This baseline selection scheme is the so-called *Threshold* algorithm; it is the one used in the development of the L1 τ_h algorithm for HGCAL presented here. Other data reduction strategies could be used instead, for instance selecting a fixed number of the highest energetic cells or reading out large areas of TCs with reduced granularity. Details on the front-end algorithms under study are given in Sect. 5.4.2, along with their impact on the the L1 τ_h trigger performance.

The core of the TPG runs in the FPGAs of the back-end electronics, which receive the TCs and module sums and creates the trigger primitives. The process is done in two stages, illustrated in Fig. 5.4. In Stage-1, each FPGA receives the TCs from a section corresponding to 1.4% of the detector, and implements repacking and energy calibration. The data are sent to the Stage-2 in a time-multiplexing fashion: each FPGA of the second stage processes one out of 18 bunch crossings and covers a section of the detector corresponding to 1/3 of one endcap, or 120°. This architecture allows the TCs to be processed over large regions covering the full depth of the

[2] MIP stands for Minimum Ionizing Particle, and expresses the energy deposition corresponding to the minimum energy loss rate of the particle through the material.

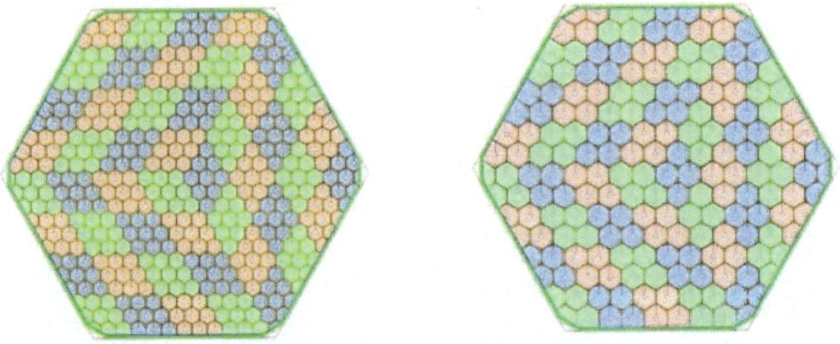

Fig. 5.3 Sketch of the trigger cell (TC) arrangement in the silicon modules of the HGCAL detector. Different TCs are indicated with different colours. They are formed by grouping 3 × 3 hexagonal silicon cells of size 0.5 cm^2 (left) or 2 × 2 hexagonal silicon cells of size 1.0 cm^2 (right). This results in and low density modules (LDM, right), each containing 48 TCs, with a total length of ∼20 cm

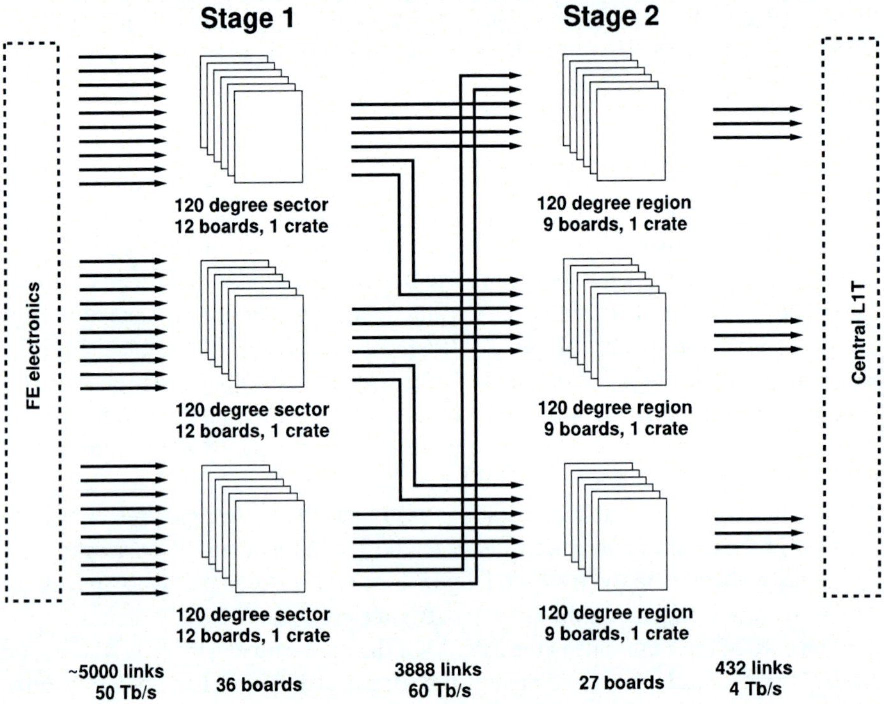

Fig. 5.4 Sketch of the HGCAL trigger primitive system. The Stage-1 receives the trigger data from the on-detector electronics and sends it to Stage-2 in a time-multiplexed fashion. The latter is connected to the central Level-1 trigger system [8]

detector. The TCs are clustered dynamically in both the transverse and longitudinal directions to form the *3D-clusters* in two steps:

1. **Seeding**: the seeds are identified as the local maxima above a threshold of $\sim$10 MIP_T, obtained from binned 2D-histograms of the cell energies projected into the $(r/z, \phi)$ plane. The seeding parameters are chosen to minimize the reconstruction of multiple seeds for single showers.
2. **Clustering**: Adjacent TCs are attached to the seed if they are found within a distance in the $(x/z, y/z)$ plane from 0.015 in the first layers to 0.050 in the last layers. In the presence of more than one seed, the TC is associated to the nearest one. The choice of the cluster size is optimized to maximize the shower containment and minimize the softer pileup contributions.

The positions assigned to the 3D-clusters correspond to the energy-weighted barycenters of the positions of the TC constituents. The highly-granular HGCAL information is used to derive additional properties of the 3D-cluster related to the compactness and substructure of the shower. Examples of these observables are the shower length, its transverse extension, the energy density in different detector layers, the TC multiplicity or the ratio of hadronic-to-electromagnetic deposits. Hence, the 3D-clusters generated this way carry significantly finer information about the calorimetric energy deposits compared to the trigger towers used during Phase 1. Besides presenting higher granularity, these trigger primitives encode information on the longitudinal spread and the distribution of the energy deposit within the shower. The inclusion of these advanced features is pivotal in the energy calibration of the L1 τ_h candidates, the removal of pileup and the identification of the individual decay modes at the extreme pileup levels expected in the HL-LHC era.

Alongside the 3D-cluster reconstruction, a projective *tower* map of the unclustered energy is built to ensure that no energy is lost. These objects are useful to perform global quantities such as energy sums at L1, but they are not used in the L1 τ_h algorithm. The HGCAL trigger primitives (3D-clusters and tower maps) are passed to the central L1 trigger system, presented in the next section, where the reconstruction of the τ_h and other physics objects is conducted making use of the trigger primitives received from all subdetectors.

5.3 The Level-1 Phase 2 Trigger

The goal of the L1 Phase 2 trigger of CMS is to exploit the new features provided by the upgraded subdetectors to attain the optimum physics selectivity in the extreme running conditions of the HL-LHC era. Besides the information from the calorimeters and the muon system, already used by the current Phase 1 trigger, the Phase 2 trigger incorporates the data from the new tracker and improved granularity of the calorimeter data. To provide sufficient time for the hardware track reconstruction and the matching of tracks to muons and calorimeter clusters, the latency of the L1 trigger is increased to from 3.8 μs to 12.5 μs for Phase 2 operation. The total output bandwidth of the upgraded trigger is likewise enlarged from 100 kHz to 750 kHz, so that thresholds comparable to Phase 1 can be retained in the expected pileup conditions. The upgraded system will be able to handle an input data as high as 63 Tb/s, compared to the ~2 Tb/s taken by the current trigger. Details on the upgraded architecture are given in the following; an overview of the proposed L1 τ_h trigger algorithms to date is found in Sect. 5.3.2.

5.3.1 Architecture

Figure 5.5 shows a sketch of the architecture of the Phase 2 L1 trigger. The system presents four independent data processing paths; they feature tracking, calorimetry, muon reconstruction and Particle Flow identification and selection techniques. The last of these modules is a key feature of the upgraded trigger: for the first time, higher-level object reconstruction and identification algorithms run in a new correlator layer which combines the information of several subdetectors. The objects built by the four paths are passed to the Global Trigger, that computes the final accept or reject decision based on a global detector view.

- The **calorimeter trigger** takes as inputs the trigger primitives from the barrel calorimeter (BC), the HGCAL and the hadron forward calorimeter (HF). An enhanced granularity is achieved compared to the Phase 1 trigger as a result of the fine segmentation provided by the HGCAL and the upgraded calorimeter read-out. The calorimeter trigger primitives are sent to the global calorimeter trigger (GCT), where the L1 e/γ, L1 τ_h, L1 jets and L1 sums are built. The L1 τ_h algorithm described in this chapter will be potentially implemented in this trigger module.
- The **muon trigger** receives inputs from various subdetectors, with an extended coverage ($|\eta| < 2.8$) provided by the upgraded muon spectrometer. Similarly to Phase 1, muon track finder algorithms reconstruct the tracks in three separate pseudorapidity regions (BMTF, OMTF and EMTF). The outputs are sent to the global muon trigger (GMT), where the combination with the tracker data is possible.
- The **track trigger** takes as input the tracks from the outer tracker and reconstructs them in the track finder (TF). A global track trigger (GTT) is included to reconstruct the primary vertices of the event and to build the tracker-only based objects. The inclusion of this module for the first time at L1 results in an enhancement of the particle identification and a better handle of the trigger rate.
- The **correlator trigger** (CT) occupies a central role in the design. It reconstructs higher-level objects from the several subdetectors, following the Particle Flow [12] offline reconstruction at CMS (see Section 3.4). The first layer of the system produces the trigger candidates by matching calorimeter clusters and tracks; the second layer applies additional identification criteria on top. The CT also hosts a simplified version of the Pileup Per Particle Identification (PUPPI) algorithm [13] at CMS; it relies on the usage of vertexing information at the particle level to remove or downweight the softer pileup contributions during reconstruction.

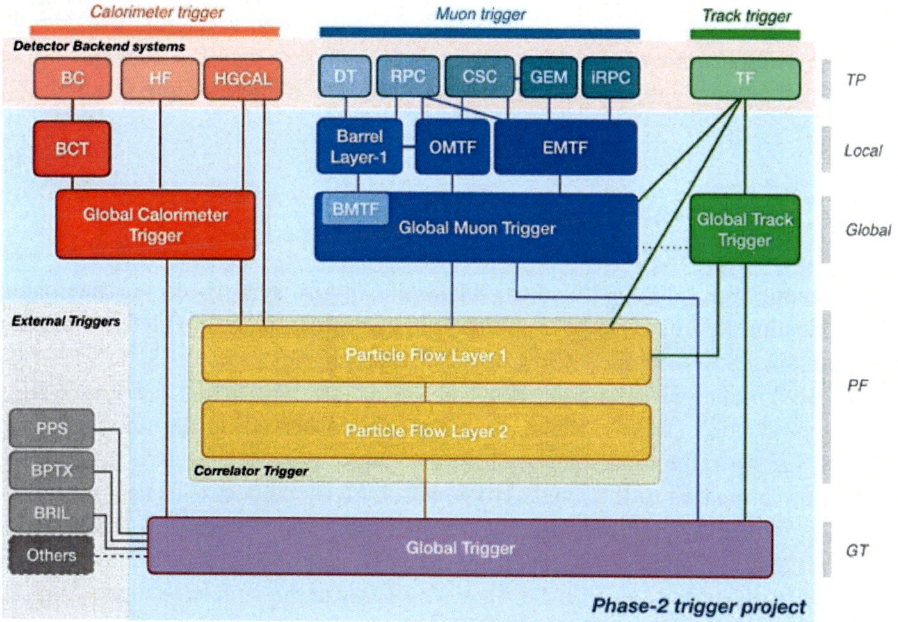

Fig. 5.5 Sketch of the Phase 2 upgraded Level-1 trigger system. The Global Calorimetry trigger (GCT) receives the trigger primitives from the barrel calorimeter (BC), the hadron forward calorimeter (HF) and the HGCAL. The Global Muon Trigger (GMT) receives the trigger primitives from the track finders (BMTF, OMTF, EMTF) in the muon subdetectors. The Track Trigger (TT) receives the trigger primitives from the track finder (TF). The Particle Flow trigger processes high-level trigger objects from multiple subdetectors in the Correlator Trigger (CT). The Global Trigger (GT) collects the outputs from the GCT, GMT, GTT and CT and computes a trigger decision [8]

- The **global trigger** collects the outputs from the GCT, GMT, GTT and CT and computes a trigger decision based on a menu of algorithms. Similarly to Phase 1, it evaluates correlation variables among various types of objects to design analysis-targeted seeds that maximize the physics extractions.

Each data processing path applies independent trigger selections targeted to specific physics requirements. The complementarity between them results in an enhanced physics acceptance, a controlable rate, a safe early commissioning and a robust system during data-taking. The data transfer amongst modules is provided by high-speed optical links (up to 28 Gb/s) that facilitate the aggregation of information from the entire detector. A global detector view is pivotal for pileup removal and for the precise evaluation of global quantities, such as the missing transverse energy and other energy sums. Instead of making use of simpler subsytem variables as is done in the current trigger, the upgraded system will more closely replicate the full offline object reconstruction. A large variety of objects are reconstructed, ranging from standalone objects with single detector information, possibly matched to L1 tracks, to full Particle Flow objects. These sophisticated objects and the correlations amongst them are used to define event selection strategies which were unthinkable with the Phase 1 trigger. New signal triggers are designed to capture the rarer and more exotic phenomena making use of object configurations that include soft and correlated muons, light mesons and displaced muons and jets, amongst others [8].

5.3.2 Level-1 τ_h Trigger Algorithms

The benefit of developing a dedicated hadronically decaying τ_h reconstruction and identification algorithm has been demonstrated during Run 2 of the LHC; its excellent performance has been shown in Chapter 4. The physics case for the L1 τ_h trigger in Phase 2 is centred around the Higgs boson sector, in particular the double Higgs boson production (HH $\rightarrow$ bb$\tau\tau$), foreseen to be observed during this operational phase. The τ_h in the final state of the HH $\rightarrow$ bb$\tau\tau$ process present a softer energy spectrum compared to the single Higgs boson H $\rightarrow \tau\tau$ decay. Hence, the former is most optimally captured with lower trigger thresholds ($\sim$ 20 GeV) compared to the ones appropriate for the latter ($\sim$ 60 GeV). Other non-Higgs boson processes are targeted by the trigger, such as the W $\rightarrow \tau\nu_\tau$ decay or the search for supersymmetric particles; these result in boosted τ_h with energies larger than $\sim$100 GeV. The wide energy range of the processes calls for a large variety of L1 τ_h algorithms, each making use of different detector inputs and reconstruction techniques, possibly deployed simultaneously for maximal physics selectivity. A summary of the algorithms and performance of the main L1 τ_h triggers currently under study is given in the following.

- The **tower** algorithm uses calorimeter towers from the barrel, the HF and the HGCAL following a similar approach to the Phase 1 τ_h trigger. The τ_h are built from trigger towers found in a 3×5 window, corresponding to dimensions of $\Delta\eta \times \Delta\phi = 0.261 \times 0.435$. The calibration is modified with respect to Phase 1 by considering the different τ_h decays, identified via the e/γ clusters contained in the τ_h core. A window of 7×7 trigger towers around the object defines the isolation region that is used to identify the τ_h and reject the jet background. The cutoff on this value is defined to maintain a 99% efficiency at 100 GeV while achieving a factor 2 rate reduction compared to the nonisolated case. The L1 candidate resulting from this algorithm is denoted as `CaloTau` in what follows.
- The **HGCAL** algorithm, studied in this thesis, is based solely on the calorimeter inputs from the HGCAL. It follows a novel approach with respect to Phase 1 and takes maximal advantage of the highly-granular available trigger information provided by the upgraded detector. The L1 τ_h candidate is seeded by a single 3D-cluster; a 3-step calibration procedure is applied on top to correct for the pileup contributions, the response of the different decay modes and the p_T-dependent resolution. Machine learning techniques scrutinize the properties of the τ_h showers to disambiguate the pileup-induced deposits and resolve the individual decay modes; the latter is performed in view of a future isolation scheme to reject the QCD-induced jet background. Details on the algorithm and performance are given in Sect. 5.4.

- The **track+calo** algorithm is seeded by high-quality track objects corresponding to the charged hadrons originating from τ lepton decays. Neighbouring tracks are added to the seed if they are found within a cone around the signal and they satisfy predefined kinematic criteria. An isolation annulus is constructed around the cone to evaluate the vertex isolation; it results in two working points (loose and tight) with increasing efficiency and decreased purity. The e/γ clusters expected from the neutral pions ($\pi^0 \rightarrow \gamma\gamma$) of the τ decay are matched to the tracks if they are found within the signal cone and they satisfy certain quality criteria. The L1 candidate constructed with this algorithm is denoted as `tkEGTau`.
- Two implementations using **Particle Flow** algorithms are under study. The first constitutes a novel L1 approach that uses a neural network (NN) to identify the τ_h candidate. It takes as inputs the Particle Flow or PUPPI trigger candidates and relies on event level quantities to improve the overall efficiency, at the expense of a more complicated firmware implementation. To reduce the rate of the lower-p_T particles, a lower limit on the NN output can be applied, resulting in two distinct working points (loose and tight) with increasing threshold values. The τ_h candidates reconstructed this way are denoted as `NNPFTau` or `NNPuppiTau` depending on the input. The second algorithm is based on the hadrons-plus-strips (HPS) reconstruction algorithm of CMS (see Section 3.4.6), but in this case it is applied in restricted $\eta - \phi$ regions. Multivariate τ_h identification techniques similar to the offline procedure at CMS follow. The output τ_h candidate of this algorithm is denoted as `HPSTau`.

Particle Flow and track+calo-based triggers encode more complex reconstruction techniques able to tackle the fake rates present in the low energy region, thus being more suited for triggers with lower thresholds in high luminosity and pileup scenarios. The robustness and simplicity of the calorimeter-based algorithms allow for stable efficiencies to be maintained at higher energies; they are more adequate for triggers with tighter thresholds. A comparison of the performance of some of these algorithms is presented in Fig. 5.6. The figure shows the selection efficiencies of the single- and double-τ_h triggers measured in HH $\rightarrow$ bb$\tau\tau$ simulated events at pileup 200 for a fixed output rate of 6.6 kHz. As expected, higher selection efficiencies are attained by the NN and track+calo algorithms at low energy, the former performing slightly better below 30 GeV. The reconstruction efficiency in the high energy regime is recovered with calorimeter-based objects; a 99% efficiency is reached at plateau.

The single-τ_h trigger rates as a function of the offline thresholds for each algorithm are shown in Fig. 5.7. In the absence of definite offline reconstruction techniques, the online-to-offline threshold projection is derived from simulated HH $\rightarrow$ bb$\tau\tau$ events by selecting the generator-level visible p_T point at which a 90% of the plateau efficiency is achieved for each L1 threshold. The NN algorithms result in the lowest rates, while the highest rates are given by the track+calo approach at energies larger than 40 GeV. This behaviour drives the definition of the benchmark τ_h seeds in a preliminary version of the L1 menu for Phase 2, shown in Table 5.1 for different algorithms. The offline thresholds are computed at 90% and 50% of the plateau efficiency; the latter is adequate for physics analysis willing to increase their acceptance

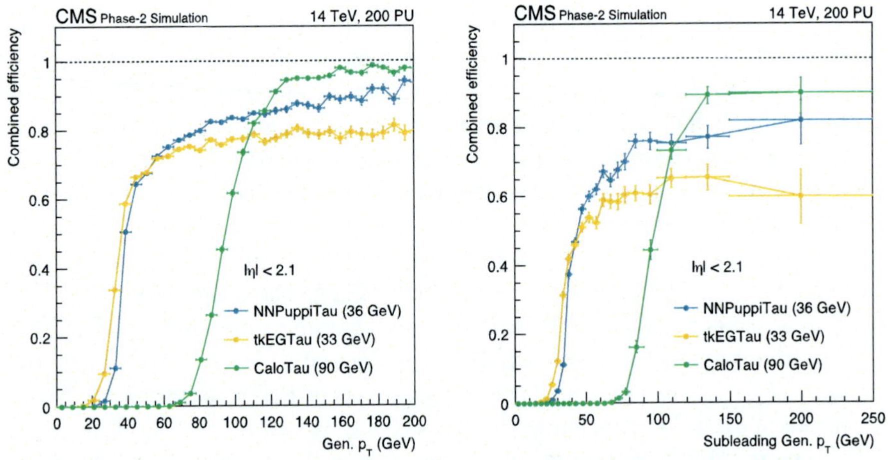

Fig. 5.6 Single-τ_h efficiency as a function of the generated τ_h p_T (left) and double-τ_h efficiency as a function of the subleading generated τ_h p_T (right). The efficiencies are evaluated in HH $\rightarrow$ bb$\tau\tau$ simulated events for various Level-1 τ_h algorithms at a fixed rate of 6.6 kHz [8]. The efficiency of the HGCAL-based algorithm is shown in Fig. 5.17a

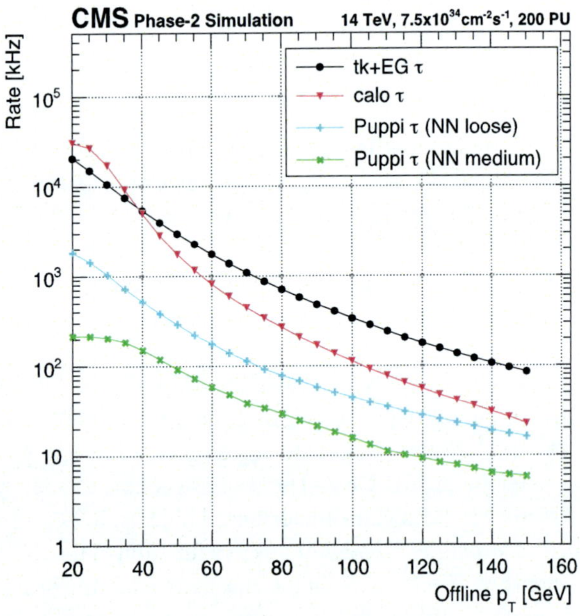

Fig. 5.7 Single-τ_h trigger rates as a function of the offline visible τ_h p_T thresholds for various Level-1 τ_h algorithms [8]. The rate of the HGCAL-based algorithm is shown in Fig. 5.18b

Table 5.1 Benchmark Level-1 τ_h seeds in a preliminary version of the trigger menu for Phase 2 [8]. The offline thresholds at the 90% (50%) plateau efficiency for a given L1 threshold are shown. The quoted rates correspond to a $\langle \mathrm{PU} \rangle = 200$ scenario. An additional requirement on the τ_h candidate of $|\eta| < 2.1$ is applied. Details about the `TkMuon`, `TkIsoElectron` and `PuppiMET` reconstruction algorithms can be found in Ref. [8]

L1 τ_h seed	Offline threshold at 90% (50%) [GeV]	Rate [kHz] $\langle \mathrm{PU} \rangle = 200$
Single `CaloTau`	150 (119)	21
Double `CaloTau`	90/90 (69/ 69)	25
Double `NNPuppiTau`	52/52 (36/36)	7
`NNPuppiTau` + `TkMuon`	36 (27), 18	2
`NNPuppiTau` + `TkIsoElectron`	39 (29), 22	13
`NNPuppiTau` + `PuppiMET`	55 (38)/190 (118)	4

selecting the events far from the plateau. For tower-based (NN-based) calorimeter objects, the di-τ_h trigger would encode an offline threshold of 90 GeV (52 GeV) for a 90% plateau efficiency. For comparison, during 2018 operations, a L1 threshold of 34 GeV resulted in a 90% efficiency for an offline p_T of ~55 GeV (see Section 4.3). However, the total rate allocated to the upgraded seeds meets the bandwidth requirement with a 50% margin, and thus looser thresholds could in principle be envisioned.

5.4 Development of a Level-1 τ_h Trigger for HGCAL

In this section, a proposal for a L1 τ_h algorithm relying on the calorimeter information from the HGCAL subdetector is presented. In contrast with the Phase 1 trigger described in Chapter 4, the algorithm faces the experimental challenges of the forward τ_h reconstruction in a pileup environment that can reach values up to 200 in the ultimate HL-LHC configuration. However, it benefits from the highly-granular data of the HGCAL detector and the large computational power of the programmable FPGAs mounted in the upgraded L1 system. The version of the algorithm described in this chapter has been derived for the HGCAL geometry v10, based on the baseline Threshold compression algorithm of the TPG, with a threshold set at 1.35 $\mathrm{MIP_T}$ (see Sect. 5.4.1). The algorithm has been developed with simulated di-τ_h events with p_T ranging from 20 to 100 GeV within the endcap region of $1.6 < |\eta| < 2.9$ in a pileup environment of 200. It should be noted that the study presented here is far from being final, but it already hints the potential of high-granularity triggering in the HGCAL during the HL-LHC.

5.4.1 Baseline Algorithm and Performance

The L1 τ_h algorithm in HGCAL takes as inputs the 3D-clusters generated by the TPG, in particular the information concerning their energy, their position and several shower shape variables, and builds the L1 τ_h candidates which are sent to the central L1 trigger. The algorithm is composed of four main steps applied in the following order:

1. **seeding** of the relevant energy deposits of the L1 τ_h candidate,
2. three-step **calibration** of the L1 τ_h candidate to remove the pileup, correct the response of the different decay modes and improve the resolution across all energy ranges,
3. rejection of the softer **pileup**-induced energy deposits for rate reduction,
4. identification of the τ_h **decay mode** as a starting point for a better isolation procedure to discriminate against jets.

Details on the different steps are given in what follows.

Seeding
Being the τ_h a collimated jet, the L1 τ_h candidates are seeded from 3D-clusters satisfying $E_T > 4$ GeV. In contrast to the Phase 1 trigger, where a seed of $E_T > 2$ GeV is identified and neighbouring clusters are aggregated, the L1 τ_h candidates in Phase 2 are built from single clusters. It is possible since the clustering procedure in the HGCAL electronics is able to capture efficiently the longitudinal and transverse extension of the shower induced by the τ_h: a single 3D-cluster contains the complete τ_h energy deposit in ~90% of the cases. Moreover, the selection of a single cluster leads to a better pileup rejection, as the softer contributions surrounding the genuine τ_h are removed.

The position assigned to the L1 τ_h candidate is the position of the 3D-cluster, computed in the HGCAL TPG as the energy-weighted position of the TC constituents. The resolution in position resulting from this assignment is illustrated in Fig. 5.8, where the η and ϕ coordinates of the L1 τ_h candidate are compared to the ones of the generator-level τ_h. For this purpose, a L1 τ_h is matched to its generator-level counterpart when it is found within $|\Delta\eta| < 0.1$ and $|\Delta\phi| < 0.2$ of the generated τ_h; the later must fulfil $p_T > 20$ GeV. If several candidates are found, the L1 τ_h with the highest energy is selected. The figures show full widths at half maximum of 0.02 rad in both the η and ϕ directions. Given the high granularity and the three-dimensional clustering of the HGCAL TPG, the position resolution of the L1 τ_h is improved by a factor ~5 with respect to Run 2 (see Fig. 4.10), where coarser granularity trigger towers were used, in a pileup environment about 4 times larger.

Energy Calibration
The energy contained in the 3D-cluster constituting the L1 τ_h candidate is denoted as raw energy (E_T^{raw}). Despite the TC calibration in the front-end electronics of the HGCAL, the raw energy does not reflect precisely the real energy carried by the τ_h. The effect is illustrated in Fig. 5.9; it shows the uncalibrated response, defined

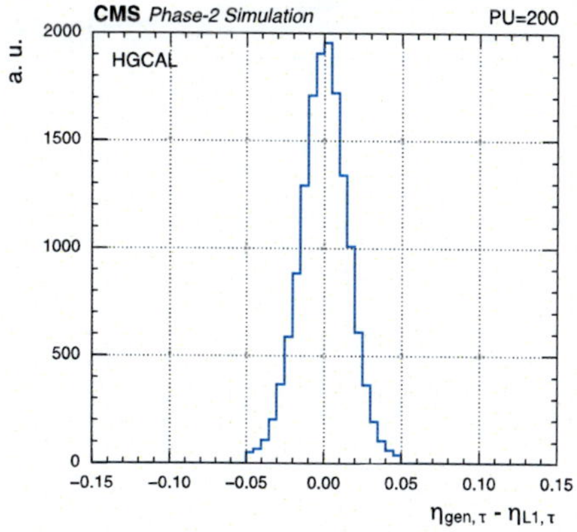

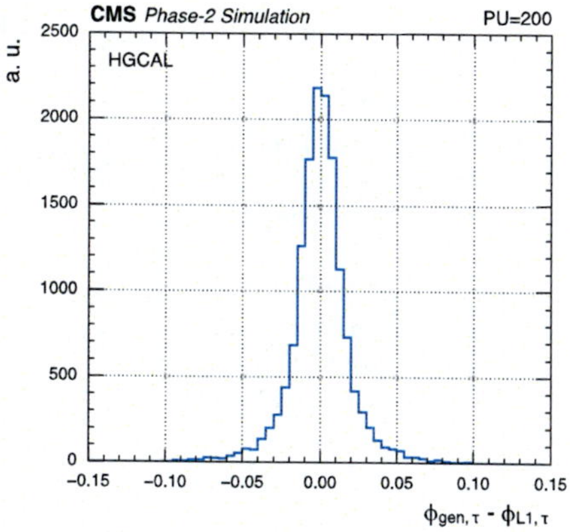

Fig. 5.8 Difference in η (left) and ϕ (right) positions between the generated visible τ_h and the Level-1 τ_h candidate, obtained from di-τ_h simulated events with 200 average pileup interactions

Fig. 5.9 Response distributions, defined as the ratio between the energy of the Level-1 τ_h candidate and the visible p_T of the generated τ_h, in the different steps of the energy calibration, obtained from di-τ_h simulated events with 200 average pileup interactions

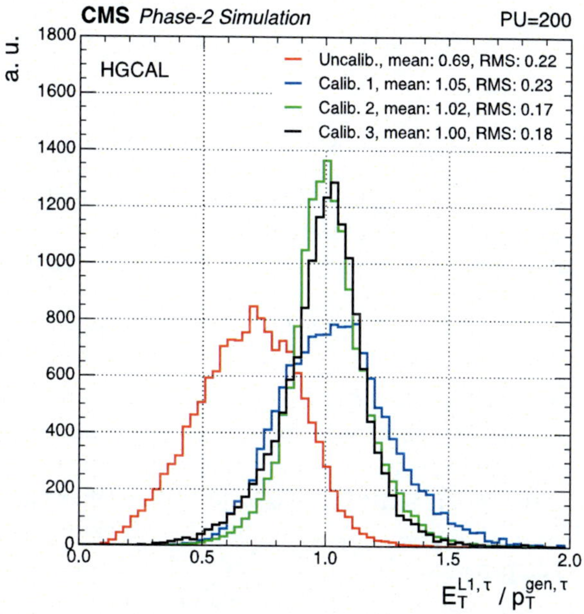

as the ratio between the raw energy and the visible p_T of the matched generator-level τ_h. Uncalibrated τ_h candidates present a mean response of 0.69, far from the expected value of 1, and a relative resolution of $\sim$31%. The suboptimal resolution is a result of the clustering procedure inducing residual losses in the energy recollection, as well as potential biases introduced from the difference in response between the electromagnetic and hadronic showers. Hence, the energy of L1 τ_h candidates is calibrated with a 3-step procedure aimed at subsequently improving different effects affecting the resolution and response.

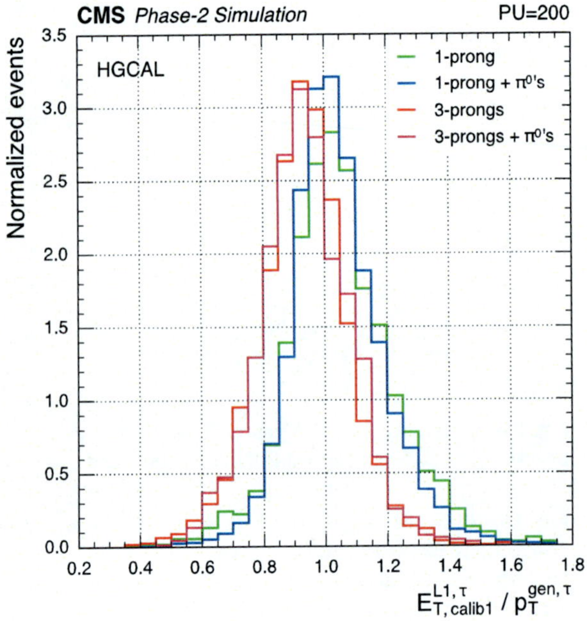

Fig. 5.10 Response distribution, defined as the ratio between the energy of the Level-1 τ_h candidate after the first calibration step (see text) and the visible p_T of the generated τ_h, for the different generated decay modes, obtained from di-τ_h simulated events with 200 average pileup interactions

The first step of the calibration consists in applying η-dependent pileup subtraction to the raw energy of the candidate to account for the pileup contributions collected during the clustering. The calibrated energy is expressed as

$$E^{L1,\tau}_{\mathrm{T,calib1}} = E^{L1,\tau}_{\mathrm{T,raw}} - (a_1 + b_1 \cdot |\eta^{L1,\tau}|) \ . \tag{5.1}$$

The coefficients a_1 and b_1 are obtained from a linear regressor. The input to the regressor is the $|\eta|$ value of the 3D-clusters; the target is the quantity $(p_{\mathrm{T}}^{\mathrm{gen}} - E_{\mathrm{T}}^{\mathrm{raw}})$. The result of the first calibration step is shown in Fig. 5.9 (blue line). The mean response has shifted from 0.69 to 1.05 and the relative resolution has been improved from 31 to 22%. The resolution achieved in the first calibration step alone is already comparable to the one observed during Run 2 operation.

The second step of the calibration is aimed at improving the energy resolution of the L1 τ_h candidate by correcting for the difference in response observed in the τ_h decay modes, illustrated in Fig. 5.10. The figure shows how the single prong decays (1-prong and 1-prong+π^0's), are slightly overcalibrated, while multiple prong decays (3-prongs and 3-prongs+π^0's) are significantly undercalibrated, a trend which has also been observed in Phase 1. It is explained from inhomogeneities in the shower containment during the clustering process, as multiple-hadron decays result in larger showers with a higher particle multiplicity. Compared to Run 2, information on the τ_h decay modes can be retrieved at L1 with more advanced shower shape variables associated to the 3D-cluster, such as the longitudinal and transverse spread and the

Table 5.2 List of input variables used in the BDTs for energy calibration, pileup rejection and decay mode identification. The ranking of variable importance is shown for each case. A hyphen (–) indicates that the variable is not considered

	Variable ranking		
Input variable	Energy BDT	Pileup BDT	Decay mode BDT
Energy	–	–	15
$\lvert\eta\rvert$-coordinate	1	8	18
Energy weighted mean of z-coordinates of TCs ($\langle z\rangle$)[1]	9	13	3
Energy-weighted RMS of z-coordinates of TCs (σ_{zz})[2]	10	14	7
Energy-weighted RMS of η-coordinates of TCs ($\sigma_{\eta\eta}$)[2]	4	2	16
Energy-weighted RMS of ϕ-coordinates of TCs ($\sigma_{\phi\phi}$)[2]	3	4	12
Energy-weighted RMS of r-coordinates of TCs (σ_{rr})[2]	7	3	8
Energy-weighted average of σ_{rr} of layers[3]	8	9	13
Total shower length[4]	15	12	10
Core shower length[5]	14	5	14
First layer	18	17	17
Layer with maximum energy	17	15	6
Number of layers at which 10% of energy is deposited	6	7	2
Number of layers at which 50% of energy is deposited	11	16	1
Number of layers at which 90% of energy is deposited	12	11	4
Number of TCs at which 67% of energy is deposited	2	1	9
Number of TCs at which 90% of energy is deposited	5	6	11
Energy in CE-H over energy in CE-E	13	10	5
Number of 3D-clusters within $\lvert\eta\rvert < 0.1$ and $\lvert\eta\rvert < 0.2$	16	–	–

[1] Computed as $\langle x\rangle = \dfrac{\sum_i^{\mathrm{TC}} E_i\, x_i}{\sum_i^{\mathrm{TC}} E_i}$

[2] Computed as $\sigma_{xx} = \sqrt{\dfrac{\sum_i^{\mathrm{TC}} E_i \cdot (x_i - \langle x\rangle)^2}{\sum_i^{\mathrm{TC}} E_i}}$

[3] σ_{rr} is computed for the layers within 5 cm of the maximum energy layer

[4] Computed as (Last layer − First layer) + 1

[5] Computed as the maximum number of consecutive layers

energy density distribution within the shower. These features are optimally combined in a BDT to derive the calibration constants of the second step. The input variables are listed in Table 5.2; a detailed characterization of the decay modes with these variables is provided in the decay mode identification step of the algorithm later on. In this case, 18 variables are used as input to the training, compared to the 4 variables

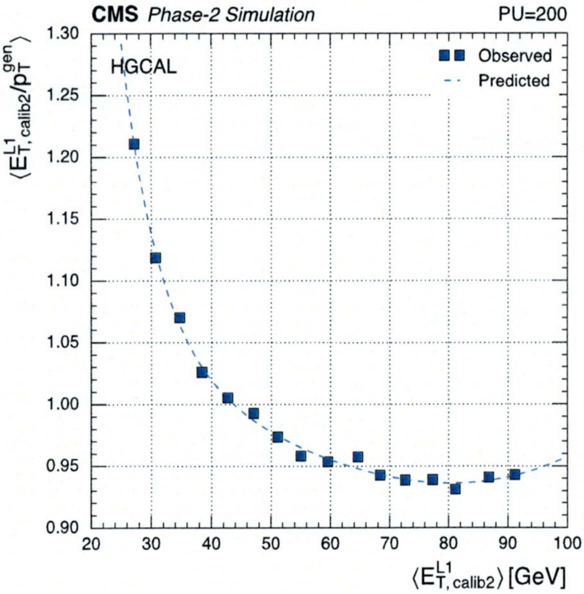

Fig. 5.11 Mean Level-1 τ_h energy response as a function of the mean Level-1 τ_h energy after the second calibration step (see text), obtained from di-τ_h simulated events with 200 average pileup interactions

employed in the calibration in Phase 1. The target of the training is the ratio between the generator visible p_T of the τ_h and the calibrated L1 energy from the first step, i.e. $p_T^{gen}/E_{T,calib1}^{L1,\tau}$; the BDT output is applied as a multiplicative correction c_2, namely

$$E_{T,calib2}^{L1,\tau} = c_2 \cdot E_{T,calib1}^{L1,\tau} . \tag{5.2}$$

The response after the second calibration step is shown in Fig. 5.9 (green line). The resolution of the L1 τ_h candidates is improved to from 22% to 17%, while the response remains unchanged with a mean value of 1.02.

A third and last calibration is applied to correct the response of the L1 τ_h candidate as a function of its energy. As illustrated in the points in Fig. 5.11, the response of the L1 τ_h candidates after the second calibration lies significantly above 1 for $E_T < 40$ GeV and remains ~0.95 for $E_T > 50$ GeV. Hence, a E_T-dependent correction c_3 is applied, obtained from a linear regressor. The inputs are the successive powers of the logarithm of the L1 energy after the second calibration, i.e.

$$1/c_3 = \sum_{i=0}^{4} k_{3,i} \cdot [\, \log(E_{T,calib2}^{L1,\tau}) \,]^i . \tag{5.3}$$

The $k_{3,i}$ coefficients are derived by the model. The predicted corrections are shown with the dashed line in Fig. 5.11. These are applied to the L1 τ_h candidate as a multiplicative correction to obtain the final calibrated energy

$$E_{T,calib3}^{L1,\tau} = c_3 \cdot E_{T,calib2}^{L1,\tau} . \tag{5.4}$$

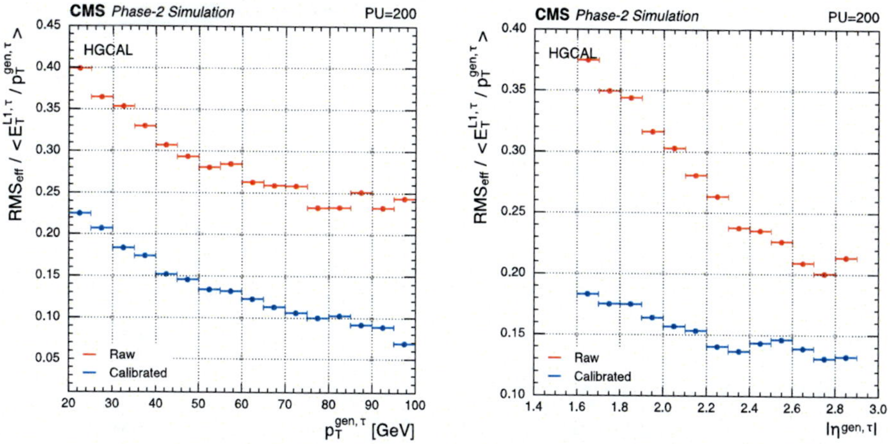

Fig. 5.12 Relative Level-1 τ_h energy resolution, defined as the effective RMS over the mean of the calibrated response distributions, as a function of the generator τ_h visible p_T (left) and $|\eta|$ (right), obtained from di-τ_h simulated events with 200 average pileup interactions

The response after this third calibration step is shown in Fig. 5.9 (black line). A more symmetric distribution is attained, with a mean value exactly at 1.00 and a relative resolution of 19%.

Owing to the increased sophistication of the τ_h calibration, the inclusive energy resolution of the L1 τ_h candidates with the new algorithm is ~15% better than the one observed during Run 2 (see Fig. 4.11) despite the harsher experimental conditions. The improvement is more evident in bins of p_T and η, as shown in Fig. 5.12 for the new algorithm (to be compared to Fig. 4.12 during Run 2). The resolution is computed as the ratio between the effective RMS and the mean of the response distributions, where the effective RMS is defined as the half-width of the shortest interval containing 68% of the events. The improvement of the resolution after calibration is clearly visible across all energy ranges; a stable resolution in all detector regions is likewise achieved.

Pileup Rejection

The calorimetric deposits due to pileup are initially removed in the first step of the calibration, where an η-dependent subtraction to the raw energy is applied to the L1 τ_h candidates. Residual pileup contributions are rejected in the next step of the algorithm with a dedicated BDT trained to discriminate the τ_h-induced clusters from the pileup-induced clusters. The pileup rejection procedure is applied after the calibration step, as it was found that this results in more stable and lower rates compared to the inverted scheme. The signal efficiency remains comparable after the change, since an early calibration only results in an homogeneous change of energy scale across all pileup clusters, leading to similar populations entering the BDT training.

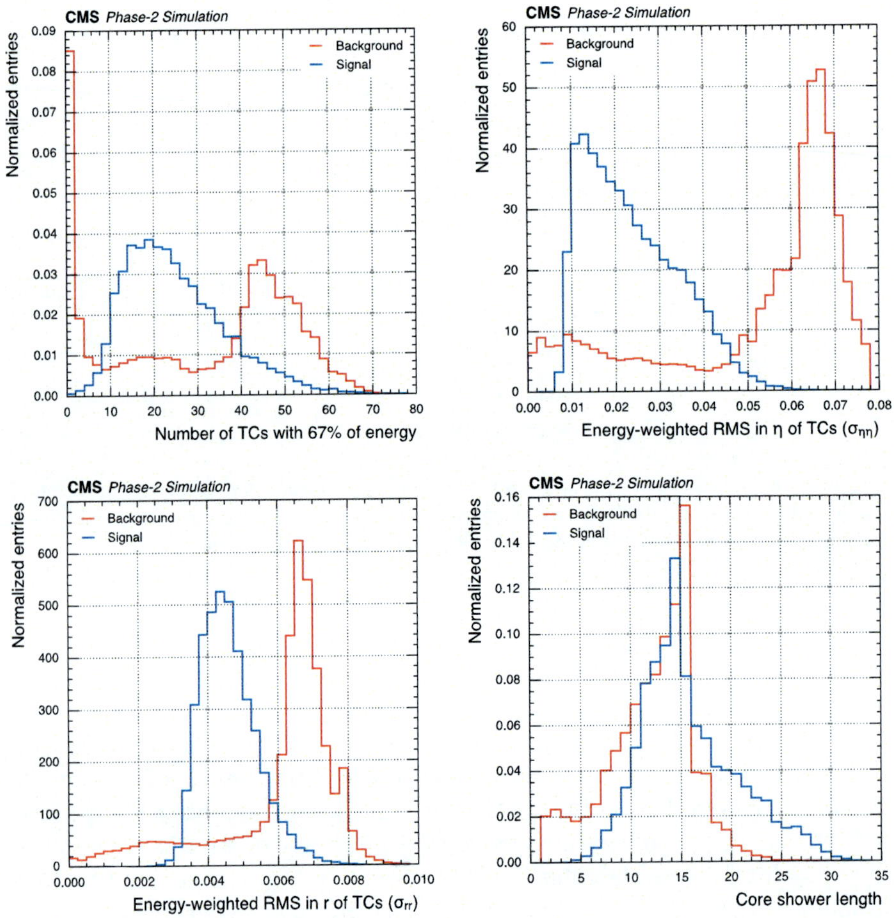

Fig. 5.13 Distributions of various input variables to the BDT used for pileup rejection, obtained from simulated events with 200 average pileup interactions. The signal corresponds to 3D-clusters ($E_{\mathrm{T}}^{\mathrm{calib}} > 4$ GeV) matched to generator-level τ_h ($p_{\mathrm{T}}^{\mathrm{vis.}} > 20$ GeV). The background corresponds to 3D-clusters ($E_{\mathrm{T}}^{\mathrm{calib}} > 20$ GeV) in neutrino events

The pileup rejection is performed by setting a cutoff on the output score of the BDT. It takes as input 17 observables associated to the 3D-cluster, listed in Table 5.2. The method is significantly more advanced than the one used during Phase 1, where the pileup was estimated based on a single variable representing the calorimetric activity in the central part of the detector. In the training of the BDT, the signal corresponds to 3D-clusters with $E_{\mathrm{T}}^{\mathrm{calib}} > 4$ GeV matched to generator-level τ_h with $p_{\mathrm{T}}^{\mathrm{vis.}} > 20$ GeV within a window $|\Delta\eta| < 0.1$ and $|\Delta\phi| < 0.2$, obtained from simulated samples of di-τ_h events with pileup 200. The background corresponds to 3D-clusters with $E_{\mathrm{T}}^{\mathrm{calib}} >$ 20 GeV from simulated neutrino events with pileup 200. The algorithm provides a predicted probability associated to each 3D-cluster ranging from 0 to 1: values closer to 0 indicate that the cluster originates from pileup particles, while values

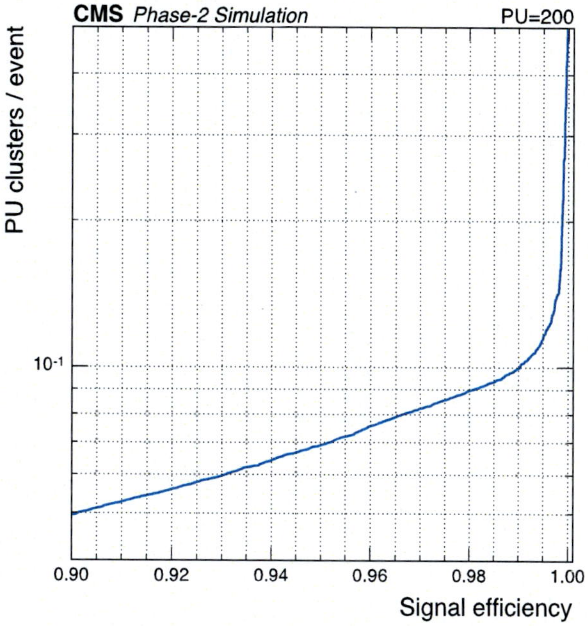

Fig. 5.14 Number of pileup clusters per event as a function of the τ_h selection efficiency resulting from the BDT used for pileup rejection, obtained from neutrino and τ_h simulated events with 200 average pileup interactions. A working point of 99% signal efficiency is applied in the algorithm

closer to 1 indicate that it originates from a τ_h candidate. The distributions of the most discriminating input variables are found in Fig. 5.13. The pileup footprint is more spread in space, with less dense energy deposits and shorter shower length. The excellent separation power of the BDT is illustrated in Fig. 5.14, which shows the signal efficiency as a function of the number of pileup clusters per event. Only L1 τ_h candidates passing the 99% signal efficiency working point are selected in the algorithm; a contamination of $\sim$0.1 pileup clusters per event remains after the selection.

Decay Mode Identification

The decay mode identification is an innovative feature in the L1 τ_h trigger and it is intended as a starting point towards the development of an advanced isolation procedure for discrimination against jets. The prediction is performed with a multi-classifier BDT which captures the shower features associated to each decay mode. Compared to the three input variables used for the derivation of the isolation cut in the Phase 1 trigger, the BDT used here combines 18 input variables related to the 3D-cluster, listed in Table 5.2. The identified decay modes (classes) are the 1-prong, 1-prong+π^0's and 3-prongs, the latter possibly with additional π^0's in the final state. The reason for merging the 3-prongs+π^0 decay into the 3-prongs decay is justified by the lower BR of the former and the more efficient overall identification achieved when classified jointly.

The distributions of some of the most discriminating observables in the decay mode identification are shown in Fig. 5.15. The plots show a very good discrimination

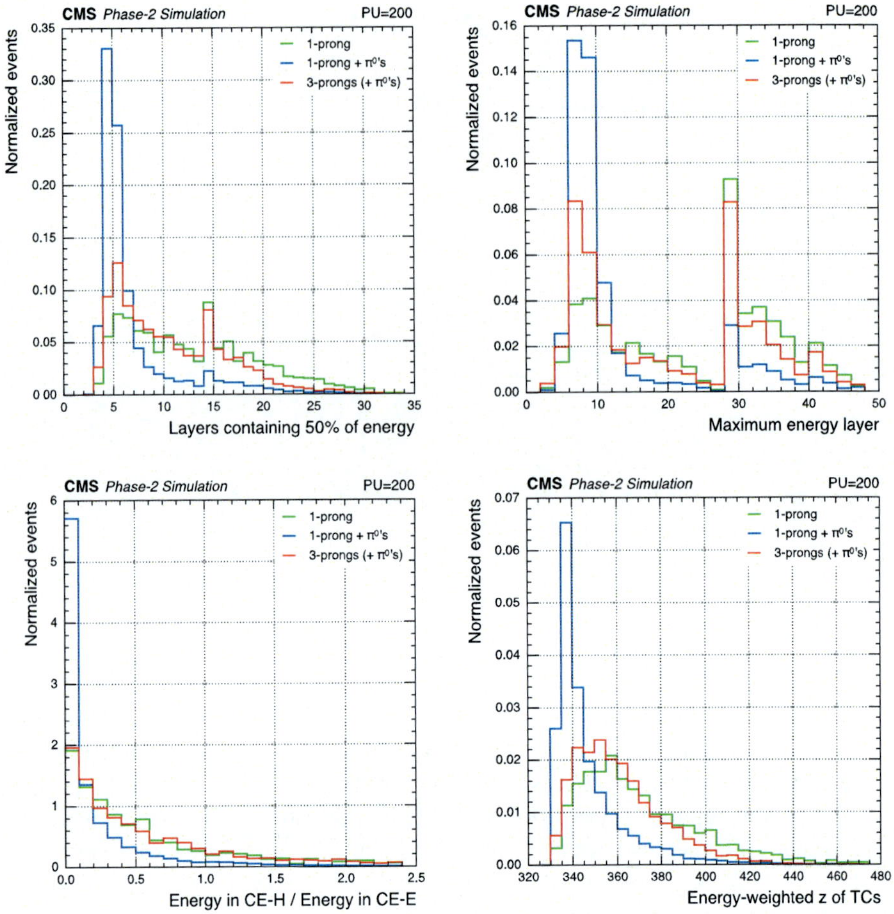

Fig. 5.15 Distributions of various input variables to the BDT used for decay mode identification, for different generated τ_h decay modes, obtained from di-τ_h simulated events with 200 average pileup interactions

of the 1-prong+π^0's decay mode, which presents higher electromagnetic deposits and a more compact shower extension than the 1-prong and 3-prongs decays. The multiclassifier BDT assigns to each L1 τ_h candidate a probability associated to each decay mode; the one presenting the highest value is taken as the predicted decay mode for a given τ_h. The correctness of the predictions is represented in the normalized confusion matrix of Fig. 5.16. A correct 1-prong+π^0's identification is achieved in 69% of the cases. Overall, the matrix shows a diagonal assignment for all decay modes, with a global success rate of 61%.

The decay mode identification algorithm presented here is an indicator of the gain that can be achieved by using multivariate techniques for particle identification at L1. In future developments, this discriminator can be adapted to separate the τ_h signal from QCD-induced jets by introducing an additional class targeted to these type of

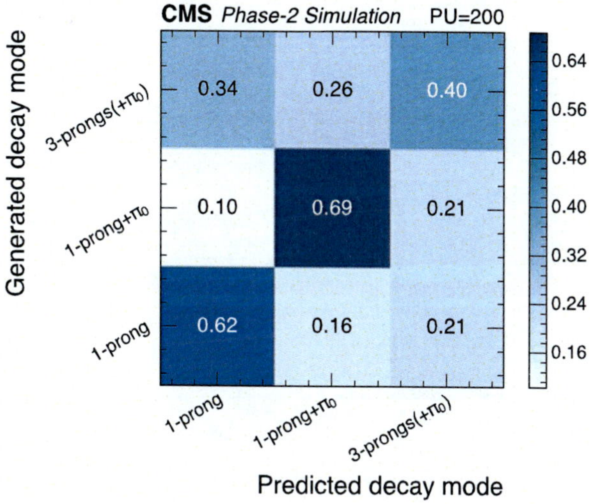

Fig. 5.16 Normalized confusion matrix relating the predicted and generated τ_h decay modes resulting from the BDT used for decay mode identification, obtained from di-τ_h simulated events with 200 average pileup interactions

jets, which typically develop larger and more diffuse showers. This overwhelming background could be efficiently removed by setting a cutoff on the probability that a candidate originates from a QCD-induced jet, additionally serving as a handle for rate reduction. Other improvements to the algorithm can be the inclusion of the information on the calorimeter activity around the L1 τ_h as input to the BDT. In the L1 τ_h Phase 1 isolation algorithm, the isolation energy is evaluated in a fixed square size surrounding the candidate, but no information on the longitudinal or transverse distribution of the energy density is extracted. This BDT—or a dedicated one—could be instrumental to evaluate the correlations amongst the shower shape variables associated to the neighbouring clusters, resulting in a more knowledgeable extraction of the signal topology. Ultimately, L1 τ_h candidates could be matched to the track candidates in the correlator to achieve the optimum jet background rejection, possibly making use of the isolation scheme already implemented in the correlator [8].

5.4.1.1 Performance

The selection efficiency of a single L1 τ_h as a function of the generated visible τ_h p_T is shown in Fig. 5.17a, obtained from simulated di-τ_h events with average pileup of 200. The efficiency is presented for typical L1 thresholds expected in the double-τ_h seeds during the HL-LHC operation. The excellent response ensures a sharp rise of the efficiency, that reaches a flat plateau of 100%. For a 50 GeV L1 threshold, 90% efficiency is reached for a visible p_T of 64 GeV at generator level. The single L1 τ_h efficiencies for the different τ_h decay modes are shown in Fig. 5.17b for a L1 threshold of 50 GeV. An enhanced selection of the 1-prong(+π^0) is achieved compared to the 3-prongs(+π^0) decays, ascribed to the improved energy recollection and response of the former. The selection efficiency of multiple prong decays could be improved by

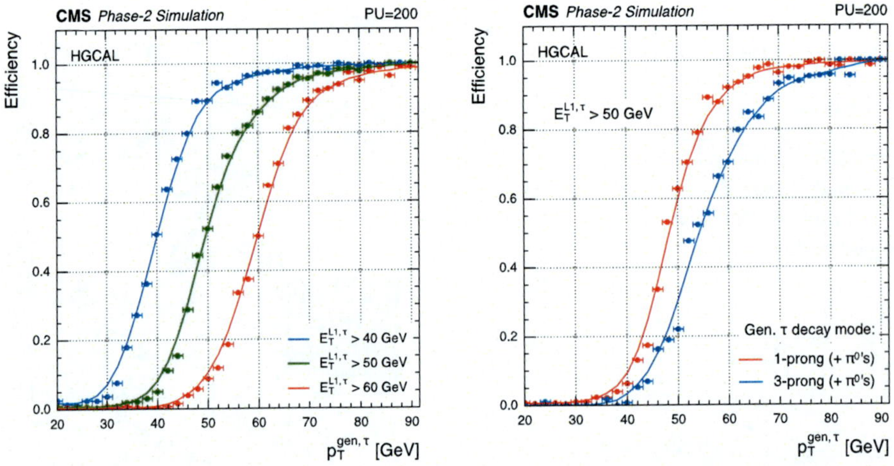

(a) Inclusive efficiency (b) Per decay-mode efficiency

Fig. 5.17 **a** Single-τ_h efficiency as a function of the generator-level τ_h visible p_T for different Level-1 thresholds. **b** Single-τ_h efficiency as a function of the generator-level τ_h visible p_T for a Level-1 threshold of 50 GeV for the different generated decay modes. The efficiencies are evaluated in di-τ_h simulated events with 200 average pileup interactions

merging secondary 3D-clusters present in the region surround the seed, as is done in the Phase 1 algorithm, and/or by matching the calorimetric clusters in the HGCAL to the L1 tracks.

In the absence of recorded data, the rates associated to the single L1 τ_h trigger in a pileup environment of 200 are derived from simulated neutrino events. They are evaluated as a function of the hypothetical offline threshold, defined as the value of the visible p_T of the generator-level τ_h at which a 95% selection efficiency is achieved for a given L1 threshold. The mapping between the L1 and offline thresholds can be found in Fig. 5.18a. The single-τ_h rates as a function of the offline threshold are presented in Fig. 5.18b for the inclusive L1 τ_h candidates and for each of the reconstructed decay modes of the algorithm. Given the limited p_T range of the samples used to derive the offline thresholds, no information on the rate for offline thresholds larger than 90 GeV is accessible. Hence, a direct comparison with the rate associated to the single L1 τ_h trigger with the tower-based approach is not possible, since the rate allocated in the menu in Table 5.1 corresponds to an offline threshold of 150 GeV. Naturally, the rates resulting from the HGCAL algorithm are expected to be higher than the ones with the tower-based approach, as the version of the algorithm presented here misses pivotal rate reduction methods such as the evaluation of the isolation and the QCD-jet background rejection. Additionally, the rate of the algorithm could be significantly reduced by matching the 3D-clusters to the L1 tracks within a predefined cone size, possibly applying certain quality and isolation criteria. Such developments are foreseen for future iterations of the algorithm.

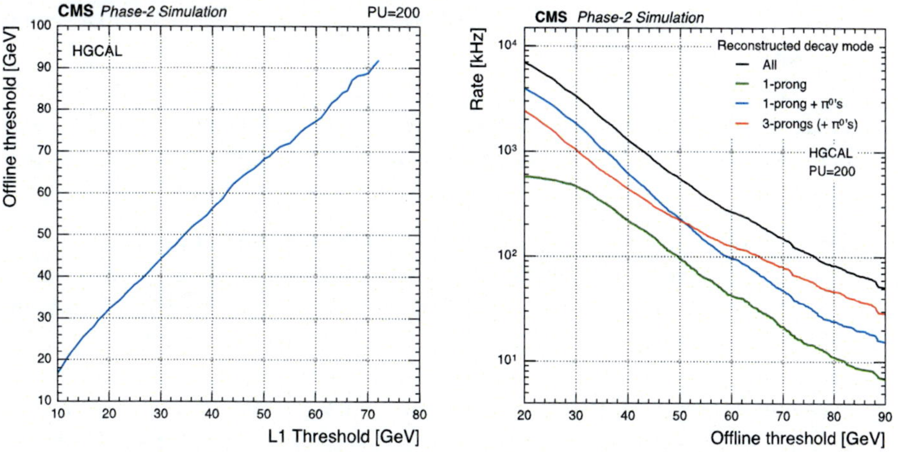

(a) L1-to-offline map (b) Rates

Fig. 5.18 **a** Mapping between the Level-1 and offline threshold, defined as the 95% efficiency point for a given Level-1 threshold. **b** Single-τ_h rate as a function of the offline threshold, evaluated in simulated neutrino events with 200 average pileup interactions

5.4.2 *Impact of the HGCAL Front-End Algorithms*

The 3D-clusters used as input to the L1 τ_h algorithm are aggregates of trigger cells (TC) built in the HGCAL front-end electronics. These TCs are formed by summing the energy deposits in the silicon sensor cells and the scintillator tiles, as described in Sect. 5.2. The affordable trigger bandwidth is insufficient to read out all of the TCs in the event: the volume of the trigger data has to be reduced before being sent to the back-end electronics for clustering. Several compression algorithms are under study for this purpose:

- The **Threshold** algorithm is the baseline algorithm used to derive the results in this chapter. The TCs are selected if they have an energy above a threshold currently set at 1.35 MIP_T. The selection results in variable data size and variable channel content sent to the back-end electronics per bunch-crossing.
- The **Best Choice** (BC) algorithm consists in selecting the N highest energy TCs per module, where N is a value ranging from 4 to 13 depending on the detector region. The selection leads to a fixed data size but variable channel content.
- The **Best Choice Coarse** (BC Coarse) is similar to the BC, but in this case the sorting and selection is done on coarsened groups of 2×2 TCs. As for the BC algorithm, the selection leads to a fixed data size but variable channel content.
- The **Super Trigger Cell** (STC) algorithm consists in summing the energy over groups of TCs, read out as a block. The grouping consists of 2×2 TCs in the high density modules and 4×4 TCs in the low density modules and the scintillator. The position assigned to these super trigger cells (STC) corresponds to the position of

the maximum energy TC. The method leads to a fixed data size and fixed channel content, as all the STC are sent out.

- The **Mixed Best Choice and Super Trigger Cell** (BC+STC) algorithm is a combination of the BC and STC algorithms. The former is applied in the CE-E, as it is better suited for smaller objects such as electrons and τ_h, while the latter is applied in the CE-H, as it is better suited for larger objects such as jets.

As each of the compression algorithms results in a different balance of energy and position resolution, their impact on the reconstruction of the L1 objects has to be evaluated to make an educated choice. The performance for the L1 τ_h is presented in the following; it is briefly compared to the performance observed for L1 e/γ and L1 jets.

The mean energy response of the calibrated L1 τ_h candidates as a function of the visible p_T and η of the generated τ_h can be found in Fig. 5.19a for the five front-end algorithms. An approximately flat response across all energy ranges and detector regions is observed for all options, meaning the L1 τ_h calibration can compensate for the biases introduced with different compression schemes. The effective RMS of the energy response is shown in Fig. 5.19b as a function of the visible p_T and η of the generated τ_h. Being the τ_h deposits very dense and collimated, the best energy resolution is provided by the Threshold algorithm. The most degraded energy resolution ($\sim$15% worse than Threshold) is observed for the BC Coarse algorithm, which is better suited for showers with lower number of hits. Similar studies performed with L1 e/γ objects favour the BC and BC+STC algorithms in terms of energy resolution; they both have equal performance since only the CE-E deposits are used as input to the L1 e/γ algorithm. The best energy resolution for L1 jets is found with the STC algorithm, as these objects present a higher number of lower energy hits.

In an environment of 200 average pileup, the STC algorithm leads to an average of 4.5 reconstructed pileup clusters per event. It almost doubles the value of 2.2 resulting from the Threshold option. Higher pileup contributions are expected from the STC algorithm because the algorithm merges the pileup and signal energies during the reconstruction of larger areas of TCs. The rest of the front-end options present an average of $\sim$1 reconstructed pileup cluster per event; the lower contamination comes from the fact that these algorithms select the TCs with the highest energy deposits. The performance of the pileup removal BDT with each of the front-end options is illustrated in the ROC curves in Fig. 5.20. The worst pileup rejection is achieved with the STC algorithm; it presents $\sim$60% more clusters per event than in the other algorithms at the 99% signal efficiency working point used by the algorithm. The presence of pileup is more significant for the STC algorithm also in the L1 e/γ trigger reconstruction, while the best performance is achieved by the Threshold option in this case.

The single-τ_h selection efficiency as a function of the visible p_T of the generated τ_h is shown in Fig. 5.21a for the five algorithms. At energy scales above 40 GeV, a slightly better efficiency is granted by the Threshold algorithm, leading to lower offline thresholds associated to each L1 threshold. The single-τ_h rates as a function of the offline reconstructed p_T, defined as the 95% efficiency point for a given L1

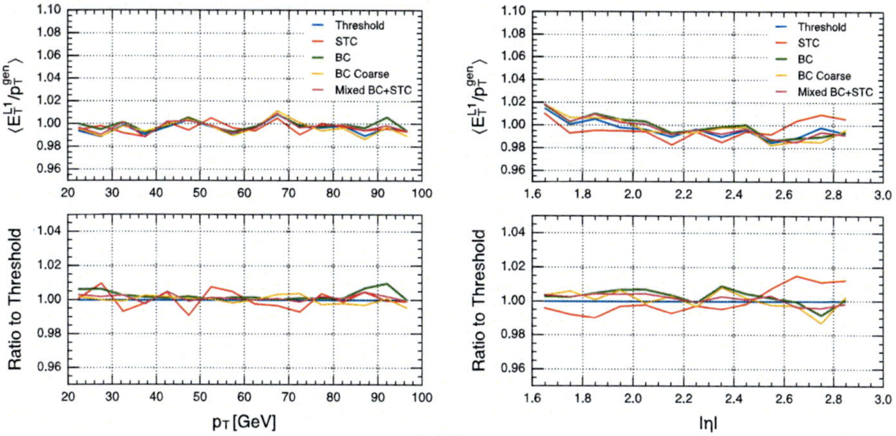

(a) Response

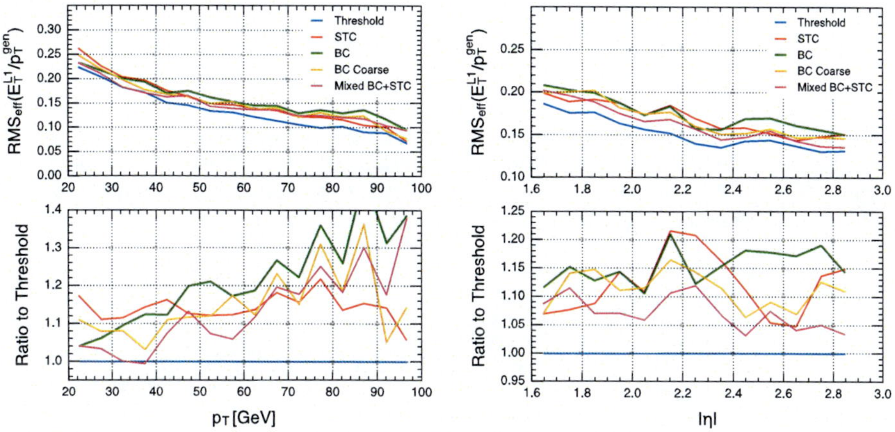

(b) Resolution

Fig. 5.19 **a** Mean Level-1 τ_h energy response as a function of the generator-level τ_h visible p_T (left) and $|\eta|$ (right). **b** Effective RMS of the Level-1 τ_h energy response distribution as a function of the generator-level τ_h visible p_T (left) and $|\eta|$ (right). The results are shown for the different front-end algorithms considered for the HGCAL trigger primitive generator

threshold, are presented in Fig. 5.21b. For an offline threshold of 50 GeV, the lowest rates are generally given by the BC+STC algorithm, whereas the BC and Threshold options yield ~30% higher rates. Owing to the higher pileup particles reconstructed, the STC option results in a ~60% rate increase with respect to the other algorithms at a 50 GeV threshold; however, it approaches the same output rate for thresholds larger than 70 GeV. For e/γ objects, the lowest rates are achieved with the Threshold algorithm, while for L1 jets the Threshold and the BC+STC are the most appropriate.

Towards the choice of the most suited algorithm for all L1 objects, the figure of merit to evaluate is the rate as a function of the offline threshold. It takes into account

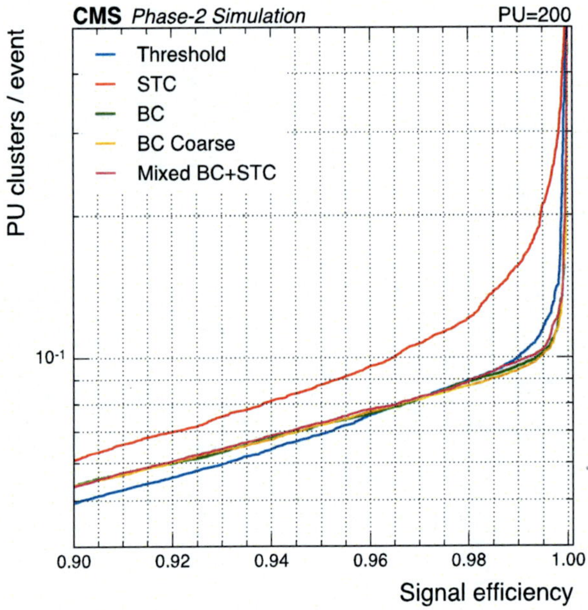

Fig. 5.20 Number of pileup clusters per event as a function of the Level-1 τ_h selection efficiency resulting from the BDT used for pileup rejection, obtained from neutrino and τ_h simulated events with 200 average pileup interactions, for the different front-end algorithms considered for the HGCAL trigger primitive generator

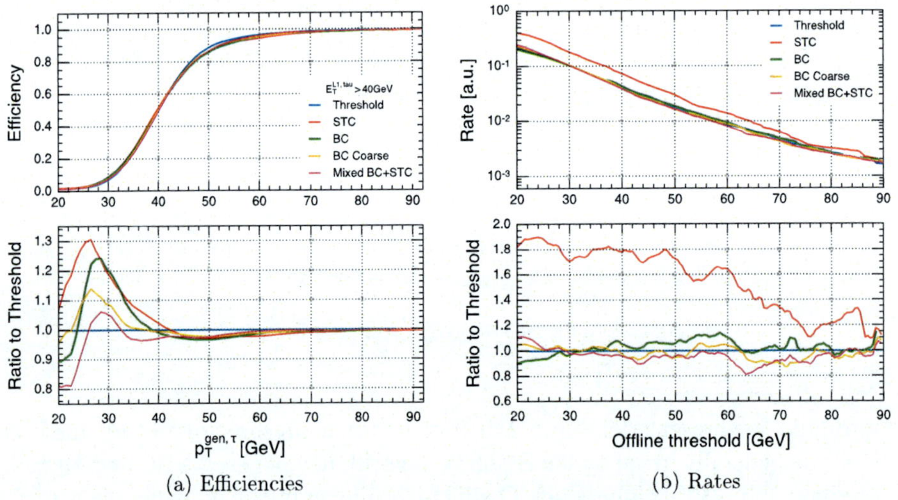

Fig. 5.21 **a** Single-τ_h efficiency as a function of the generator-level τ_h visible p_T for a L1 threshold of 50 GeV. **b** Single-τ_h rate as a function of the offline threshold, defined as the 95% efficiency point for a given Level-1 threshold, evaluated in simulated neutrino events with pileup 200. Results are shown for the different front-end algorithms considered for the HGCAL trigger primitive generator

the background rejection, the signal efficiency and the resolution of the algorithms; in addition, it is the quantity that is used in the definition of the trigger menus for Phase-2. Overall, the Threshold and mixed BC+STC algorithms are the most suited options for all calorimetric objects, as they provide an adequate performance in both the low and high offline threshold regions. The final choice of the front-end algorithm to be deployed in the HGCAL TPG is still to be made as this manuscript is being written.

5.5 Conclusions

Significant changes in the LHC operations are foreseen during the upcoming HL-LHC era, namely the increase in instantaneous luminosity of up to a factor 4 with respect to the current configuration and average pileup values reaching 200 in the ultimate running conditions. Very high particle occupancy and radiation doses are expected; to achieve the high signal sensitivities required for a rich physics program, the CMS Level-1 trigger will be upgraded along with several subdetector components and electronics. The HGCAL is a brand new calorimeter that will replace the highly damaged ECAL and HCAL detectors in the forward region; it will provide trigger primitives with an unprecedented granularity. Making use of the expertise acquired during Run 2 operations, I single-handedly developed the first L1 τ_h algorithm in the HGCAL exploiting the full triggering potential of the HGCAL detector. The efficient τ_h selections resulting from the algorithm will be pivotal for the physics case of the HL-LHC, in particular in the search for lower cross section processes, such as the double Higgs boson production, but also in the physics reach of the $t\bar{t}H$ and tH processes.

Profiting from the clustering of the on-detector electronics, the algorithm builds upon the 3-dimensional shower information of the HGCAL trigger primitives. The fine granularity results in a five-fold improvement in the position resolution with respect to Phase 1. Compared to the limited input variables used for calibration in the Phase 1 trigger, a more detailed knowledge of the shower features allows for sophisticated BDT-based energy corrections to be developed. They result in an energy resolution of 19% ($\sim$14% better than during Run 2) and a sharp selection efficiency that reaches 100% at plateau. The fine trigger primitive generation also results in an enhanced pileup removal, as the diffuse shower features of the pileup-induced particles can be more accurately resolved. In contrast with the coarse pileup estimator of Phase 1, the upgraded trigger implements a multivariate BDT able to capture the genuine signal with a 99% efficiency while suffering from a pileup contamination as low as 0.1 pileup clusters per event. The identification of the τ_h candidate is improved with a multiclassifier BDT which learns the specific features of the different τ_h final states and can predict the correct decay mode with a success rate of 61%.

The full firmware implementation of the L1 τ_h trigger in HGCAL is still to be conducted and tested in the menu along with the other physics objects. Despite being at a preliminary stage, the algorithm already presents a remarkable performance and certainly hints the potential of high-granularity triggering in the forward region. Several improvements to the trigger, mainly concerning the reduction of its rate, have been identified and will be deployed during the subsequent versions of the algorithm. The decay mode multiclassifier will be enlarged to include a dedicated QCD-induced jet identification for background rejection, complemented with the addition of an isolation scheme. The L1 τ_h candidates reconstructed at HGCAL can be matched to the charge and neutral hadrons reconstructed in the L1 track trigger. Additional identification criteria can be applied to reduce the rate at the correlator trigger, where reconstruction à la Particle Flow can be envisioned.

References

1. Cepeda M et al (2018) Report from working group 2: higgs physics at the HL-LHC and HE-LHC. Technical report CERN-LPCC-2018-04, CERN, Geneva, Dec 2018
2. Maltoni F, Pagani D, Shivaji A, Zhao X (2017) Trilinear Higgs coupling determination via single-Higgs differential measurements at the LHC. Eur Phys J C 77(12)
3. CMS Collaboration (2018) Sensitivity projections for Higgs boson properties measurements at the HL-LHC. Technical report CMS-PAS-FTR-18-011, CERN, Geneva
4. CMS Collaboration (2017) The Phase-2 upgrade of the CMS tracker. Technical report CERN-LHCC-2017-009. CMS-TDR-014, CERN, Geneva, Jun 2017
5. CMS Collaboration (2017) The Phase-2 upgrade of the CMS Barrel Calorimeters. Technical report CERN-LHCC-2017-011. CMS-TDR-015, CERN, Geneva, Sep 2017
6. CMS Collaboration (2017) The Phase-2 upgrade of the CMS Muon Detectors. Technical report CERN-LHCC-2017-012. CMS-TDR-016, CERN, Geneva, Sep 2017
7. CMS Collaboration (2017) Technical proposal for a MIP timing detector in the CMS experiment Phase 2 upgrade. Technical report CERN-LHCC-2017-027. LHCC-P-009, CERN, Geneva, Dec 2017
8. CMS Collaboration (2017) The Phase-2 upgrade of the CMS Endcap Calorimeter. Technical report CERN-LHCC-2017-023. CMS-TDR-019, CERN, Geneva, Nov 2017
9. CMS Collaboration (2020) The Phase-2 upgrade of the CMS Level-1 Trigger. Technical report CERN-LHCC-2020-004. CMS-TDR-021, CERN, Geneva, Apr 2020
10. CMS Collaboration (2015) Technical proposal for the Phase-II upgrade of the CMS Detector. Technical report CERN-LHCC-2015-010. LHCC-P-008. CMS-TDR-15-02, CERN, Geneva, Jun 2015
11. PICMG (2008) Advanced TCA base specification: advanced TCA. PICMG, Wakefield, MA
12. CMS Collaboration (2017) Particle-flow reconstruction and global event description with the CMS detector. JINST 12(10):P10003–P10003 Oct
13. Bertolini D, Harris P, Low M, Tran N (2014) Pileup per particle identification. J High Energy Phys 2014(10):59

Chapter 6
Search for the $t\bar{t}H$ and tH Processes in Multileptonic Final States

This chapter describes the analysis strategy aimed at the measurement of the cross section of the Higgs boson produced in association with one or two top quarks, denoted respectively as tH and $t\bar{t}H$, where the Higgs boson decays to WW^*, ZZ^* or $\tau^+\tau^-$. The subsequent decays of the top quarks and the Higgs boson products result in final states with several electrons, muons and hadronically decaying τ leptons (τ_h), commonly referred to as *multileptonic* final states. The physics analysis is performed with the proton-proton collision data recorded at a centre-of-mass energy $\sqrt{s} = 13$ TeV and 25 ns bunch crossing period by the CMS experiment during Run 2. The analyzed dataset corresponds to integrated luminosities of 35.9 fb^{-1}, 41.5 fb^{-1} and 59.7 fb^{-1} collected during 2016, 2017 and 2018, respectively, yielding a total of 137 fb^{-1}, an unprecedented dataset size in high energy physics. Cross section measurements of these production modes have been conducted by the ATLAS and CMS experiments with smaller datasets [1, 2]; a summary of those results is provided in Sect. 2.2.4.

The tH and $t\bar{t}H$ processes have very low cross sections at the LHC, amounting to around 89 fb and 506 fb [3, 4] at $\sqrt{s} = 13$ TeV, respectively. The multileptonic final states targeted in this thesis originate primarily from the Higgs boson decaying into WW^* (BR $\approx 22\%$), followed by the $\tau^+\tau^-$ (BR $\approx 6\%$) and ZZ^* (BR $\approx 3\%$) decays. The products of these decays result in multiple leptons ($\ell = e, \mu$) and τ_h via the processes

- $WW^* \rightarrow \ell\nu_\ell\, qq'$ or $WW^* \rightarrow \ell\nu_\ell\, \ell\bar{\nu}_\ell$,
- $ZZ^* \rightarrow \ell\ell\, qq'$ or $ZZ^* \rightarrow \ell\ell\, \nu\bar{\nu}$,
- $\tau\tau \rightarrow \ell\nu_\ell\bar{\nu}_\tau\, \ell\bar{\nu}_\ell\nu_\tau$, $\tau\tau \rightarrow \ell\nu_\ell\bar{\nu}_\tau\, \tau_h\nu_\tau$ or $\tau\tau \rightarrow \tau_h\bar{\nu}_\tau\, \tau_h\nu_\tau$.

The decays $H \rightarrow ZZ^* \rightarrow 4\ell$ are not included in the search, as they profit from a dedicated analysis [5]; in any case, the BR of this process is too small for the measurement of the tH and $t\bar{t}H$ processes with the available datasets. The top quark decays either

C. Martin Perez, *Search for the Higgs Boson Produced in Association with Top Quarks with the CMS Detector at the LHC*, Springer Theses,
https://doi.org/10.1007/978-3-030-90206-3_6

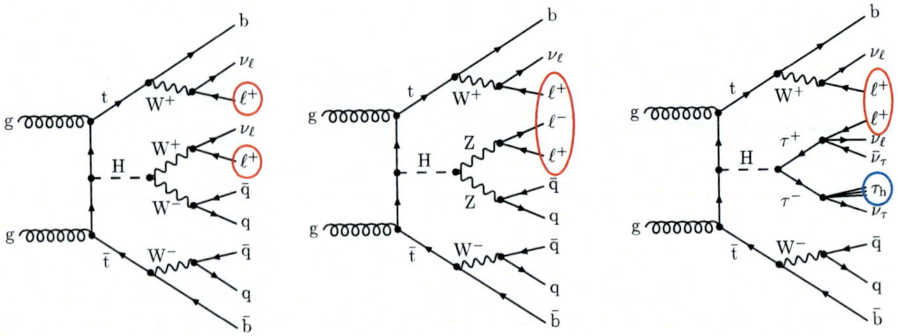

Fig. 6.1 Examples of leading-order Feynman diagrams of the tt̄H process with multileptonic final states in hadron colliders. The categories in the analysis are defined according to the number of leptons ($\ell = e, \mu$) and hadronically decaying τ (τ_h) present in the final state

leptonically ($t \to bW^+ \to b\ell^+\nu_\ell$) or hadronically ($t \to bW^+ \to bqq'$), and analogously for the antitop quarks. Examples of leading order Feynman diagrams of the tt̄H production are shown in Fig. 6.1. A graphical display of a tt̄H event candidate recorded by the CMS experiment in 2018 is presented in Fig. 6.2; it results from the decay of the Higgs boson into a pair of W bosons. Besides the presence of multiple electrons, muons and τ_h, the experimental signatures of the analysis include one (tH) of two (tt̄H) b-jets from the top quark decays, light-quark jets from the Higgs boson or the top quark(s) decays and E_T^{miss} due to the neutrinos.

The exploration of the tt̄H and tH productions in multileptonic final states requires the experimental capability to identify and reconstruct the multiple objects of the final states, the methods used for this being presented in Sect. 6.1. The reconstructed objects are used in the selection of the signal-like events, which enter the so-called *signal regions* (SR), described in Sect. 6.2. These are defined according to the detector signature of the targeted decays, in such way that the events arising from other background processes are rejected as much as possible while keeping a reasonable signal selection efficiency. The main background contributions in the signal regions arise from the irreducible tt̄W, tt̄Z and diboson processes, along with the reducible backgrounds from tt̄+jets in which non-prompt leptons are misidentified as prompt leptons or quark and gluon jets are misidentified as τ_h. An accurate understanding of the expected signal and background yields is required. Monte Carlo (MC) simulations are used to estimate the signal and irreducible backgrounds; the reducible backgrounds, imperfectly modelled by the simulation, are estimated from signal-free data regions. The estimation of the backgrounds is explained in Sects. 6.3 and 6.5.

The observed data events are compared to the predicted signal and background yields in the signal regions. Various imprecisely-known experimental or theoretical effects can affect the yields of the data and the simulation; they are evaluated and taken into account as systematic uncertainties associated to the result, included in the statistical interpretation as explained in Sect. 6.6. The ultimate goal is to measure the cross section of the tt̄H and tH processes; it is done by means of a maximum-likelihood (ML) fit to the distributions of sensitive observables in the signal regions

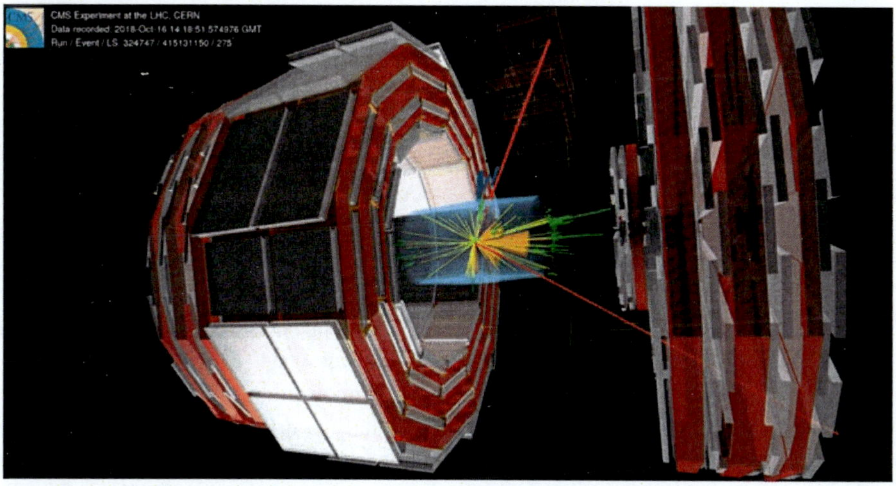

Fig. 6.2 Graphical display of a $t\bar{t}H$ event candidate recorded by the CMS detector. The Higgs boson decays into a pair of W bosons, while each top quark decays to a W boson and a *b*-quark. Two of these W bosons decay into two same-sign muons (red lines) and neutrinos (undetected). The other two W bosons produce four light quarks. The orange cones correspond to jets resulting from the hadronization of the quarks

and background-enriched control regions. These observables are built with multivariate (MVA) techniques so that the maximal separation is achieved in the shape of the distributions of the signals and the backgrounds, as explained in Sect. 6.7. The value of the top Yukawa coupling y_t is extracted taking into account that the $t\bar{t}H$ and tH production rates vary differently as a function of this coupling (see Sect. 2.2). For this purpose, both SM and BSM interpretations are included.

The analysis presented in this manuscript constitutes the first in which the $t\bar{t}H$ and tH production processes are studied in a combined search, motivated by the high degree of overlap between their experimental signatures. The analysis strategies are inspired from the previous searches (see Sect. 2.2.4); however, innovative tools that improve the sensitivity are introduced in the search. The physics object reconstruction profits from optimized methods, in particular from the DNN-based τ_h and *b*-jets identification and a dedicated BDT-based prompt lepton identification. The separation of the $t\bar{t}H$ and tH signals from the background processes is attained with advanced multivariate techniques, following two independent approaches. The first approach, referred to as the *main analysis*, constitutes the public result within the CMS collaboration [6]. It employs machine learning techniques consisting of BDTs [7] and DNNs [8] to separate the signals between themselves and from the backgrounds. The second approach, referred to as *control analysis*, is my personal contribution to the analysis; it is developed as a cross check of the methods employed in the main analysis. It is based on the Matrix Element Method (MEM) [9], along with suitably chosen single-variable distributions, in both cases targeting the $t\bar{t}H$ signal alone. The two approaches were developed in parallel and share the same strategies described in this chapter. Comprehensive explanations of the signal extraction techniques of the main analysis and the control analysis are given in Chaps. 7 and 8, respectively.

The physics search presented in this thesis is the result of the collaborative work between scientists from several universities and institutes in CMS [6]. Within this team, I have had the chance to participate in all steps of the analysis chain, from beginning to end. This includes, in particular, the technical implementation and optimization of the particle reconstruction techniques, the definition of the trigger selections and the signal regions, as well as the processing of all the data and simulated events. The core of my contribution is the development of the control analysis, namely the MEM used for signal extraction, together with the statistical interpretation of the results in the main analysis and in the control analysis. Moreover, I have had a leading role in the coordination and technical synchronization between the different institutes, as well as in the development of the analysis documentation, as I have been the main author of the internal technical note describing the analysis. I have presented the work included in this thesis in two international conferences: the Higgs Hunting conference in 2018 [10] and the High Energy Physics conference of the European Physical Society (EPS-HEP) in 2019 [11]. The results of the analysis were made public during the 40th International Conference on High Energy Physics (ICHEP) in 2020 [12] and released in a dedicated CMS publication after that [13].

6.1 Physics Objects Selection

One of the key ingredients of the analysis is the object identification, in particular the leptons, as they constitute a handle to reject the QCD and non-prompt lepton backgrounds. The selection of the analysis objects is designed to achieve a maximal purity in the signal regions; the objects are individually reconstructed by combining the information of the subdetectors with the standard CMS algorithms (see Sect. 3.4) and, on top of these, analysis-specific identification and quality criteria are applied to enhance the sensitivity. Different levels of object selections are defined and used to construct the non-overlapping regions of the analysis, namely the signal regions and the signal-free data regions where the backgrounds are estimated and validated.

Muons and Electrons

Muons are reconstructed in CMS by linking track segments from the silicon tracker to compatible hits in the muon chambers (see Sect. 3.4.2). Only muons within the muon system acceptance ($|\eta| < 2.4$) and with $p_T > 5$ GeV are considered. The muons used in this analysis pass the loose working point of the muon PF identification, designed to discriminate the genuine muons from those arising from decays of kaons or pions according to the quality of the reconstructed tracks. Additionally, upper limits are set on the longitudinal and transverse impact parameters (d_z, d_{xy}) of the muon track with respect to the primary vertex, as well as on the significance of the 3-dimensional impact parameter (SIP_{3D}), which is defined as the impact parameter divided by its uncertainty. These requirements serve to reject badly reconstructed tracks or tracks originating from pileup vertices.

Electrons are reconstructed by linking a track in the tracker to a supercluster in the ECAL (see Sect. 3.4.3). They are required to be within the tracker acceptance ($|\eta| < 2.5$) and to have $p_{\mathrm{T}} > 7\,\mathrm{GeV}$. The electron identification is based on the loose working point of the BDT algorithm (*Fall17noIsoV2*), which combines observables related to the matching of tracks and clusters, the compactness of the shower and the amount of bremsstrahlung radiation along the electron trajectory. Additionally, electrons must satisfy quality criteria regarding the width of the corresponding ECAL supercluster ($\sigma_{i\eta i\eta}$), the ratio of the HCAL and ECAL deposited energies (H/E) and the difference between the reciprocal of the cluster energy and the reciprocal of its track momentum (1/E-1/p). Similarly to the muons, upper limits on the impact parameters of the electron track with respect to the primary vertex are applied. Electron candidates that arise from photon conversions are removed by requiring that the associated track leaves a hit in each layer of the pixel detector that it crosses except for at most one.

The muons and electrons satisfying the criteria above are referred to as *loose leptons*; they are used for overlap cleaning between the different objects, as well as for the computation of the leptons invariant mass to reject background processes containing leptonically-decaying Z bosons (mainly $\mathrm{t\bar{t}Z}$). In addition to the loose selection, two tighter levels of selection are defined. The most stringent one, resulting in the so-called *tight leptons*, is applied to select the leptons produced from W, Z or τ decays, while rejecting as much as possible the contributions from non-prompt leptons from the decays of *B* hadrons or in-flight decays of pions and kaons, typically present in $\mathrm{t\bar{t}}$+jets background events. Tight leptons are used to define the signal regions of the analysis. Even after applying the tight selection, there is a residual contribution of non-prompt leptons in the signal regions, which is estimated from data control regions. These control regions are defined with the same selections as the signal regions but relaxing the lepton identification from the tight to the *fakeable lepton* selection. Fakeable leptons are additionally used to compute global quantities used for event selection. The complete list of cuts applied to the three levels of muon and electron selections (loose, fakeable, tight) can be found in Tables 6.1 and 6.2.

Prompt leptons are expected to be isolated, while non-prompt leptons are often reconstructed within jets. In contrast with the standard method used in CMS, where the isolation of the leptons is evaluated in a fixed cone size, the leptons in this analysis are identified by means of their *mini-isolation* (I_ℓ). This quantity evaluates the scalar p_{T} values of charged particles, neutral hadrons and photons reconstructed within a cone centred on the lepton direction that shrinks with increasing values of the lepton p_{T}. The mini-isolation enhances the selection efficiency of isolated leptons in events with boosted hadronic activity. The contribution of pileup to the isolation sum is reduced by considering only charged particles originating from the lepton production vertex and by applying corrections to the neutral particles that enter the sum. The mini-isolation is computed as

$$I_\ell = \sum_{\text{charged}} p_{\mathrm{T}} + \max\left(0, \sum_{\text{neutral}} p_{\mathrm{T}} - \rho\mathcal{A}\left(\frac{R}{0.3}\right)^2\right), \tag{6.1}$$

Table 6.1 Loose, fakeable and tight selection criteria for muons (see text). The observables with † indicate that the selection is applied when the muons fails (passes) the requirement lepton-MVA > 0.85. A hyphen (−) indicates selection criteria that do not apply

	Muons		
Observable	Loose	Fakeable	Tight
p_T	>5 GeV	>10 GeV	>10 GeV
$\|\eta\|$	<2.4	<2.4	<2.4
$\|d_{xy}\|$	<0.05 cm	<0.05 cm	<0.05 cm
$\|d_z\|$	<0.1 cm	<0.1 cm	<0.1 cm
SIP_{3D}	<8	<8	<8
Muon identification	Loose	Loose	Medium
Mini-isolation I_ℓ	$<0.4 \times p_T$	$<0.4 \times p_T$	$<0.4 \times p_T$
b-tagging of nearby jet	–	Loose (Medium)†	Medium
$p_T^{jet}/p_T^\ell - 1$	–	<0.5 (–)†	–
Lepton-MVA score	–	<0.85 (>0.85)†	>0.85

Table 6.2 Loose, fakeable and tight selection criteria for electrons (see text). The observables with † indicate that the selection is applied when the electron fails (passes) the requirement lepton-MVA > 0.80. A hyphen (−) indicates selection criteria that do not apply

	Electrons		
Observable	Loose	Fakeable	Tight
p_T	>7 GeV	>10 GeV	>10 GeV
$\|\eta\|$	<2.5	<2.5	<2.5
$\|d_{xy}\|$	<0.05 cm	<0.05 cm	<0.05 cm
$\|d_z\|$	<0.1 cm	<0.1 cm	<0.1 cm
SIP_{3D}	<8	<8	<8
$\sigma_{i\eta i\eta}$	–	<{0.011/0.030}[a]	<{0.011/0.030}[a]
H/E	–	<0.10	<0.10
1/E - 1/p	–	>−0.04	>−0.04
Track missing hits	≤ 1	0	0
Electron identification	Loose	Eff-80 (Loose)†	Loose
Mini-isolation I_ℓ	$<0.4 \times p_T$	$<0.4 \times p_T$	$<0.4 \times p_T$
b-tagging of nearby jet	–	Medium	Medium
$p_T^{jet}/p_T^\ell - 1$	–	<0.7 (–)†	–
Lepton-MVA score	–	<0.80 (>0.80)†	>0.80

[a]Barrel/endcaps

where ρ represents the energy density of neutral particles reconstructed within the geometric acceptance of the tracking detectors and $\mathcal{A}$ are the effective area corrections, obtained from the simulation by studying the correlations of I_ℓ and ρ separately for muons and electrons. The size of the cone is given by

$$R = \begin{cases} 0.05 & \text{if } p_\mathrm{T} > 200\ \text{GeV} \\ 10\ \text{GeV}/p_\mathrm{T} & \text{if } 50 < p_\mathrm{T} < 200\ \text{GeV} \\ 0.20 & \text{if } p_\mathrm{T} < 50\ \text{GeV} \end{cases}, \tag{6.2}$$

compared to the 0.4 (0.3) cone size used for muons (electrons) in the standard CMS isolation.

The properties of the reconstructed jet closest to the lepton (within $\Delta R < 0.4$) are instrumental to select prompt leptons. The most suitable observables for such purpose are the ratio of the lepton and the jet p_T values ($p_\mathrm{T}^\ell/p_\mathrm{T}^\mathrm{jet}$), the component of the lepton momentum in the direction transverse to the jet ($p_\mathrm{T}^\mathrm{rel}$), the number of charged constituents in the jet and the b-tagging score of the jet, the latter being evaluated with the DeepJet discriminator [14]. The mini-isolation and jet properties are used to determine the different levels of lepton selection, as shown in Tables 6.1 and 6.2.

In order to enhance the separation power between prompt and non-prompt leptons, the aforementioned variables are combined in a BDT-based multivariate discriminator (*lepton-MVA*) specifically developed for this analysis. The discriminator is trained with simulated events separately for electrons and muons. The signal leptons are those arising from the prompt decay of the W boson or the τ in a $\mathrm{t\bar{t}H}$ simulated sample; the background leptons arise from semileptonic $\mathrm{t\bar{t}}$ events which are not matched to a prompt W boson or τ decay. The variables used in the training are related to the kinematics, isolation, identification and impact parameters of the leptons; they are combined with variables related to the closest jet ($p_\mathrm{T}^\ell/p_\mathrm{T}^\mathrm{jet}$, $p_\mathrm{T}^\mathrm{rel}$, b-tagging score and charged constituents). The excellent performance of the discriminator is illustrated in the ROC curve of Fig. 6.3. The output of this BDT drives the tight lepton selection (see Tables 6.1 and 6.2), where the optimal working point was carefully tuned to maximize the physics sensitivity to the $\mathrm{t\bar{t}H}$ process; it corresponds to a prompt-lepton selection efficiency of ~85% (~60%) for muons (electrons).

The contribution of the non-prompt lepton background in the signal regions is estimated from data in dedicated control regions defined according to the multiplicity of fakeable leptons (see Sect. 6.3.3). The fakeable lepton selection is shown in Tables 6.1 and 6.2; it includes relaxed identification criteria with respect to the tight leptons. The selection cuts are defined to maintain a high acceptance rate of non-prompt leptons, while minimizing the systematic uncertainties associated to the data-driven method. The latter are estimated in dedicated closure tests that compare the data-driven non-prompt background estimation to that obtained from $\mathrm{t\bar{t}}$+jets and QCD multijet simulated events (see Sect. 6.3.3). In order to improve the accuracy of the non-prompt background estimation, the p_T of the fakeable leptons (which do not pass the tight selection) is replaced by the so-called *cone-p_T*. This variable is

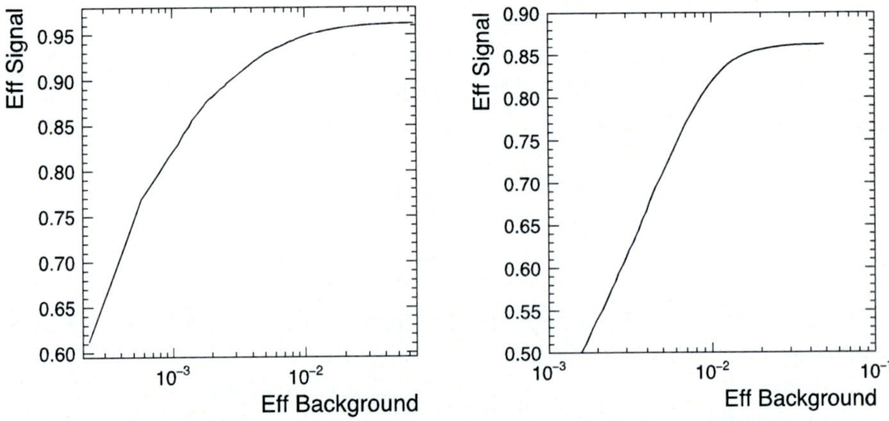

Fig. 6.3 Prompt-lepton (signal) efficiency as a function of the non-prompt lepton (background) efficiency for muons (left) and electrons (right) resulting from the lepton-MVA developed within the $t\bar{t}H$ multileptons analysis. The tight leptons used in the signal regions of the analysis correspond to the working points of 85% signal efficiency for muons and 60% for electrons [15]

designed as a proxy of the p_T of the parton hadronizing into the jet in which the lepton is found. It is defined as 0.9 times the p_T of the nearest jet if the jet is within $\Delta R < 0.4$, and $p_T^\ell/(p_T^\ell + I_\ell)$ otherwise. This specific computation of the fakeable lepton p_T generally exceeds the standard reconstruction p_T.

Hadronic τ Leptons

Hadronically decaying τ leptons are reconstructed with the hadrons-plus-strips (HPS) algorithm [16] (see Sect. 3.4.6). The algorithm reconstructs the individual decay modes by matching the PF jet charged constituents to the cluster strips in the calorimeters expected from the neutral pion decays. Only τ_h which satisfy the criteria $p_T > 20$ GeV and $|\eta| < 2.3$ are considered in the analysis, provided they do not overlap with a loose electron or muon within $\Delta R < 0.3$. As for electrons and muons, an upper cut is applied on the longitudinal impact parameters of the τ_h track relative to the primary vertex.

Similarly to the leptons, three levels of τ_h identification criteria are defined in the analysis (loose, fakeable, tight), the corresponding selections given in Table 6.3. Loose τ_h are used for the overlap removal of the jets and to veto the presence of τ_h in the purely leptonic categories. The fakeable τ_h are employed to define the data control regions where to contribution of the background arising from jets misidentified as τ_h is estimated (see Sect. 6.3.3). Finally, the tight τ_h are used for the selection of events in the signal regions.

The three levels of selection are defined based on the output of the DeepTau [17] algorithm, trained to discriminate genuine τ_h against quark and gluon jets, electrons and muons. Each channel in the analysis requires different τ_h multiplicities and hence suffers from a different background contribution arising from jets misidentified as τ_h. Hence, different working points of the corresponding discriminator are required for the tight τ_h selection in the different categories (see Sect. 6.2); they are chosen

Table 6.3 Loose, fakeable and tight selection criteria for τ_h (see text). A hyphen (–) indicates selection criteria that do not apply.

	Hadronic τ leptons		
Observable	Loose	Fakeable	Tight
p_T	>20 GeV	>20 GeV	>20 GeV
$\|\eta\|$	<2.3	<2.3	<2.3
$\|d_z\|$	<0.2 cm	<0.2 cm	<0.2 cm
Identification versus jets	VVLoose	VVLoose	Per channel†
Identification versus muons	–	VLoose	VLoose
Identification versus electrons	–	VVVLoose	VVVLoose

†See Tables 6.6 and 6.7

to maximize the signal efficiency in each case. They range from the very-loose to the very-tight working points, corresponding to signal efficiencies from 90% to 50% and jet misidentification probabilities from 4% to 0.3%. After overlap removal of the leptons, the presence of background arising from muons and electrons misidentified as τ_h is much less significant: the very loose (very-very-very loose) working points are used in the discriminant against muons (electrons) in all channels.

Jets and b-Tagging

Jets are reconstructed using the anti-k_T algorithm [18] with a distance parameter $R = 0.4$, taking as input the charged and neutral hadrons reconstructed by the PF algorithm (see Sect. 3.4.5). The energy of the jets is corrected for residual pileup effects and calibrated as a function of p_T and η. Jets considered in the analysis are required to pass the loose (tight) PF identification working points in 2016 (2017 and 2018). The ones that share a PF candidate with any fakeable electron, muon or τ_h within $\Delta R < 0.4$ are rejected to avoid misidentification.

The nominal jets (or simply jets) in the analysis satisfy $p_T > 25$ GeV and are found within the tracker acceptance ($|\eta| < 2.4$). An additional jet collection, denoted as *forward jets*, is defined to maximize the acceptance to the spectator quarks in tH events, expected to be emitted in the forward direction. Forward jets have the same p_T requirement as nominal jets but are found within $2.4 < |\eta| < 5.0$. They are subject to a special treatment due to the problematic noise in the forward region resulting from the ageing of the ECAL and the ineffective pileup mitigation with the unforeseen LHC filling scheme in 2017. This noise results in a bias in the estimation of the p_T of soft jets, which is not compensated with the application of the energy corrections. Hence, the p_T requirement is tightened to $p_T > 60$ GeV for forward jets in the region $2.7 < |\eta| < 3.0$. This leads to an improvement of the data-simulation agreement throughout all the detector acceptance, as demonstrated in Fig. 6.4, which shows the $|\eta|$ position of jets and forward jets in a control region enriched in the $t\bar{t}Z$ background (see Sect. 6.5).

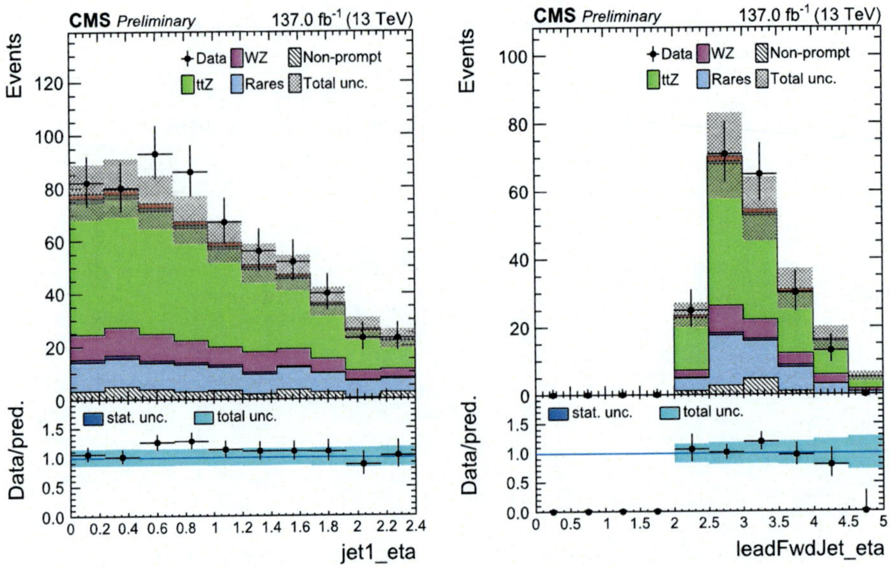

Fig. 6.4 Distributions of $|\eta|$ of the leading jets (left) and leading forward jets (right) in a control region enriched in the $t\bar{t}Z$ background. The data and the simulation agree within uncertainties

Jets that originate from the hadronization of b-quarks (b-jets) are identified from the nominal jet selection making use of the DeepJet algorithm [14]. Two different working points of the algorithm are used: the loose and the medium; they correspond to b-jet selection efficiencies of 84% and 70%, and mistag rates of light quark and gluon jets of 11% and 1.1%, respectively. The tH topology can be extracted making use of the number of *light jets* in the event, defined as

$$N_{\text{light jets}} = (N_{\text{jets}} - N_{\text{loose b-jets}}) + N_{\text{forward jets}} \,, \tag{6.3}$$

where $(N_{\text{jets}} - N_{\text{loose b-jets}})$ represents the number of nominal jets ($|\eta| < 2.4$) that fail loose b-tagging working point and $N_{\text{forward jets}}$ is the number of forward jets ($|\eta| > 2.4$), which are not subject to any b-tagging requirement.

Missing Transverse Energy

The missing transverse energy ($E_{\text{T}}^{\text{miss}}$) accounts for the presence of neutrinos in the event; it is computed as the negative of magnitude of the vectorial transverse momentum sum of all particles reconstructed by the PF algorithm (see Sect. 3.4.7). To mitigate the influence of pileup, the $E_{\text{T}}^{\text{miss}}$ is complemented with the observable $H_{\text{T}}^{\text{miss}}$. The latter is defined as the magnitude of the vectorial transverse momentum sum of the reconstructed jets and fakeable electrons, muons and τ_h in the event, i.e.

$$H_{\text{T}}^{\text{miss}} = |\sum_{\text{leptons}} \vec{p}_{\text{T},\ell} + \sum_{\tau_h} \vec{p}_{\text{T},\tau} + \sum_{\text{jets}} \vec{p}_{\text{T},j}| \,. \tag{6.4}$$

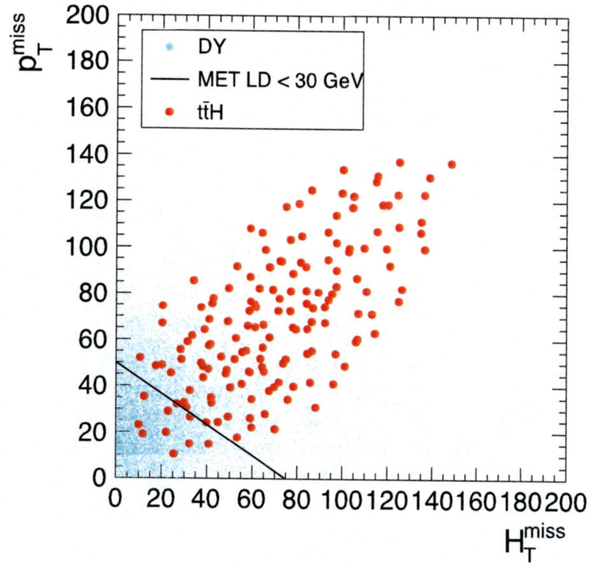

Fig. 6.5 Correlation between E_T^{miss} and H_T^{miss} in the $t\bar{t}H$ and Drell-Yan processes for simulated events with at least 4 reconstructed jets. A requirement of $L_D > 30$ GeV (see Eq. 6.5) is applied in the event selection of the analysis to reject the Drell-Yan background

The H_T^{miss} variable has worse resolution than E_T^{miss} and is therefore less sensitive to undetectable particles; however, it is more robust against pileup, energy mismeasurements and spurious signals. A linear combination of the E_T^{miss} and H_T^{miss} variables is used for event selection in the analysis, defined as

$$L_D = 0.6 \times E_T^{miss} + 0.4 \times H_T^{miss} . \quad (6.5)$$

The coefficients in Eq. 6.5 were optimized in the context of the Run 1 analysis to maximize the separation between the $t\bar{t}H$ signal and the Z+jets background [19]. The combination of the two observables exploits the fact that E_T^{miss} and H_T^{miss} are more correlated in events with genuine missing energy compared to events with instrumental missing energy. Figure 6.5 illustrates the high level of correlation observed for the $t\bar{t}H$ signal compared to the Drell-Yan background for events with at least 4 jets. A typical lower threshold of 30 GeV on the value of L_D is applied in the signal regions of the analysis to remove the contribution from this background.

6.2 Event Selection

The signal events arising from the $t\bar{t}H$ and tH production modes with $H \rightarrow WW^*$, $H \rightarrow ZZ^*$ and $H \rightarrow \o^+\o^-$ are targeted in the 10 mutually exclusive final states (*categories* or *channels*), defined according to the multiplicity of light leptons (electrons or muons) and τ_h. Three of the channels are purely leptonic and contain no τ_h; they require the presence of two leptons ($2\ell ss + 0\tau_h$) with charge of the same sign, three leptons ($3\ell + 0\tau_h$) or four leptons ($4\ell + 0\tau_h$). Four of the channels require

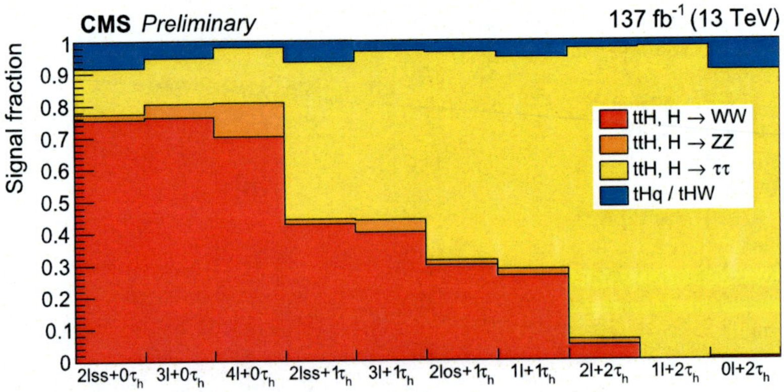

Fig. 6.6 Relative signal contributions present in the 10 signal regions of the analysis, defined according to the number of light leptons ($\ell = e, \mu$) and hadronically decaying τ leptons (τ_h) in the event, obtained from the simulation

one τ_h, either with one lepton ($1\ell + 1\tau_h$), with two leptons with charge of the same ($2\ell ss + 1\tau_h$) or opposite ($2\ell os + 1\tau_h$) sign, or with three leptons ($3\ell + 1\tau_h$). Two channels require two τ_h, along with one lepton ($1\ell + 2\tau_h$) or with two leptons ($2\ell + 2\tau_h$). The last channel is purely hadronic and requires the presence of two τ_h and no leptons ($0\ell + 2\tau_h$).

The relative signal yields in the different categories is shown in Fig. 6.6, considering the inclusive tH process and the $t\bar{t}H$ process split by decay mode. As far as the $t\bar{t}H$ process is concerned, the $H \rightarrow WW^*$ decay is mostly targeted with the three purely leptonic categories ($2\ell ss + 0\tau_h$, $3\ell + 0\tau_h$ and $4\ell + 0\tau_h$), while the $H \rightarrow ZZ^*$ decay is targeted with the $3\ell + 0\tau_h$ and $4\ell + 0\tau_h$ categories. Categories containing one τ_h have comparable yields of the $H \rightarrow WW^*$ and $H \rightarrow ø^+ø^-$ decays, while most of the yield in categories requiring two τ_h arises from the $H \rightarrow ø^+ø^-$ decay. The highest tH signal yield -in any decay mode configuration- is found in channels where the total multiplicity of leptons and τ_h is lower.

Triggers and $E_{\mathrm{T}}^{\mathrm{miss}}$ Filters

The events used in the analysis have been collected with a combination of single, double and triple lepton triggers, along with triggers based on the presence of a lepton and a τ_h or two τ_h. Events recorded in any data-taking period and in the simulated samples are selected if they pass the HLT paths required in each category, defined consistently with the minimum lepton and τ_h multiplicities of the channel in question. Single, double and triple lepton triggers are used in the categories requiring at least one, two or three leptons, respectively. Lepton+τ_h triggers are used in the $1\ell + 1\tau_h$ and $1\ell + 2\tau_h$ categories. In the $0\ell + 2\tau_h$ category, only double-τ_h triggers are used. The p_{T} thresholds of these triggers vary slightly during data-taking periods depending on the experimental conditions; they are shown in Table 6.4. To achieve the highest efficiency, the analysis employs a combination of HLT paths with different p_{T} thresholds when multiple are available. Muon paths encoding requirements on

Table 6.4 HLT paths used to record events, subdivided according to the flavour, and their corresponding p_T thresholds. Thresholds indicated as X-Y refer to the ranges used across the HLT paths of the same kind

HLT path	Flavour	p_T thresholds (GeV)
Single-lepton	e	23–35
	μ	22–27
Double-lepton	ee	23 (leading), 12 (subleading)
	$e\mu$	23 (leading), 8–12 (subleading)
	$\mu\mu$	17 (leading), 8 (subleading)
Triple-lepton	eee	16 (leading), 12 (subleading), 8 (third)
	$ee\mu$	12 (leading), 12 (subleading), 8 (third)
	$e\mu\mu$	9 (leading), 9 (subleading), 9 (third)
	$\mu\mu\mu$	12 (leading), 10 (subleading), 5 (third)
Lepton+τ_h	$e\tau_h$	24 (e), 20–30 (τ_h)
	$\mu\tau_h$	19–20 (μ), 20–27 (τ_h)
Double-τ_h	$\tau_h\tau_h$	35–40

the impact parameters (d_z) or the dimuon invariant mass ($m_{\mu\mu} > 3.8$ GeV) are also included.

In addition, the selected events are required to pass a series of E_T^{miss}-related filters; they are aimed at rejecting events with a large unphysical or uninteresting E_T^{miss}. These typically relate to spurious signals in the detector, originating from particles produced from the interactions of protons in the low density tails of the beam (*beam halo*), cosmic rays, miscalibrations and/or detector noise. These events are identified with dedicated algorithms based on the timing, the pulse shape and the topology of the signals from the subdetectors. The application of such filters guarantees a better agreement between the data and the simulation.

Signal Regions

The 10 signal regions of the analysis are designed to target the $\mathrm{t\bar{t}H}$ production with a specific decay mode configuration. Three of these ($2\ell ss + 0\tau_h$, $3\ell + 0\tau_h$, $2\ell ss + 1\tau_h$) are also expected to be sensitive to the tH process. The list of targeted decays and tailored kinematic selections of each signal region is shown in Tables 6.5, 6.6 and 6.7. The p_T and η requirements applied to electrons, muons and τ_h are driven by the trigger thresholds and the detector acceptance. Electrons and muons in the signal regions are required to pass the tight selection criteria (see Tables 6.1 and 6.2). The same applies to the τ_h (see Table 6.3), but in this case the tight selection is defined with a channel-dependent working point chosen to maximize the sensitivity, also outlined in the tables. The selected jets and b-jets follow the criteria described in Sect. 6.1. The 10 channels are mutually exclusive: lepton and τ_h vetoes are applied to ensure that no selected event enters two categories simultaneously.

The charge of the leptons and τ_h is required to be consistent with the expectation in each decay mode. Some categories implement specific requirements on the charge

Table 6.5 Event selections applied in the $2\ell ss + 0\tau_h$, $3\ell + 0\tau_h$ and $4\ell + 0\tau_h$ categories

Selection	$2\ell ss + 0\tau_h$	$3\ell + 0\tau_h$	$4\ell + 0\tau_h$
$t\bar{t}H$ decays	$t \to b\ell\nu$, $t \to bqq'$, $H \to WW^* \to \ell\nu qq'$	$t \to b\ell\nu$, $t \to b\ell\nu$, $H \to WW^* \to \ell\nu qq'$; $t \to b\ell\nu$, $t \to bqq'$, $H \to WW^* \to \ell\nu\ell\nu$; $t \to b\ell\nu$, $t \to bqq'$, $H \to ZZ^* \to \ell\ell qq'/\ell\ell\nu\nu$	$t \to b\ell\nu$, $t \to b\ell\nu$, $H \to WW^* \to \ell\nu\ell\nu$; $t \to b\ell\nu$, $t \to b\ell\nu$, $H \to ZZ^* \to \ell\ell qq'/\ell\ell\nu\nu$
Triggers	Single/double-lepton	Single/double/triple-lepton	
Lepton p_T	$p_T > 25, 15$ GeV	$p_T > 25, 15, 10$ GeV	$p_T >$ 25, 15, 15, 10 GeV
Lepton η	$\lvert\eta\rvert < 2.5(e)/2.4(\mu)$		
Charge	Same-sign leptons, charge quality	–	–
Jets	≥ 3 jets*		≥ 3 jets
b-tagging	≥ 1 medium b-tagged jets or ≥ 2 loose b-tagged jets		
E_T^{miss}	$L_D > 30$ GeV	$L_D > 30$ (45) GeV if $n_{jet} < 4$ (and SFOS†)	
Dilepton mass	$m_{\ell\ell} > 12$ GeV, $\lvert m_{\ell\ell} - m_Z\rvert > 10$ GeV		
4-lepton mass	–	$m_{4\ell} > 140$ GeV	

†SFOS: same-flavour opposite-charge leptons
*The jet multiplicity selections are relaxed to target the tH signal (see text)

of the leptons; they must present the same sign in the $2\ell ss + 0\tau_h$ and $2\ell ss + 1\tau_h$ categories and opposite sign in the $2\ell os + 1\tau_h$ category. The separation is motivated by the fact that the former has a more favourable signal-to-background ratio, since the same-sign requirement removes the large background arising from $t\bar{t}$+jets production with dileptonic decays of the top quarks. The purity of the events in the $2\ell ss + 0\tau_h$ and $2\ell ss + 1\tau_h$ categories is increased by requiring a high-quality charge measurement: the relative uncertainty of the muon track curvature has to be lower than 20% and the independent measurements of the electron charge in the tracker and the ECAL have to be consistent.

The selections on the reconstructed number of (nominal) jets exploit the fact that the jet multiplicity is generally larger in the signal events compared to the background. The requirements follow the expected jet multiplicities in each decay mode, but they are also aimed at selecting events in which some jets fall outside of the detector acceptance or do not pass the kinematic cuts, thereby increasing the signal efficiency. The total number of expected jets in a $t\bar{t}H$ event at LO is $N_{jet} = 10 - 2N_\ell - 2N_\tau$, where N_ℓ and N_τ are the number of leptons and τ_h, respectively. The requirements on the jet multiplicity in the signal regions are relaxed to allow for up to two of these jets to fail the selection. Additionally, a selection based on the multiplicity of b-jets is applied. Two of the jets present in the $t\bar{t}H$ signal originate from b-quarks and should in principle pass the loose working point of the b-tagging discriminator.

Table 6.6 Event selections applied in the $1\ell + 1\tau_h$, $2\ell ss + 1\tau_h$, $2\ell os + 1\tau_h$ and $3\ell + 1\tau_h$ categories

Selection	$1\ell + 1\tau_h$	$2\ell ss + 1\tau_h$	$2\ell os + 1\tau_h$	$3\ell + 1\tau_h$
$t\bar{t}H$ decays	$t \to bqq'$, $t \to bqq'$ $H \to \tau\tau \to$ $\ell\nu\nu\tau_h\nu$	$t \to b\ell\nu$, $t \to bqq'H \to \tau\tau \to \ell\nu\nu\tau_h\nu$		$t \to b\ell\nu$, $t \to$ $b\ell\nu$ $H \to \tau\tau \to$ $\ell\nu\nu\tau_h\nu$
Triggers	Single-lepton, lepton+τ_h	Single/double-lepton		Single/double/ triple-lepton
Lepton p_T	$p_T >$ $30(e)/25(\mu)$ GeV	$p_T > 25, 15(e)/10(\mu)$ GeV		$p_T >$ 25, 15, 10 GeV
Lepton η	$\|\eta\| < 2.1$	$\|\eta\| < 2.5(e)/2.4(\mu)$		
τ_h p_T	$p_T > 30$ GeV	$p_T > 20$ GeV		
τ_h η	$\|\eta\| < 2.3$			
τ_h identification	Medium	Very-loose	Very-tight	Very-loose
Charge	–	Same-sign ℓ, charge quality, $\sum_{\ell,\tau_h} q = \pm 1$	Opposite-sign ℓ	$\sum_{\ell,\tau_h} q = 0$
Jets	≥ 4 jets	≥ 3 jets*	≥ 3 jets	≥ 2 jets
b-tagging	≥ 1 medium b-tagged jets or ≥ 2 loose b-tagged jets			
E_T^{miss}	–	$L_D > 30$ GeV		$L_D > 30$ (45) GeV if $n_{jet} < 4$ (and SFOS[†])
Dilepton mass	$m_{\ell\ell} > 12$ GeV	$m_{\ell\ell} > 12$ GeV, $\|m_{\ell\ell} - m_Z\| > 10$ GeV		

*The jet multiplicity selections are relaxed to target the tH signal (see text)

This selection is likewise relaxed to maintain the signal efficiency, allowing for one of those two jets to fail the reconstruction cuts or the b-tagging loose working point, provided that the other one passes the medium b-tagging working point.

In the categories of the main analysis where a reasonable yield of tH events is expected and where the backgrounds are not dominated by non-prompt leptons ($2\ell ss + 0\tau_h$, $3\ell + 0\tau_h$ and $2\ell ss + 1\tau_h$), the jet multiplicity requirements are modified to accommodate the tH signature. The jet selection in this case is targeted at the tHq production and requires the presence of at least one jet passing the medium b-tagging working point and at least one light jet (see Sect. 6.1); they are applied taking into account that the spectator quark may be emitted in the forward region and that only one b-tagged jet is expected. As both the tH and $t\bar{t}H$ signals are targeted in these categories, events passing either of the targeted jet selections are accepted. The distinction between the two signals is attained at a later stage with dedicated nodes in the DNN algorithm used to extract the signal in the main analysis (see Chap. 7). Since only the $t\bar{t}H$ cross section is measured in the control analysis, the tH-targeted jet selection does not apply.

Table 6.7 Event selections applied in the $0\ell + 2\tau_h$, $1\ell + 2\tau_h$ and $2\ell + 2\tau_h$ categories

Selection	$0\ell + 2\tau_h$	$1\ell + 2\tau_h$	$2\ell + 2\tau_h$
$t\bar{t}H$ decays	$t \to bqq', t \to bqq'$ $H \to \tau\tau \to \tau_h\nu\tau_h\nu$	$t \to b\ell\nu, t \to bqq'$ $H \to \tau\tau \to \tau_h\nu\tau_h\nu$	$t \to b\ell\nu, t \to b\ell\nu$ $H \to \tau\tau \to \tau_h\nu\tau_h\nu$
Triggers	Double-τ_h	Single-lepton, lepton+τ_h	Single/double-lepton
Lepton p_T	–	$p_T >$ $30(e)/25(\mu)$ GeV	$p_T >$ $25, 15(e)/10(\mu)$ GeV
Lepton η	–	$\|\eta\| < 2.1$	$\|\eta\| < 2.5(e)/2.4(\mu)$
τ_h p_T	$p_T > 40$ GeV	$p_T > 30, 20$ GeV	$p_T > 20$ GeV
τ_h η	$\|\eta\| < 2.1$		$\|\eta\| < 2.3$
τ_h identification	Loose	Medium	
Charge	$\sum_{\tau_h} q = 0$		$\sum_{\ell,\tau_h} q = 0$
Jets	≥ 4 jets	≥ 3 jets	≥ 2 jets
b-tagging	≥ 1 medium b-tagged jets or ≥ 2 loose b-tagged jets		
E_T^{miss}	–	–	$L_D > 30$ (45) GeV if $n_{jet} < 4$ (and SFOS†)
Dilepton mass	–	–	$m_{\ell\ell} > 12$ GeV, $\|m_{\ell\ell} - m_Z\| > 10$ GeV

†SFOS: same-flavour opposite-charge leptons

The contamination from Drell-Yan (DY) production with $Z/\gamma^* \to \ell\ell$ in the categories with at least two leptons is reduced by applying a lower threshold on the linear discriminant L_D, which accounts for the presence of neutrinos in the final state. The nominal selection in these channels is set to $L_D > 30$ GeV (see Fig. 6.5); however, this cutoff is modified in some categories to increase the signal efficiency and improve the background rejection. In the $2\ell ss + 0\tau_h$ and $2\ell ss + 1\tau_h$ channels, the cut is only applied if the two reconstructed leptons are electrons, as the probability to mismeasure the charge of the electrons is significant ($\sim$0.1%), while it is negligible for muons (see Sect. 6.3.3). Owing to the steeply falling distribution in N_{jet} of the DY background in the $3\ell + 0\tau_h$, $4\ell + 0\tau_h$, $3\ell + 1\tau_h$ and $2\ell + 2\tau_h$ categories, the L_D requirement is only applied in events that contain less than 4 jets. If the event contains a same-flavour opposite-sign (SFOS) fakeable lepton pair, the requirement is tightened to $L_D > 45$ GeV in these categories.

The background contributions of processes involving Z bosons ($t\bar{t}Z$, tZq, WZ and DY) in the signal regions are suppressed by vetoing events that contain a pair of loose leptons of SFOS with an invariant mass close to the Z boson mass ($|m_{\ell\ell} - m_Z| >$ 10 GeV). In the $2\ell ss + 0\tau_h$ and $2\ell ss + 1\tau_h$ categories, this Z veto is applied only to electron pairs of the same charge due to their higher charge misidentification rate. Moreover, events containing a pair of loose leptons with $m_{\ell\ell} < 12$ GeV are not considered in the analysis; they typically originate from cascade decays of heavy flavour hadrons or low-mass DY processes, which are not well modelled in the simulation. Finally, in the $3\ell + 0\tau_h$ and $4\ell + 0\tau_h$ categories, events containing four loose leptons with $m_{4\ell} < 140$ GeV are rejected to avoid overlap with the $t\bar{t}H$ and tH production modes of the dedicated $H \rightarrow ZZ^* \rightarrow 4\ell$ analysis [5].

6.3 Signal and Background Estimations

Multiple sources of background affect the $t\bar{t}H$ and tH searches. These are initially rejected by putting into place the dedicated selection strategies detailed in Sect. 6.2, but the yields of these background processes still significantly exceed the $t\bar{t}H$ and tH yields. The signal-to-background ratio is particularly disfavoured in channels with low lepton and τ_h multiplicities, as more SM processes are expected present such topologies; however, these final states profit from a higher signal acceptance, and therefore improve the global sensitivity of the analysis. The background processes can be qualitatively classified as irreducible or reducible.

- The **irreducible backgrounds** refer to those processes that are indistinguishable from the signal, as they result in the same final state and have a high probability of entering the signal regions. The leading contribution to the irreducible backgrounds originates from the associated production of a Z or W boson with a top quark pair ($t\bar{t}V$). Along with the tH and $t\bar{t}H$ signals, these backgrounds are also interesting themselves as studies of the SM consistency; their production cross section is comparable to that of the signals. Irreducible background are estimated from the Monte Carlo (MC) simulation.
- The **reducible backgrounds** instead arise from erroneous object identification, mostly suppressed with tight object selection criteria. The most common source of reducible background comes from non-prompt leptons or misidentified τ_h. Additionally, the events where the charge of the leptons is mismeasured or where the electrons arise from photon conversions are considered reducible backgrounds. Although the misidentification probability for these objects is usually small, the cross sections of the background processes from which they originate is many orders of magnitude larger than the signals, and thus yield a sizeable contribution in the signal regions. Reducible backgrounds are estimated from data, expect for the photon conversions, which are estimated from the simulation.

The signal and irreducible background processes relevant to this analysis are listed in Table 6.8, together with the corresponding cross sections. The same table shows the MC generators used to simulate the processes and the associated perturbative QCD accuracy employed in the modelling. More details on the estimation of the signal and background processes are given in Sects. 6.3.1–6.3.3. For all simulated samples, the NNPDF3.0 [20] set of parton distribution functions (PDF) is used. The generated events are interfaced with Pythia [21], that simulates the quark hadronization and fragmentation effects, the underlying events and the multiple parton interactions. The generated events are passed through a detailed simulation of the CMS detector response based on the Geant4 software [22]; they are then processed using the

Table 6.8 List of signals and irreducible backgrounds used in the analysis, estimated with MC simulation. The generators and the perturbative QCD accuracy used in the modelling are indicated together with the cross sections

Process	Generator/qQCD accuracy	Cross section [pb]
Signals		
$t\bar{t}H$ ($H \nrightarrow b\bar{b}$)	MADGRAPH5_AMC@NLO / NLO	0.2118
tHq	MADGRAPH5_AMC@NLO / LO	0.07096
tHW	MADGRAPH5_AMC@NLO / LO	0.01561
$t\bar{t}$+X backgrounds		
$t\bar{t}W$ ($W \rightarrow \ell\nu$)	MADGRAPH5_AMC@NLO / NLO	0.1960
$t\bar{t}WW$	MADGRAPH5_AMC@NLO / LO	0.006981
$t\bar{t}Z$ / $t\bar{t}\gamma^*$ ($t\bar{t} \rightarrow 2\ell 2\nu 2b$)		
$1 < m_{\ell\ell} < 10$ GeV	MADGRAPH5_AMC@NLO / NLO	0.0822
$m_{\ell\ell} \geq 10$ GeV	MADGRAPH5_AMC@NLO / NLO	0.2814
$t\bar{t}$ + jets[†]		
$t\bar{t} \rightarrow 2\ell 2\nu 2b$	POWHEG / NLO	88.40
$t\bar{t} \rightarrow \ell\nu 2q2b$	POWHEG / NLO	366
$t\bar{t} \rightarrow 4q2b$	POWHEG / NLO	378
Electroweak backgrounds		
$W^{\pm}W^{\mp}$ ($\rightarrow \ell\ell\nu\nu$)	POWHEG / NLO	12.2
WZ		
$WZ \rightarrow \ell\ell\ell\nu$	MADGRAPH5_AMC@NLO / NLO	4.43
$WZ \rightarrow \ell\ell qq$[†]	MADGRAPH5_AMC@NLO / NLO	5.60
ZZ		
$ZZ \rightarrow \ell\ell\ell\ell$	POWHEG / NLO	1.256
$ZZ \rightarrow \ell\ell qq$[†]	MADGRAPH5_AMC@NLO / NLO	5.52
Z/γ^* + jets ($Z/\gamma^* \rightarrow \ell\ell$)[†]		
$10 \leq m_{\ell\ell} < 50$ GeV	MADGRAPH5_AMC@NLO / LO	18610
$m_{\ell\ell} \geq 50$ GeV	MADGRAPH5_AMC@NLO / NLO	6077.22
Rare backgrounds		
tW+jets	POWHEG / NLO	35.8
tZq ($Z \rightarrow \ell\ell$)	MADGRAPH5_AMC@NLO / NLO	0.07358
$W^{\pm}W^{\pm}$ + 2 jets	MADGRAPH5_AMC@NLO / LO	0.04926
$W^{\pm}W^{\pm}$ double scattering	PYTHIA / LO	0.2232
WWW	MADGRAPH5_AMC@NLO / NLO	0.2086
WWZ	MADGRAPH5_AMC@NLO / NLO	0.1676
WZZ	MADGRAPH5_AMC@NLO / NLO	0.05701
ZZZ	MADGRAPH5_AMC@NLO / NLO	0.01473
$t\bar{t}t\bar{t}$	MADGRAPH5_AMC@NLO / NLO	0.008213
Higgs boson backgrounds		
ggH	POWHEG / NLO	48.58
VBF	POWHEG / NLO	3.782
WH	MADGRAPH5_AMC@NLO / NLO	1.373
ZH	MADGRAPH5_AMC@NLO / NLO	0.8839
$t\bar{t}WH$	MADGRAPH5_AMC@NLO / LO	0.001582
$t\bar{t}ZH$	MADGRAPH5_AMC@NLO / LO	0.001535
HH (gluon fusion)	MADGRAPH5_AMC@NLO / LO	0.03105

[†]Considered irreducible background only in $1\ell + 1\tau_h$ and $0\ell + 2\tau_h$ categories

same versions of the CMS event reconstruction algorithms as the data. Appropriate corrections are applied to the simulated events in order to account for residual differences observed with respect to the data (see Sect. 6.4). The accuracy of the modelling of the main irreducible backgrounds by the MC simulation is validated by checking the data-simulation agreement in dedicated control regions enriched in the corresponding background (see Sect. 6.5).

6.3.1 Signal Modelling

Samples of the $t\bar{t}H$, tHq and tHW processes produced by MC simulation are used for the purpose of estimating the signal yields in the signal regions and control regions of the analysis. These are simulated using the MADGRAPH5_AMC@NLO generator [23]. The $t\bar{t}H$ process is modelled at NLO accuracy in perturbative QCD (pQCD); it is normalized according to its cross section computed at NLO with the electroweak corrections presented in Sect. 2.2.1. The tH processes are simulated at LO accuracy in qQCD; the estimation of their yield is improved by normalizing the simulated events at NLO accuracy. The tHq sample is generated in the 4FS, as it describes better the additional b-quark in the gluon splitting. The tHW sample is generated in the 5FS to avoid interferences with $t\bar{t}H$ at LO. The s-channel tH process is not considered in the signal simulation, as its cross section is only $\sim$4% of the tHq cross section; its contribution is negligible with the integrated luminosity and sensitivity of the analysis.

The signal samples are generated assuming all couplings of the Higgs boson to have the values expected in the SM. Deviations from the SM prediction of the Higgs-W coupling (λ_{HWW}) and the Higgs-top coupling (y_t), parametrized with the κ_{W} and κ_t coupling modifiers, induce a variation in the cross sections of these processes. In the case of $t\bar{t}H$, variations of κ_t scale the cross section by a factor $|\kappa_t|^2$. In the case of the tH signal, deviations of κ_v and κ_t from the SM prediction follow the parametrization in Eq. 2.45. Additionally, the variations induce changes on the event kinematics of the tH process, uniquely defined by the ratio κ_t/κ_v, as illustrated in Fig. 6.7. These are accounted for by applying appropriate weights to the simulated events. Variations of κ_t are considered in the range $-3 \leq \kappa_t \leq 3$ in steps of 0.25, while κ_v takes values of 0.5, 1.0 and 1.5. The changes in the kinematic properties of the tH events affect the detector acceptance and the efficiency of the analysis, as shown in Fig. 6.8.

6.3.2 Irreducible Background Modelling

The dominant irreducible background in most channels is due to the $t\bar{t}W(W)$ and $t\bar{t}Z$ processes, collectively referred to as $t\bar{t}V$. The $t\bar{t}V$ events that enter the signal regions typically have genuine prompt leptons and b-jets arising from the hadronization of

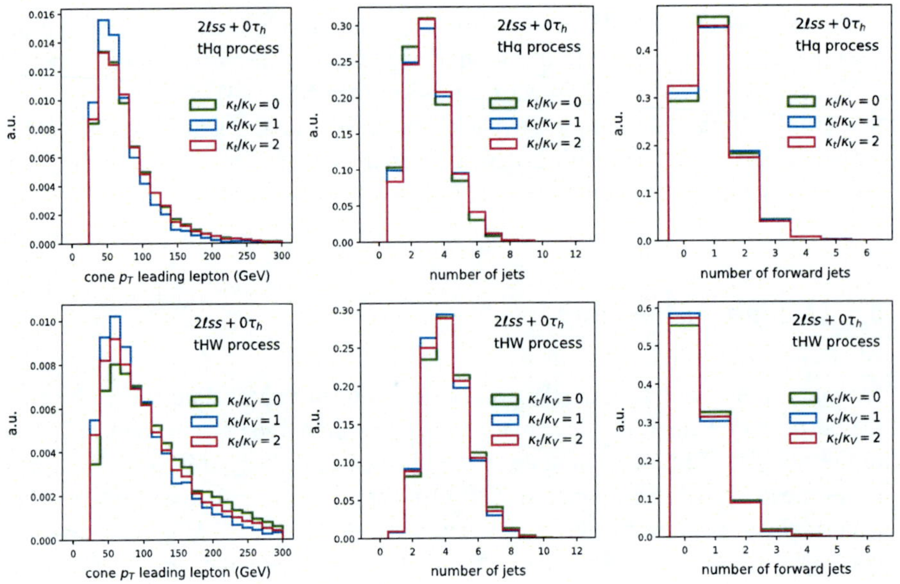

Fig. 6.7 Normalized distributions of various observables for the tHq (upper row) and tHW (bottom row) processes in the $2\ell ss + 0\tau_h$ signal region for different values of the ratio between the Higgs-top (κ_t) and Higgs-W (κ_V) coupling modifiers, obtained from the simulation

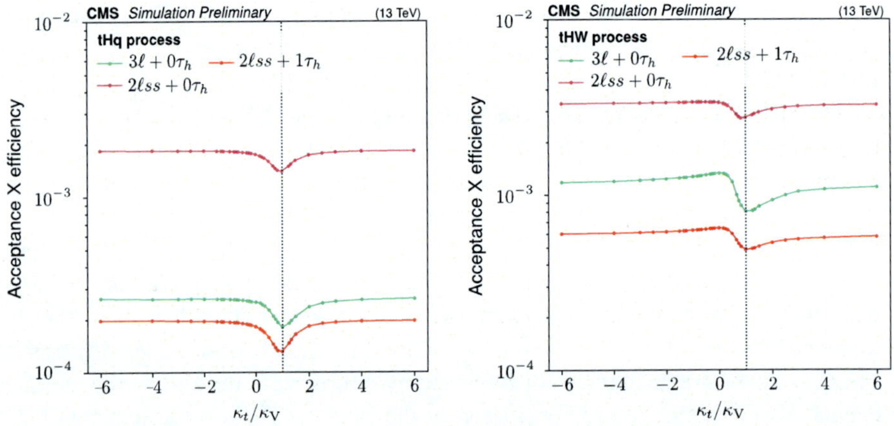

Fig. 6.8 Probability for tHq (left) and tHW (right) simulated events to pass the selection criteria of the $2\ell ss + 0\tau_h$, $3\ell + 0\tau_h$ and $2\ell ss + 1\tau_h$ categories as a function of the ratio between the Higgs-top (κ_t) and Higgs-W (κ_V) coupling modifiers

b-quarks, whereas the reconstructed τ_h are a mixture of genuine τ_h and misidentified quark and gluon jets. Background events from $t\bar{t}Z$ production can pass the Z veto in the event selection when the one of the leptons resulting from the Z boson decays fails to be reconstructed, or when the Z boson decays to τ leptons that decay leptonically, shifting $m_{\ell\ell}$ to lower values due to the presence of neutrinos. Additional background contributions arise from off-shell $t\bar{t}\gamma^*$ production, included in the simulated $t\bar{t}Z$ sample. The $t\bar{t}W$ and $t\bar{t}Z/t\bar{t}\gamma^*$ backgrounds are simulated at NLO in pQCD using MADGRAPH5_AMC@NLO [23], while the $t\bar{t}WW$ background is simulated at LO with the same program. In the case of the $t\bar{t}W$ background, additional $\mathcal{O}(\alpha_s\alpha^3)$ corrections are included to improve the modelling, as recently derived in Ref. [24]. The contribution of this correction is sizeable in this process, amounting to a 12% increase with respect to leading order. The simulated $t\bar{t}V$ samples are normalised to their cross sections computed at NLO accuracy; however, dedicated control regions enriched in $t\bar{t}Z$ and $t\bar{t}W$ events allow for the normalization of these processes to be left freely floating in the maximum likelihood fit, constrained by the data (see Sect. 6.5).

The cross section for $t\bar{t} + jets$ events is around three orders of magnitude larger than the cross section of the $t\bar{t}W$ and $t\bar{t}Z$ processes; however, the $t\bar{t}$+jets background is strongly reduced by the lepton and τ_h identification criteria in most categories. Consequently, the $t\bar{t}$+jets background contribution manifests when a non-prompt lepton or a jet is misidentified as a prompt lepton, or when a quark or gluon jet is misidentified as a τ_h. The estimation of this reducible background is performed with a data-driven method (see Sect. 6.3.3). Nonetheless, a significant irreducible contribution of the $t\bar{t}$+jets events is expected in the $1\ell + 1\tau_h$ and $0\ell + 2\tau_h$ channels. In these channels, the process is modelled with the POWHEG generator at NLO accuracy; it is normalized to the cross section computed at NNLO. The p_{T} spectrum of the top quark is found to be harder in the simulation than in the data: the effect is corrected by applying p_{T}-dependent weights to the simulation (see Sect. 6.4).

Another relevant background in most channels arises from *electroweak* processes, namely the diboson productions WZ and ZZ (and, less notably, $W^{\pm}W^{\mp}$) with two or more jets. The generator MADGRAPH5_AMC@NLO is used at NLO to model the WZ and semileptonic ZZ contributions, while POWHEG at NLO is used to model $W^{\pm}W^{\mp}$ and the purely leptonic decays of ZZ. The processes are normalized to their cross sections computed at NNLO accuracy. In the $1\ell + 1\tau_h$ and $0\ell + 2\tau_h$ channels, the DY process ($Z/\gamma^* \rightarrow \tau\tau$) constitutes another relevant background. It is modelled with MADGRAPH5_AMC@NLO at LO and NLO accuracies in the low and high dilepton invariant mass regimes, respectively, and normalized to its cross section computed at NNLO. The agreement between the data and simulated DY events is improved by applying weights to the simulation derived from dedicated $Z \rightarrow \ell\ell$ control regions (see Sect. 6.4).

Due to the low rate of the signals, smaller background contributions must be taken into consideration, denoted collectively as *rares*. These include processes such as tW+jets, tZq, same-sign $W^{\pm}W^{\pm}$ pairs, triboson (WWW, WWZ, WZZ, ZZZ) and $t\bar{t}t\bar{t}$. The tW+jets process is modelled with POWHEG at NLO accuracy. The $W^{\pm}W^{\pm}$ production is simulated using the program MADGRAPH5_AMC@NLO in

LO accuracy, except for the contribution from double-parton interactions, which is simulated with PYTHIA at LO accuracy. The rest of the rare backgrounds are simulated at NLO using MADGRAPH5_AMC@NLO. All rare processes are normalized to their cross section computed at NLO.

Finally, the *Higgs* boson production via non-$t\bar{t}H$ mechanisms contributes with small yields to all channels. It includes ggH, VBF, WH, ZH, the production of the Higgs boson with W and Z bosons plus pairs of top quarks ($t\bar{t}WH$, $t\bar{t}ZH$) and non-resonant Higgs boson pair production via the gluon fusion (HH). The modelling of the ggH and VBF processes is done with NLO accuracy using the POWHEG generator, while the rest of the processes are modelled using MADGRAPH5_AMC@NLO, at NLO accuracy for WH and ZH, and at LO accuracy for $t\bar{t}WH$, $t\bar{t}ZH$ and HH. The normalization of the Higgs boson processes is applied as computed in Ref. [3].

In order to avoid double-counting of the background estimates obtained from the simulation and from the data, the reconstructed electrons, muons and τ_h in the irreducible backgrounds are required match their generator level equivalent within $\Delta R < 0.3$; upper bounds on their relative transverse momentum are also imposed.

6.3.3 Reducible Background Estimation

The reducible backgrounds, arising from the erroneous object identification, are categorized in three different sources:

- The **fakes background** refers to events in which at least one reconstructed lepton is due to the misidentification of a non-prompt lepton or hadron, or at least one reconstructed τ_h arises from the misidentification of a jet. The main contribution to the fake background originates from $t\bar{t}$+jet production.
- The **charge flip background** results from events in which the charge of a reconstructed lepton is mismeasured. This background is specific to the $2\ell ss + 0\tau_h$ and $2\ell ss + 1\tau_h$ channels; it originates mainly from $t\bar{t}$+jet events in which both top quarks decay leptonically.
- The **conversions background** is due to events in which one or more reconstructed electrons are due to the conversion of a photon (real or virtual) into an electron-positron pair. One converted electron or positron typically carries most of the energy of the photon, while the other fails to be reconstructed.

The fakes and charge flip backgrounds are estimated from the data, while the conversions background is estimated from the simulation. The method used in the estimation are detailed in what follows.

Fakes Background

The fakes background is estimated from data using the *fake factor* (FF) method, developed within the $t\bar{t}H$ multilepton analysis group. The method consists in selecting a sample of events which satisfies the selection criteria of the signal regions (see Sect. 6.2), except that muons, electrons and τ_h are demanded to pass the fakeable selection criteria instead of the tight (see Tables 6.1, 6.2 and 6.3). The events in which the leptons and τ_h satisfy the tight selection are vetoed to avoid overlap with the signal regions. The event sample selected this way is called *application region* (AR). This region is enriched in fake leptons and τ_h; the contribution of the fake background in the signal regions is estimated with a suitable transfer factor, or weight w, applied to events in the AR. The weights are given by

$$w = (-1)^n \prod_{i=1}^{n+1} \frac{f_i}{1 - f_i} , \tag{6.6}$$

where the product extends over all electrons, muons and τ_h which pass the fakeable but fail the tight selection criteria, and n is the total number of such leptons or τ_h. The symbol f_i represents the *fake rate*; it is the probability for an electron, muon or τ_h that passes the fakeable selection to also pass the tight one. The product $(-1)^n$ is introduced in order to avoid double counting in the signal region.

The fake rates f_i are measured separately for leptons and τ_h in dedicated *measurement regions* (MR). In the case of the leptons, the sample is enriched in multijet data events, and the fake rates are measured separately for electrons and muons, parametrized by the lepton cone-p_T and η. Only events containing exactly one fakeable electron or muon and at least one jet further than $\Delta R = 0.7$ from the lepton are considered. The fake rates f_i for τ_h are measured in $t\bar{t}$+jets data events in which the W boson pair from the top quarks decays to an opposite-sign electron-muon pair. The measurement is done separately for the barrel and the endcap regions as a function of the p_T and identification working point. Selected events contain at least one fakeable τ_h and at least two jets, which cannot overlap with the leptons or τ_h within $\Delta R < 0.3$.

To compute the value of the fake rates, the events selected in the MR are subdivided into *pass* or *fail* sub-samples according to whether the fakeable lepton or τ_h passes or fails the tight selection. In the case of the τ_h, the tight selection is defined according to the identification working points of each category (see Tables 6.5, 6.6 and 6.7), for which the fake rates are derived separately. The fake rate f_i is obtained as

$$f_i = \frac{N_{\text{pass}}}{N_{\text{pass}} + N_{\text{fail}}} , \tag{6.7}$$

where N_{pass} (N_{fail}) is the number of events in the pass (fail) sample. The measured electron and muon fake rates in the barrel region with the data collected during 2018 are found in Fig. 6.9. The measured τ fake rates in the barrel for the loose and medium identification points with the data collected in 2018 are found in Fig. 6.10. Depending on the detector region, the fake rates range from 3 to 13% for electrons, from 2 to 20% for muons and from 10 to 50% for τ_h.

In the $2\ell ss + 1\tau_h$ and $3\ell + 1\tau_h$ categories, a modified version of the FF method is used. In this case, the AR is defined by relaxing the identification criteria to the fakeable selection only for the leptons (and not for the τ_h). Correspondingly, only

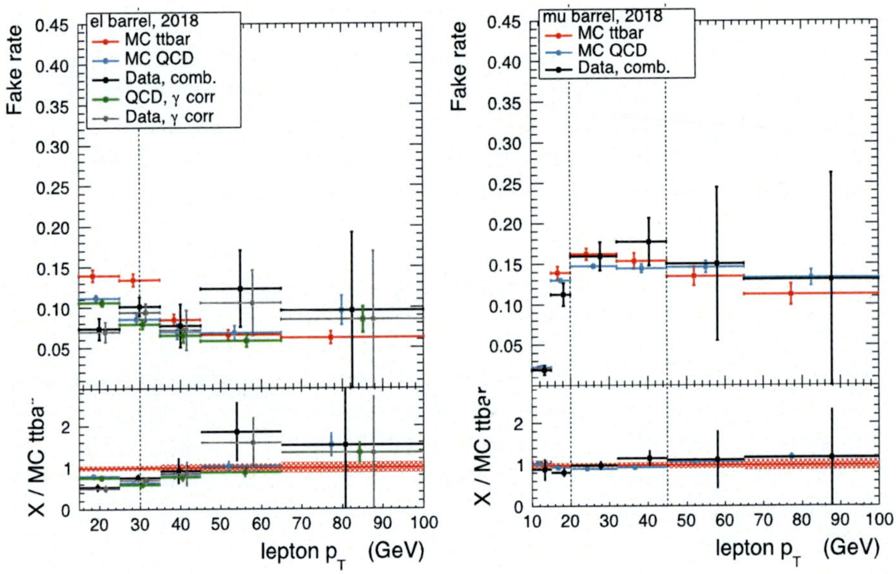

Fig. 6.9 Fake rate f_i for electrons (left) and muons (right) in the barrel, measured in 2018 data, as a function of the lepton p_T. The corresponding fake rates measured in multijet and $t\bar{t}$+jet simulated events are also shown for comparison. In the case of electrons, the measured rates applying photon conversion veto ("γ corr.") are shown

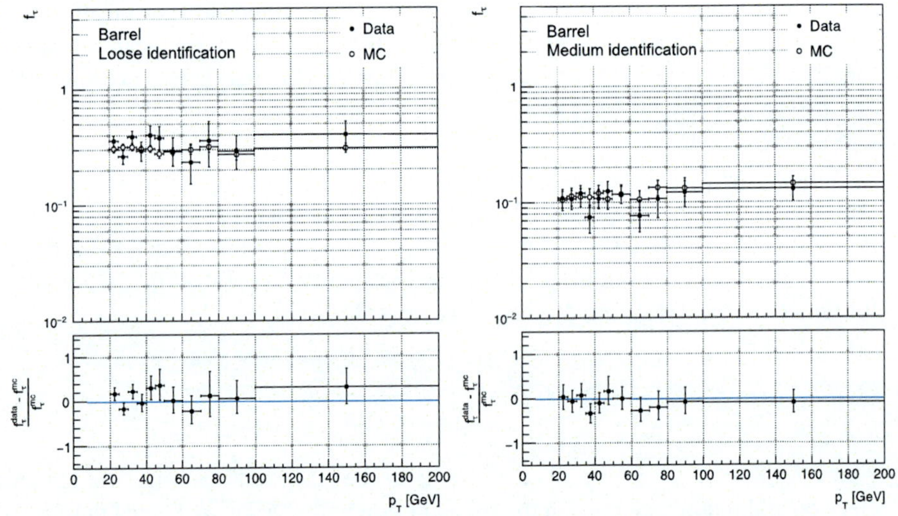

Fig. 6.10 Fake rate f_i for τ_h in the barrel for the loose (left) and medium (right) identification working points, measured in 2018 data, as a function of the τ_h p_T. The fake rates for the loose working point are measured requiring the double-τ_h trigger selection of the $1\ell + 1\tau_h$ and $0\ell + 2\tau_h$ categories. The corresponding fake rates measured in $t\bar{t}$+jet simulated events are also shown for comparison

the lepton fake rates are considered when computing the extrapolation weights in Eq. 6.6. The background contribution where the reconstructed leptons are prompt but the reconstructed τ_h is a misidentified jet is modelled using the simulation. The reason for the modification is that in these categories ~30% of the $t\bar{t}$H yield corresponds to misidentified τ_h from jets. If these events are used for the fake rate measurement, they cannot be used for the purpose of inferring the signal rates, reducing the sensitivity of these categories by ~1/3.

In order to check the validity of the fake background estimation, a closure test is performed in simulated events. It ensures that no potential biases are introduced due to the differences in flavour composition or lepton kinematics between the MR and the signal region. The closure test for the fake lepton background estimation is illustrated in Fig. 6.11 for the $2\ell ss + 0\tau_h$ category. In the figure, the grey line corresponds to the event yield in the signal regions obtained from simulated $t\bar{t}$ events. It is compared to the yield estimated by the method applied to $t\bar{t}$ simulated events in the AR. The latter is done with the fake rates measured both in $t\bar{t}$ (blue line) and multijet (red line) simulated events. The comparison between these two allow to evaluate the dependency of the fake rates between these two samples. The comparison between the $t\bar{t}$ yield in the signal region and the estimation using the fake rate obtained from a multijet sample is an estimator of the bias introduced by the method. Very good closure is observed for muons, but a ~30% non-closure is observed for electrons. The difference is accounted for by applying a multiplicative correction to the fake rates measured in data. A similar closure test is applied in the τ_h misidentification rate.

Flips Background

The charge flip background is relevant in the $2\ell ss + 0\tau_h$ and $2\ell ss + 1\tau_h$ categories of the analysis. It is estimated from data, following a similar strategy to the estimation of the fakes background. In this case, the sample of events in the AR pass the same selection criteria as the signal region, except that the two leptons are required to have opposite charge instead of the same charge. Extrapolation weights are applied to the AR to estimate the flip background in the signal region. In the $2\ell ss + 0\tau_h$ channel, the weight is given by the sum of the probabilities to mismeasure the charge of each of the two leptons; in the $2\ell ss + 1\tau_h$ channel, only the lepton that has the same charge as the τ_h is considered, as the condition $\sum_{\ell,\tau_h} q = \pm 1$ is already applied in the event selection.

The *charge misidentification rate* for electrons (muons) is measured in a MR of $Z/\gamma^* \to ee$ ($Z/\gamma^* \to \mu\mu$) events, parametrized as a function of the p_T and the η of the electron (muon). The set of events selected in these MR is subdivided into samples of same-sign or opposite-sign charge electrons (muons); the charge misidentification rate is given by

$$f_i = \frac{N_{ss}}{N_{ss} + N_{os}}, \tag{6.8}$$

where N_{ss} (N_{os}) is the number of events in the same-sign (opposite-sign) sample.

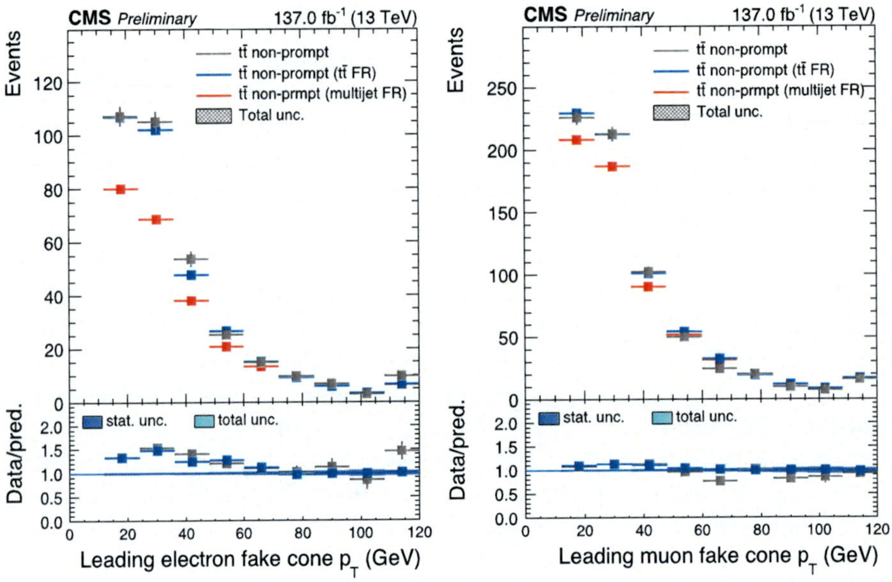

Fig. 6.11 Fake lepton background closure tests showing the distributions of cone-p_T of electrons (left) and muons (right) in the $2\ell ss + 0\tau_h$ category. The figure compares the yields of $t\bar{t}$ simulated events in the signal region (grey) to the yields obtained with $t\bar{t}$ simulated events in the AR weighted by the fake rates derived in a $t\bar{t}$ (blue) and multijet (red) simulated events

The charge misidentification rate for muons is found to be negligible as a result of the quality requirements applied to the tracks during object selection. However, electrons can radiate significantly, leading to an incorrect measurement of the sign of the curvature with a higher probability. The measured for electrons in different p_T and $|\eta|$ regions and years of data-taking are shown in Table 6.9. They vary from $\sim$0.005% in the low-p_T region of the barrel to around $\sim$0.15% in the high-p_T region of the endcap. It is noteworthy that the rates for 2017 and 2018 are significantly lower than in 2016 as a result of the pixel detector upgrade. Owing to the higher mismeasurement rate, the charge flip background in the $2\ell ss + 0\tau_h$ and $2\ell ss + 1\tau_h$ categories is considered only in events with one or two electrons.

Conversions Background

The background contribution from $t\bar{t}$ events with real photons ($t\bar{t}\gamma$) is typically due to a $\gamma \rightarrow e^+e^-$ conversion where one of the lepton carries most of the energy of the photon, and the other one fails to be reconstructed. Such events are mostly suppressed with a photon conversion veto during reconstruction and by requiring that the electron candidate tracks have hits in each of the traversed layers of the pixel detector (see Sect. 6.2). The contribution of $t\bar{t}$ events with virtual photons ($t\bar{t}\gamma^*$) typically produces low-mass electron or muon pairs which can be effectively suppressed by requiring their dilepton mass to be bigger than 12 GeV. Consequently, the yield of $t\bar{t}\gamma$ and $t\bar{t}\gamma^*$ events in the signal regions is small in the $2\ell ss + 0\tau_h$, $3\ell + 0\tau_h$, $2\ell ss + 1\tau_h$ and $2\ell os + 1\tau_h$ categories, and negligible in all other.

Table 6.9 Charge misidentification rates for electrons of different p_T and η ranges, measured in $Z/\gamma^* \rightarrow ee$ events for 2016, 2017 and 2018. The errors associated correspond to the statistical uncertainties in the measurement region

		Charge misidentification rate (%)		
Year	$\|\eta\|$	$10 \leq p_T <$ 25 GeV	$25 \leq p_T <$ 50 GeV	$p_T \geq 50$ GeV
2016	$\|\eta\| < 1.479$	0.056 ± 0.021	0.009 ± 0.001	0.015 ± 0.004
	$1.479 \leq \|\eta\| <$ 2.5	0.061 ± 0.009	0.102 ± 0.004	0.162 ± 0.010
2017	$\|\eta\| < 1.479$	0.012 ± 0.005	0.005 ± 0.001	0.011 ± 0.003
	$1.479 \leq \|\eta\| <$ 2.5	0.043 ± 0.008	0.050 ± 0.003	0.088 ± 0.010
2018	$\|\eta\| < 1.479$	0.004 ± 0.006	0.006 ± 0.001	0.011 ± 0.014
	$1.479 \leq \|\eta\| <$ 2.5	0.036 ± 0.006	0.051 ± 0.002	0.097 ± 0.008

Unlike the other reducible backgrounds, the conversions background is estimated from simulated events of the background processes considered in the analysis (see Table 6.8) along with $t\bar{t}\gamma$, $t\gamma$, $W\gamma$, $Z\gamma$ and $WZ\gamma$. Only events in which at least one reconstructed electron is matched to a generated photon within $\Delta R < 0.3$ are considered. Additionally, the p_T of the generator level photon is required to be greater than half the p_T of the reconstructed electron.

6.4 Data-to-Simulation Corrections

To correct for small disagreements between the data and the simulation, suitable weights, or *scale factors* (SF), are applied to the simulation to recover the behaviour of the data. The discrepancies, of the order of a few percent, generally concern the selection efficiencies or energy resolution of the physics objects, unforeseen detector effects which are not taken into account in the simulation or theoretical limitations that lead to imperfect modelling of some processes. The corrections applied in this analysis are described in the following.

- **Pileup**: The production of the simulated datasets of the analysis is a computing-intensive operation that extends over several weeks; hence, the MC samples are usually generated before knowing the exact pileup distribution of the data. As the detector response and the reconstruction of the physics objects are affected by the pileup, the simulated events are weighted according to the ratio between the pileup distribution in data and that in the MC, illustrated in Fig. 6.12 for the full Run 2 data.

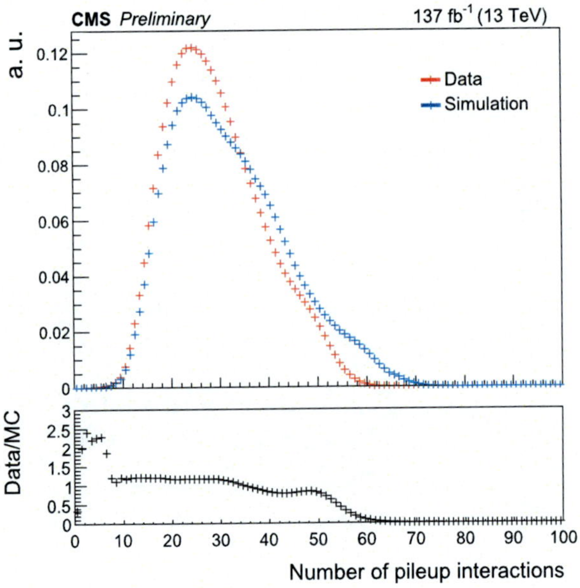

Fig. 6.12 Comparison of the number of pileup interactions observed in the Run 2 data (red) to the profile used in the MC simulations (blue). The ratio in the bottom pad is used to derive a correction weight for events in the simulation

- **Trigger efficiency**: The trigger selections defined in Sect. 6.2 are applied to the data and the simulation; systematic differences between the selection efficiencies observed in both are corrected in the simulation with a scale factor encoding the ratio of the two. The efficiencies are measured with the tag-and-probe technique, separately for each data-taking period. The efficiencies of the single and double lepton triggers are measured in $Z/\gamma^* \rightarrow ee/\mu\mu$ events as a function of the p_T, the multiplicity and the flavour (2μ, $e+\mu$, $2e$, 3ℓ) of the leptons; the corrections take values of 0.94-1.01. The efficiency of the lepton+τ_h triggers are measured in $Z/\gamma^* \rightarrow ee/\mu\mu/\tau\tau$ events in bins of the lepton properties (p_T, η, flavour) and the τ_h properties (p_T, ϕ, decay mode, identification working point). The efficiencies of the double-τ_h trigger are measured in $Z/\gamma^* \rightarrow \tau\tau$ events as a function of the aforementioned τ_h properties.
- **Electron and muon identification and isolation efficiency**: The loose and tight lepton selection criteria (see Tables 6.1 and 6.2) lead to residual differences in selection efficiencies in the data and the simulation. They are measured in $Z/\gamma^* \rightarrow \ell\ell$ events with the tag-and-probe technique as a function of the lepton p_T and η, separately for electrons and muons. It is done in two stages: the efficiency for the reconstructed leptons to pass the loose selection is measured first, followed by the efficiency for loose leptons to also pass the tight selection. The scale factors associated to the former have typical values of 0.94-1.06; the most significant corrections associated to the latter are related with low-p_T electrons, and have values of 0.52-1.05. The efficiencies measured in data and simulation for the loose leptons to pass the tight selection are compared in Fig. 6.13 for the $2\ell ss + 0\tau_h$ category with 2018 detector conditions.

- τ_h **identification efficiency and energy scale**: The efficiency for τ_h to pass the identification working points applied in the signal regions has been measured with the tag-and-probe technique in $Z/\gamma^* \to \tau\tau$ events, in bins of p_T, η and decay mode of the τ_h. The differences in the efficiency observed in the data and in the simulation have values of 0.8-1.2. Another correction is applied to account for the difference in the τ_h energy scale; it is measured in $Z/\gamma^* \to \tau\tau \to \mu\tau_h$ events separately for the different decay modes. The energy scale is generally lower in data than in the MC, with scale factors of 0.97-0.99.
- b**-tag efficiency and mistag rate**: Small differences have been observed between the data and simulation in the efficiency for b-jets to pass the loose and medium working points of the DeepJet algorithm, as well as in the mistag rate for light flavour quarks and gluon jets. The former has been measured in $t\bar{t}$+jet and multijet events and the later in $Z \to \ell\ell$ events with jets. The measurement is done as a function of the jet p_T and η, the flavour of the underlying generator-level jet the b-tagging score. The scale factors in bins of the jet p_T measured during 2017 data-taking are shown in Fig. 6.14 for the loose and the medium b-tagging working points.
- **Prefiring probability of the Level-1 ECAL trigger**: During 2016 and 2017 detector operation, a gradual timing shift of the ECAL crystal signal response in the endcap region was observed. The effect was accounted for and corrected before 2018 operation began. This timing shift was propagated to the ECAL trigger primitives, resulting in their wrong association to earlier bunch crossing, or *prefiring* [26]. The fraction of ECAL trigger primitives produced in earlier bunch crossing strongly depends on η and reaches significant values for the very forward regions of the calorimeter. As a consequence of one of the Level-1 trigger rules forbidding two consecutive bunch crossings to fire [27], the prefiring induced by the ECAL mistiming resulted in a small fractions of events lost. The effect cannot be accounted for by the simulation, so appropriate weights are applied encoding the probability of an event to prefire according to the p_T and η of the forward jets and photons in the event. The impact of prefiring is ~3% and ~4% for the $t\bar{t}H$ and $t\bar{t}Z$ processes, respectively, as shown in Fig. 6.15 for the $2\ell ss + 0\tau_h$ category with 2017 detector conditions.
- **Normalization of the Drell-Yan background**: In the $1\ell + 1\tau_h$ and $0\ell + 2\tau_h$ categories, the contribution of the irreducible $Z/\gamma^* \to \tau\tau$ background is significant. Differences in the event yield of this process in data and MC are applied as normalization scale factors to the simulation. The corrections are estimated in an orthogonal region enriched in $Z/\gamma^* \to \ell\ell$ events, in different bins of jet and b-tagged jet multiplicities. The corrections take values of 0.87-1.49; they are applied in the relevant categories.
- **Top quark p_T in $t\bar{t}$ background**: The irreducible background of $t\bar{t}$ events in the $1\ell + 1\tau_h$ and $0\ell + 2\tau_h$ categories is modelled using the POWHEG [28] generator at NLO accuracy in QCD (see Sect. 6.3.2). The p_T spectrum of the top quark is found to be harder in the simulation than in the data. It is corrected with an analytic function derived from the ratio between the distribution of the top quark p_T at NNLO QCD and NLO electroweak accuracy [29] and the distribution in top

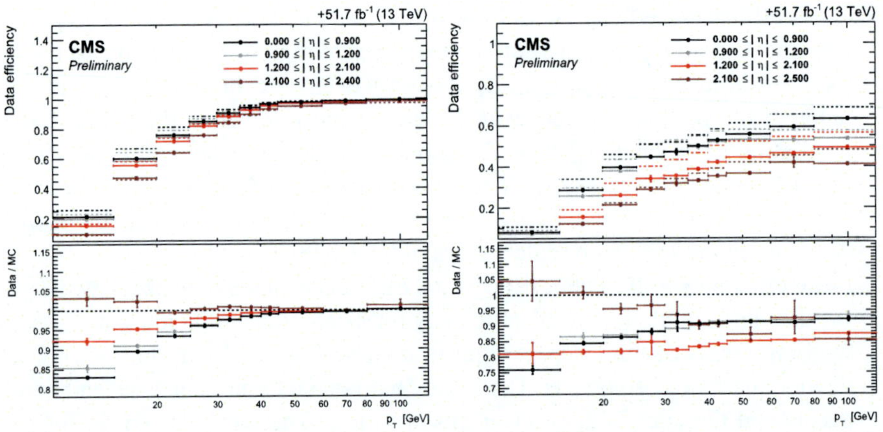

Fig. 6.13 Efficiency of loose muons (left) and electrons (right) to pass the tight selection criteria, as measured in $Z \to \ell\ell$ events using a tag-and-probe method as a function of p_T in bins of $|\eta|$ for 2018 data-taking period in the $2\ell ss + 0\tau_h$ category. The data efficiency corresponds to the dashed lines, while the simulation efficiency corresponds to the solid lines

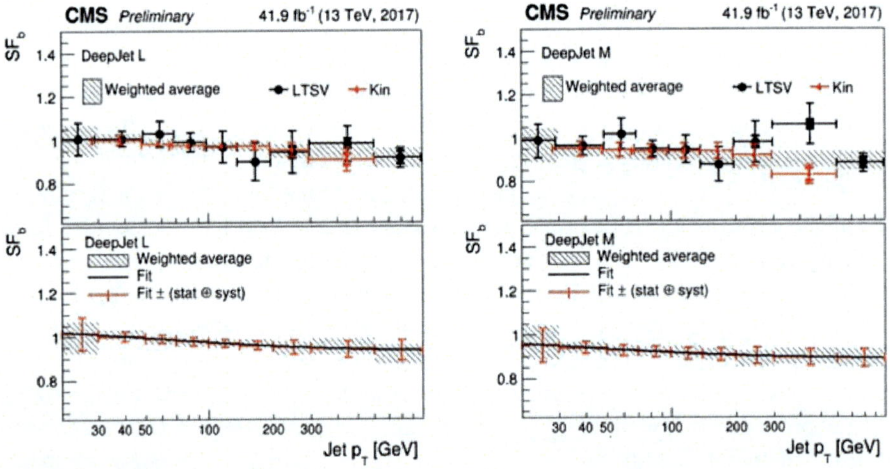

Fig. 6.14 Data-to-simulation scale factors for b-jets as a function of jet p_T for the loose (left) and medium (right) working point using the DeepJet b-tagging algorithm in the 2017 data-taking period. The top pad shows the SFs measured in dilepton top quark pair events (Kin) and in multijet events (LTSV), while the bottom pad shows the combined scale factors with the result of a fitted function superimposed [25]

quark p_T in the POWHEG sample of the analysis. The ratio is fitted with an exponential as a function of the p_T of the generator-level top quark and applied as a correction to the simulation in the relevant signal regions. The function is illustrated in Fig. 6.16 for 2018 detector conditions.

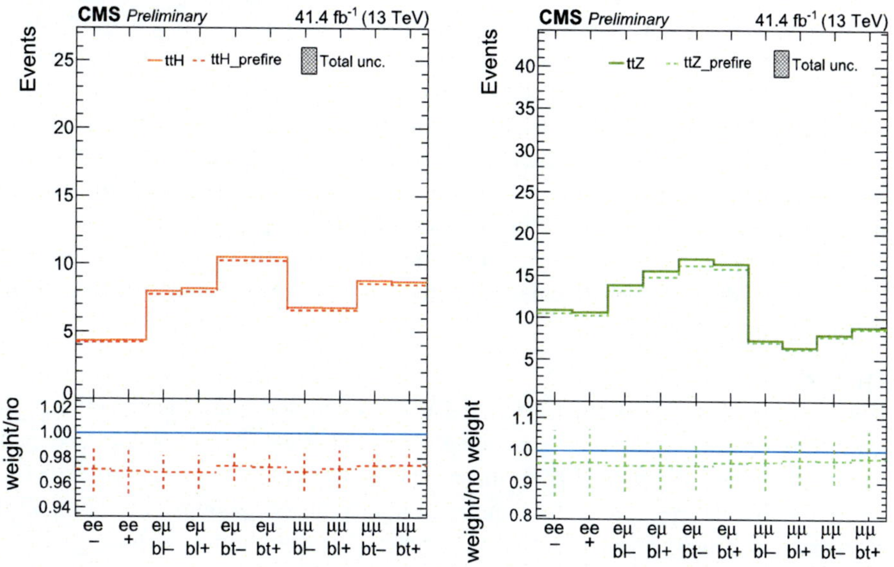

Fig. 6.15 Distributions of the $t\bar{t}H$ and $t\bar{t}Z$ yields in the $2\ell ss + 0\tau_h$ categories, split according to the lepton flavour, charge and b-jet multiplicity, obtained from the simulation in 2017 detector conditions. The dashed lines correspond to the distributions after applying the Level-1 ECAL trigger prefiring correction (see text)

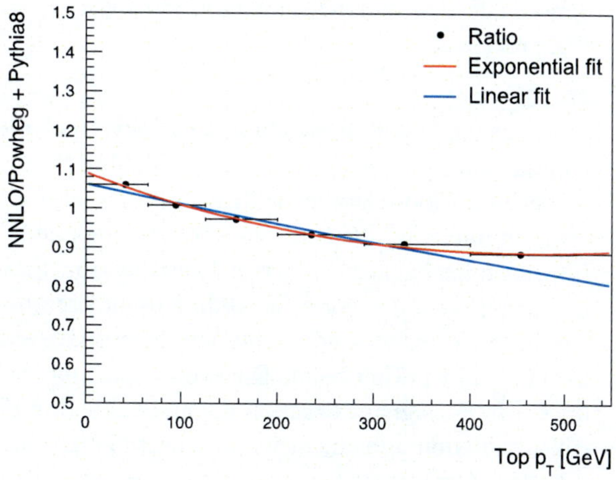

Fig. 6.16 Ratio of the top quark p_{T} spectra in $t\bar{t}$+jets events computed at NNLO QCD + NLO electroweak accuracy [29] and in the POWHEG sample used in the analysis, fitted with a function of the form $\exp(a + b \times p_{\mathrm{T}} + c \times p_{\mathrm{T}}^2)$ (red line)

6.5 Background Control Regions

The estimation of the main irreducible backgrounds in the analysis is validated in the so-called *control regions*, which are orthogonal to the signal regions defined in Sect. 6.2. These regions are enriched in specific background events, thus the modelling of such backgrounds is validated by probing the data-simulation agreement. The 3ℓ control region (3ℓ-CR) is introduced to validate the $t\bar{t}Z$ and WZ backgrounds.

The 4ℓ control region (4ℓ-CR) serves to validate the $t\bar{t}Z$ and ZZ backgrounds. To validate the $t\bar{t}W$ background, the main analysis employs a dedicated node in the DNN used for the signal extraction in the $2\ell ss + 0\tau_h$ category, while the control analysis makes use of a $2\ell ss$+3 jets control region ($2\ell ss$+3j-CR). The distributions in these regions are fitted simultaneously with the signal regions to extract the signal rates. The normalization of the $t\bar{t}W$ and $t\bar{t}Z$ backgrounds is left freely floating and is determined in situ in the signal extraction fit. This way, no assumption is made on their cross section; however, assumptions are made on their acceptance and the shapes of the distributions included in the fit, but these are expected to be less affected by higher-order corrections.

3ℓ-CR

Events in the 3ℓ-CR, enriched in $t\bar{t}Z$ and WZ backgrounds, are selected by applying the same criteria as the $3\ell + 0\tau_h$ category (see Table 6.5), except that events are required to have a pair of SFOS loose leptons with mass $|m_{\ell\ell} - m_Z| < 10$ GeV. Additionally, the requirement on the multiplicity of jets is relaxed to demand the presence of at least one jet and no requirement on the number of b-tagged jets is applied. The contributions arising from $t\bar{t}Z$ and WZ production are separated by classifying the events in 12 different subcategories according to the multiplicity of jets and b-tagged jets, separately for each lepton flavour combination (eee, $ee\mu$, $e\mu\mu$, $\mu\mu\mu$). The event yields for each subcategory before including the CR in the signal extraction fit can be found in Fig. 6.17 (left). The figure shows good agreement between the data and the simulation and a visible separation between the $t\bar{t}Z$ and the WZ processes.

4ℓ-CR

Events in the 4ℓ-CR, enriched in ZZ background, are selected by applying the same criteria as the $4\ell + 0\tau_h$ category (see Table 6.5), except that events must contain a pair of SFOS loose leptons with mass $|m_{\ell\ell} - m_Z| < 10$ GeV, and no requirement on the multiplicity of jets and b-tagged jets is applied. To separate the ZZ background from the other backgrounds, mostly arising from the $t\bar{t}Z$ process, events are classified in four different categories according to the presence of a second SFOS lepton pair within the Z mass window, the number of jets and the number of b-jets passing the medium working point. The event yields for each subcategory before including the CR in the signal extraction fit can be found in Fig. 6.17 (right), showing a good data-simulation agreement within uncertainties and a good separation between the $t\bar{t}Z$ and ZZ processes.

2ℓss+3j-CR

Events in the 2ℓss+3j-CR, enriched in $t\bar{t}W$ background, are selected by applying the same criteria as the $2\ell ss + 0\tau_h$ category (see Table 6.5), but requiring exactly 3 jets. Figure 6.18 shows the distributions of the invariant mass of the dilepton pairs in this region, separately for each lepton flavour combination (ee, $e\mu$, $\mu\mu$), before including the CR in the signal extraction fit. The binning is chosen such that the separation between $t\bar{t}W$ and the rest of the backgrounds is maximized; the good agreement between the data and the MC simulation is probed.

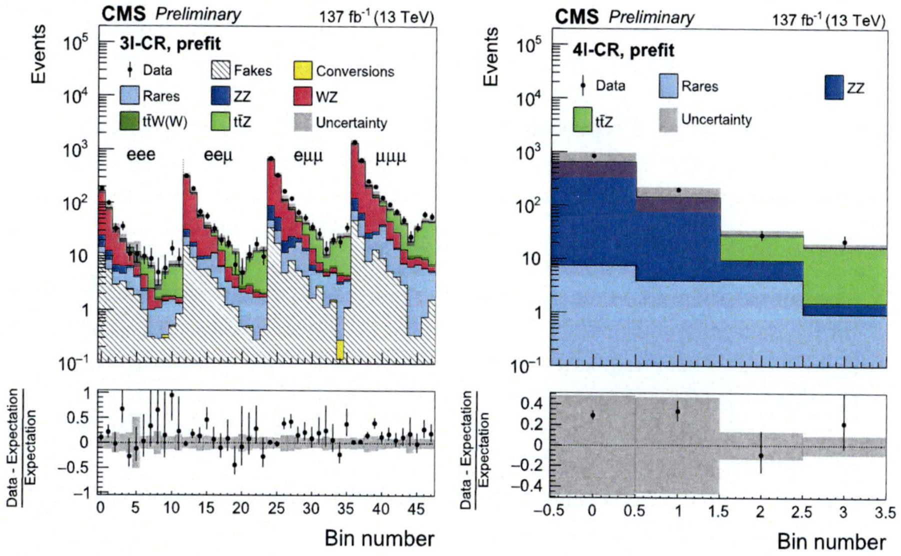

Fig. 6.17 Event yields in the (left) 3ℓ control region, enriched in the $t\bar{t}Z$-WZ backgrounds, in different bins of jet and b-jet multiplicities, and (right) 4ℓ control region, enriched in the $t\bar{t}Z$-WZ background, in different bins of SFOS and b-jet multiplicities. The distributions are included in the maximum likelihood fit of the main and control analyses

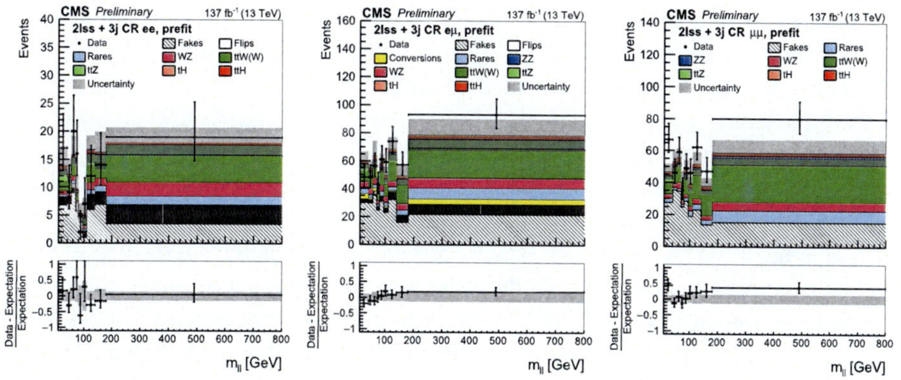

Fig. 6.18 Invariant mass of the dilepton pairs in the $2\ell ss + 3$ jets control region, enriched in the $t\bar{t}W$ background, for ee (left), $e\mu$ (centre) and $\mu\mu$ (right) pairs. The distributions are included in the maximum likelihood fit of the control analysis

6.6 Systematic Uncertainties

Various sources of systematic uncertainties may affect the precision in the measurement of the cross sections of the signals. They can affect the yields of the signal and background processes (*rate* uncertainties) or the shapes of the distributions of the observables used for signal extraction (*shape* uncertainties). The treatment of these uncertainties in the statistical methods used to measure the $t\bar{t}H$ and tH signal rates is explained in Sect. 6.7. The effects arising from the same source cause correlations across different event categories and data-taking periods, which are taken into account in the statistical analysis. Although the main analysis and the control analysis make use of different observables in the fit, the sources of uncertainty are mutual, as both share the same analysis strategy; however, they can have a different impact on the measurement of the signal rates.

Experimental Uncertainties

- **Luminosity**: The rate uncertainty on the integrated luminosity amounts to 2.6% for the data recorded in 2016 [30] and 2018 [31], and 2.4% for the data recorded in 2017 [32]. This uncertainty is obtained from dedicated Van-der-Meer scans and stability of detector response during data taking. The correlated effects across all years of data-taking amount to 1.4%, 1.3% and 2.1% in 2016, 2017 and 2018, respectively.
- **Pileup**: The shape uncertainty related to the number of PU interactions is obtained by varying the number of inelastic pp interactions superimposed in simulated events by the corresponding uncertainty, amounting to $\pm 4.6\%$. The resulting effect on the simulated signal and background yields amounts to less than 1%; it is correlated across the three data-taking periods.
- **Trigger efficiency**: The shape uncertainty due to the trigger efficiency is obtained by varying the corresponding scale factors within their uncertainties. The uncertainties relate to the statistical power of the DY samples used in the tag-and-probe method, as well as the effect of choosing other variables rather than the p_T in the parametrization. Trigger uncertainties amount to 2% for leptonic triggers and to 5% for lepton+τ_h and double-τ_h triggers; they are treated as uncorrelated across the data-taking years.
- **Lepton identification and isolation efficiency**: The uncertainty due to the electron and muon identification and isolation efficiency is obtained by varying the corresponding scale factors within their uncertainties, obtained from the tag-and-probe measurement in DY events in data and MC; they amount to 3–5% for electrons and 2% for muons. Additionally, the efficiencies are cross-checked with those obtained in $t\bar{t}$+jet events to account for the different event topology of DY events and the signal region, taken into account as a systematic uncertainty amounting to 1–2%. The uncertainties affect both the shape and the rates of the signals and backgrounds; they are correlated across the three years of data-taking.
- τ_h **identification efficiency**: The shape uncertainty related to the measurement of the τ_h selection efficiencies for the different identification working points arises mainly from statistical sources and amounts to 5%. The uncertainty is treated as uncorrelated across the data-taking periods.
- τ_h **energy scale**: The shape uncertainties associated to the energy scale of the τ_h are obtained by varying the corresponding scale factors by their uncertainties, which have a value of 1.2%; they are treated as uncorrelated across the three years of data-taking.
- **Jet energy scale**: The shape uncertainties associated to the jet energy scale and resolution are obtained by shifting the energy scale applied to the reconstructed

jets by $\pm 1\sigma$ around the central value. The impact on the event yields is evaluated by re-running the object and event selections on all simulated samples after applying the variations. The uncertainties are parametrized as a function of the jet p_T and η, and are split into 11 different sources according to the detector regions. They are treated as correlated when the measurement of the uncertainty is done in a consistent way across the data-taking periods, and uncorrelated otherwise. The effect of the uncertainties on the jet energy resolution is evaluated the same way but it is found to be negligible.

- **b-tag efficiency and mistag rate**: The shape uncertainties on the b-tagging efficiencies are evaluated by varying the scale factors within their uncertainties; they take into account the impact of the jet energy scale uncertainties and the purity and statistical power of the samples used to derive the corrections. The mistag rates for light-quark and gluon jets are measured with uncertainties of 5–10% (20–30%) for the loose (medium) b-tagging working point. The uncertainties arising from statistical effects are treated as uncorrelated across years, while the experimental ones are correlated.
- **Prefiring probability of the Level-1 ECAL trigger**: The effects due to the uncertainty in the Level-1 prefiring correction are taken into account by varying the corresponding scale factors by their uncertainty; they lead to a shape uncertainty of $\sim$1%, considered uncorrelated in 2016 and 2017.

Modeling Uncertainties

- **$t\bar{t}H$ and tH cross sections**: The signal rates are measured in units of their SM production rates, and hence the measurements are affected by the uncertainties on the cross sections. The rate uncertainty of the $t\bar{t}H$ cross section, computed at NLO accuracy, amounts to $^{+6.8\%}_{-10.0\%}$, of which $^{+5.8\%}_{-9.3\%}$ are due to missing higher orders and 3.6% corresponds to the uncertainties in the PDF and α_s [3]. The rate uncertainty of the tH cross section, computed at NLO accuracy, amounts to $^{+4.2\%}_{-6.8\%}$, of which $^{+4.1\%}_{-6.7\%}$ are due to missing higher orders and 1.0% arises from uncertainties in the PDF and α_s. Moreover, the shape uncertainties due to missing higher-order corrections to these cross sections are estimated by varying the renormalization (μ_R) and factorization (μ_F) scales up and down by a factor two with respect to their nominal values. The theoretical uncertainties associated to the rates of the signals are correlated across data-taking periods.
- **$t\bar{t}W$ and $t\bar{t}Z$ cross sections**: The rate uncertainty of the $t\bar{t}W$ cross section, computed at NLO accuracy, amounts to $^{+13.5\%}_{-12.2\%}$, of which $^{+12.9\%}_{-11.5\%}$ are due to missing higher orders and 4% arises from uncertainties in the PDF and α_s [3]. The rate uncertainty of the $t\bar{t}Z$ cross section, computed at NLO accuracy, amounts to $^{+11.7\%}_{-10.2\%}$, of which $^{+11.2\%}_{-9.6\%}$ are due to missing higher orders and 3.4% arises from uncertainties in the PDF and α_s [3]. These theoretical uncertainties are correlated across all data-taking years.
- **Higgs boson branching ratios**: The rate uncertainty in the BR of the Higgs boson amounts to 1.54% for $H \rightarrow WW^*$ and $H \rightarrow ZZ^*$ and to 1.65% for $H \rightarrow \phi^+\phi^-$ [3]; they are correlated across the years of data-taking.

- **Diboson background**: Rate uncertainties of 30% are assigned to WZ and ZZ processes resulting from the extrapolation of their rates from the control regions to the signal regions. The uncertainties relate to the limited statistical power and knowledge of the heavy flavour content of the samples; they are correlated across data-taking periods.
- **Rares background**: A conservative rate uncertainty of 50% is assigned to the normalization of the rare processes to account for the limited knowledge of their cross section and their extrapolation to the phase space of the analysis. The uncertainty is correlated across data-taking years.
- **Fakes background**: The estimation of the backgrounds arising from fake leptons and τ_h is subject to several rate and shape uncertainties. They result from the statistical power of the MR, the subtraction of the prompt lepton contamination in the MR and the differences in the background composition between the MR and the AR. The latter was verified in closure tests between the data and the simulated events (see Sect. 6.3.3). An additional 30% rate uncertainty is assigned to simulated events in the $2\ell ss + 1\tau_h$ and $3\ell + 1\tau_h$ categories in which the reconstructed leptons are genuine but the τ_h originates from a misidentified jet; the fake background is estimated from the simulation in these cases. The uncertainties associated to the fake lepton (τ_h) background estimation are treated as correlated (uncorrelated) across the years.
- **Flips background**: The rate uncertainty on the charge flip background estimation in the $2\ell ss + 0\tau_h$ and $2\ell ss + 1\tau_h$ categories amounts to 30%; it is evaluated similarly to the fakes background and it is correlated across data-taking years.

6.7 Statistical Methods

The statistical procedure to evaluate the presence or absence of the signals in the data follows the frequentist approach endorsed by the ATLAS and CMS collaborations in the context of the Higgs boson analysis combination in 2011 [33]. The method consists in interpreting the compatibility of the observed data with the background-only hypothesis (H_{B}) or with the signal plus background hypothesis ($H_{\text{S+B}}$). The estimation is done based on the binned distributions of suitably chosen observables in each category (and subcategory) of the analysis, optimized to disambiguate the signals from the backgrounds. The outputs of machine learning algorithms are used for this purpose in the main analysis, the methods being described in Chap. 7. In the control analysis, the invariant mass of the leptonic systems and the output of the Matrix Element Method are used, as explained in Chap. 8.

Maximum Likelihood

In each bin i of the distributions of the discriminating observables of each of the 10 channels considered in the analysis, the expected yields under the $H_{\text{S+B}}$ hypothesis can be parametrized as

$$\nu_i(\mu_{t\bar{t}H}, \mu_{tH}, \vec{\theta}) = \mu_{t\bar{t}H} \cdot s_i^{t\bar{t}H}(\vec{\theta}) + \mu_{tH} \cdot s_i^{tH}(\vec{\theta}) + b_i(\vec{\theta}) \,, \tag{6.9}$$

where $s_i^{t\bar{t}H}$ and s_i^{tH} are the expected yields of $t\bar{t}H$ and tH signals in the bin i, respectively, and b_i is the expected yield of background in the same bin. In the context of the Higgs boson searches, the signal yields are fixed to their SM prediction; they are scaled by the *signal strength* modifiers

$$\mu_{t\bar{t}H} = \frac{\sigma_{t\bar{t}H}}{\sigma_{t\bar{t}H}^{SM}} \,, \quad \mu_{tH} = \frac{\sigma_{tH}}{\sigma_{tH}^{SM}} \,, \tag{6.10}$$

which are the *parameters of interest* (POI) of the model. In the background-only hypothesis H_B, the model considers $\mu = 0$. For simplicity, the global naming convention $\mu s_i(\vec{\theta}) \equiv \mu_{t\bar{t}H} \cdot s_i^{t\bar{t}H}(\vec{\theta}) + \mu_{tH} \cdot s_i^{tH}(\vec{\theta})$ is adopted in what follows.

The predictions of the signal and background yields in Eq. 6.9 are subject to the systematic uncertainties described in Sect. 6.6. Each independent source of systematic uncertainty is represented by a *nuisance parameter* θ, collectively denoted $\vec{\theta}$. The rate uncertainties affect the global signal and background yields, while the shape uncertainties affect the repartition of these yields along the bins of the discriminating observable. The nominal value of these parameters $\tilde{\theta}$ are typically estimated a priori by pre-existing auxiliary measurements, to which a *probability density function* (p.d.f.) $\rho(\theta|\tilde{\theta})$ is associated, also referred to as *prior*. Nuisance parameters which are not a priori constrained by any measurement or consideration are assigned uniform priors.

In order to estimate the values of the POIs and their associated uncertainties, a binned *likelihood function* incorporating all known information regarding the parameters of the model is constructed. This function quantifies the compatibility between the observed data and the prediction, for given values of the POIs and the nuisance parameters. It is expressed as the product of the Poisson probabilities to observe n_i events in the bin i, scaled by the nuisance probability function $\rho(\theta|\tilde{\theta})$, namely

$$\mathcal{L}(\text{data}|\mu, \vec{\theta}) = \prod_i \frac{\left(\mu s_i(\vec{\theta}) + b_i(\vec{\theta})\right)^{n_i}}{n_i!} e^{-(\mu s_i(\vec{\theta}) + b_i(\vec{\theta}))} \cdot \rho(\theta|\tilde{\theta}) \,. \tag{6.11}$$

The estimation of the POIs and the nuisances parameters is done via the *maximum likelihood* method, which consists in finding the point $\mu \equiv \hat{\mu}$ and $\vec{\theta} \equiv \hat{\vec{\theta}}$ in the parameter space which makes the likelihood function $\mathcal{L}(\text{data}|\mu, \vec{\theta})$ maximal, namely the point for which the observed data are the most probable. The values $\hat{\mu}$ and $\hat{\vec{\theta}}$ are called *best estimates*. For computational simplicity, it is common to minimize $-2\ln(\mathcal{L})$ instead. The uncertainty on the best estimate $\hat{\mu}$ is obtained by determining the lower and upper bounds, μ_{min} and μ_{max}, for which the value of $-2\ln(\mathcal{L})$ exceeds the minimum by one unit, corresponding to a coverage probability of 68%. The differences $\delta_+ = \mu_{max} - \hat{\mu}$ and $\delta_- = \hat{\mu} - \mu_{min}$ correspond to the combined uncertainties associated to $\hat{\mu}$.

The statistical model above accommodates two POIs, $\mu_{t\bar{t}H}$ and μ_{tH}, that are measured simultaneously. Such configuration is used to derive the results of the main analysis, where both μ_{tH} and $\mu_{t\bar{t}H}$ float simultaneously in the maximum likelihood fit. One can also choose to measure these parameters separately by leaving one POI floating freely in the fit, while fixing the other to its SM expectation ($\mu = 1$). This is done in the control analysis, since only the $t\bar{t}H$ signal rate is measured ($\mu \equiv \mu_{t\bar{t}H}$), and the tH process is considered as backgrounds, its rate being fixed to $\mu_{tH} = 1$.

Nuisance Parameters

The sources of uncertainty included in this analysis are described in Sect. 6.6. The shape of their associated nuisance probability function in Eq. 6.11 is determined according to the type of uncertainty source. When the uncertainty is derived from an independent measurement, the log-normal distribution is used and is defined as

$$\rho(\theta|\tilde{\theta}) = \frac{1}{\sqrt{2\pi}\,\ln\kappa} \exp\left(- \frac{\left(\ln(\theta/\tilde{\theta})\right)^2}{2\,(\ln\kappa)^2} \right) \frac{1}{\theta}\,, \tag{6.12}$$

with $\kappa = 1 + \epsilon$, being ϵ the relative scale of the uncertainty. This function is more appropriate than a Gaussian for large uncertainties, such as the cross section of a process or the integrated luminosity, as it has the advantage that is described correctly the positively defined observables by going to zero at $\theta = 0$. For small uncertainties, the Gaussian distribution of width ϵ is recovered, employed to estimate parameters which can take either negative or positive values.

The uncertainties on the background estimated from data control regions, such as the 3ℓ-CR and the 4ℓ-CR, are treated with a separate approach. The estimation consists in relating the number of events in the signal region n and the number of events in the control region N by an *extrapolation factor* α, with $n = \alpha N$. The uncertainty on n is described by the gamma distribution

$$\rho(n|N) = \frac{1}{\alpha}\frac{(n/\alpha)^N}{N!} \exp(-n/\alpha)\,, \tag{6.13}$$

which represents the propagation of the statistical uncertainties on N from the control region to the signal region.

Additional uncertainties arise from the limited amount of simulated events available to model the signal and background processes. These uncertainties are treated with the Barlow-Beeston method [34], that estimates the statistical uncertainty of the total predicted number of events by assigning a single nuisance parameter to each bin of the distributions via a Poisson or a Gaussian prior depending on the statistics.

Hypothesis Testing

In order to draw conclusions from the results of the maximized likelihood, a hypothesis test is typically specified in terms of a *test statistic*. It is defined in such a way as to quantify, within observed data, behaviours that distinguish the H_{S+B} hypothesis from the H_B hypothesis. A common method used for this purpose is the *profile likelihood ratio* (PLR)

$$\lambda(\mu) = \frac{\mathcal{L}(\text{data}|\mu, \hat{\hat{\vec{\theta}}}_\mu)}{\mathcal{L}(\text{data}|\hat{\mu}, \hat{\vec{\theta}})} . \tag{6.14}$$

The numerator corresponds to the case where μ is fixed to a given value, and $\hat{\hat{\vec{\theta}}}_\mu$ is the *conditional maximum likelihood estimator* associated, namely the value of $\vec{\theta}$ that maximizes the likelihood for the fixed μ. The denominator corresponds to the maximized likelihood obtained when both μ and $\vec{\theta}$ are left floating freely. By construction, the PLR is bound between 0 and 1, and higher values indicate better compatibility between the data and the H_{S+B} hypothesis. It is common to use the logarithmic form of the PLR, defined as $q_\mu = -2\ln\lambda(\mu)$. When measuring a single parameter μ, q_μ is expressed graphically as scans of q_μ for a given μ within a certain interval, where the minimum is the best estimate. The 68% and 95% confidence level (CL) intervals are typically deduced from the conditions $q_\mu < 1$ and $q_\mu < 3.84$, respectively.

Significance

The test statistic q_μ under the background-only hypothesis H_B can be used to estimate the *significance* of an observed result. This is done via the *p-value* p_0, representing the probability of getting a result as incompatible (or more) as the observed one, assuming the background-only hypothesis H_B. It is expressed as

$$p_0 = \int_{q_{obs}}^{\infty} f(q_0|H_B)\, dq_0 , \tag{6.15}$$

where $q_0 = -2\ln\lambda(\mu = 0)$, and $f(q_0|H_B)$ is the probability density function of q_0 under the hypothesis H_B. The hypothesis H_B is excluded if p_0 lies below a predefined threshold. The p-value is typically converted into an equivalent significance Z, defined as

$$Z = \Phi^{-1}(1 - p_0) , \tag{6.16}$$

which is expressed in units of Gaussian standard deviations σ, with Φ^{-1} being the inverse of the cumulative distribution (or *quantile*) of the standard Gaussian. For a signal process such as the Higgs boson, the particle physics community claims an *evidence* when a significance of 3σ is reached ($p_0 = 1.4 \times 10^{-3}$), and an *observation* when a significance of 5σ is reached ($p_0 = 2.9 \times 10^{-7}$).

Expected Results

In order to avoid the potential bias arising from the statistical methods, the standard procedure in CMS is to develop an analysis in a *blinded* manner, meaning the strategy is defined on studies based solely on the simulation or data in signal-free regions. In this context, it is instrumental to use the *expected results* to estimate the sensitivity of the experiment. In principle, this consists in generating a large set of pseudo-data (toys) distributed according to the likelihood function, and extracting the nominal values of the POI from the median. This is computationally expensive; however, a very good approximation is provided by the *Asimov dataset* [35], a single representative dataset in which the estimates of the parameters are set to their expected values and the statistical fluctuations are suppressed. The usage of the Asimov dataset allows the impact of the systematic uncertainties and the validity of the statistical method to be verified before looking at the data, i.e., *unblinding*.

References

1. CMS Collaboration (2018) Measurement of the associated production of a Higgs boson with a top quark pair in final states with electrons, muons and hadronically decaying τ leptons in data recorded in 2017 at $\sqrt{s} = 13$ TeV. Technical Report CMS-PAS-HIG-18-019, CERN, Geneva
2. CMS Collaboration (2019) Search for associated production of a Higgs boson and a single top quark in proton-proton collisions at $\sqrt{s} = 13$TeV. Phys Rev D 99:092005
3. LHC Higgs Cross Section Working Group, de Florian D et al (2016) Handbook of LHC Higgs cross sections: 4. Deciphering the nature of the Higgs sector. Technical Report FERMILAB-FN-1025-T, CERN-2017-002-M
4. Demartin F, Maier B, Maltoni F, Mawatari K, Zaro M (2017) tWH associated production at the LHC. Eur Phys J C 77(1)
5. CMS Collaboration (2019) Measurements of properties of the Higgs boson in the four-lepton final state in proton-proton collisions at $\sqrt{s} = 13$ TeV. Technical Report CMS-PAS-HIG-19-001, CERN, Geneva
6. CMS Collaboration (2020) Higgs boson production in association with top quarks in final states with electrons, muons, and hadronically decaying tau leptons at $\sqrt{s} = 13$ TeV. Technical Report CMS-PAS-HIG-19-008, CERN, Geneva
7. Chen T, Guestrin C (2016) XGBoost. In: Proceedings of the 22nd ACM SIGKDD international conference on knowledge discovery and data mining
8. Martín A et al (2016) TensorFlow: large-scale machine learning on heterogeneous distributed systems. arXiv:1603.04467
9. Volobouev I (2011) Matrix element method in HEP: transfer functions, efficiencies, and likelihood normalization. arXiv:1101.2259
10. International conference (2018) Higgs Hunting. http://wpsist.lal.in2p3.fr/higgshunting2018/
11. International conference (2019) European physical society - high energy physics. http://eps-hep2019.eu/
12. International conference (2020) 40th international conference on high energy physics. https://www.ichep2020.org/
13. CMS Collaboration (2021) Measurement of the Higgs boson production rate in association with top quarks in final states with electrons, muons, and hadronically decaying tau leptons at $\sqrt{s} = 13$ TeV. Eur Phys J (C 81, 378)
14. CMS Collaboration (2018) Identification of heavy-flavour jets with the CMS detector in pp collisions at 13 TeV. JINST 13(05):P05011–P05011
15. Sanchez Cruz S (2020) Search for new physics in events with high transverse momentum leptons with the CMS detector at the LHC. Ph.D. thesis at Universidad de Oviedo
16. CMS Collaboration (2018) Performance of reconstruction and identification of τ leptons decaying to hadrons and ν_τ in pp collisions at $\sqrt{s} = 13$ tev. JINST 13(10):P10005–P10005
17. CMS Collaboration (2019) Performance of the DeepTau algorithm for the discrimination of taus against jets, electron, and muons. Technical Report CMS-DP-2019-033, CERN, Geneva
18. Cacciari M, Salam GP, Soyez G (2008) The anti-k_t jet clustering algorithm. J High Energy Phys 2008(04):063–063
19. CMS Collaboration (2014) Search for the associated production of the Higgs boson with a top-quark pair. Technical Report CMS-HIG-13-029. CMS-HIG-13-029. CERN-PH-EP-2014-189
20. Ball RD et al (2015) Parton distributions for the LHC run II. J High Energy Phys 2015(4)

21. Sjöstrand T et al (2015) An introduction to PYTHIA 8.2. Comput Phys Commun 191:159–177
22. Agostinelli S et al (2003) Geant4 - a simulation toolkit. Nucl Instrum Meth A 506(3):250–303
23. Alwall J et al (2014) The automated computation of tree-level and next-to-leading order differential cross sections, and their matching to parton shower simulations. J High Energy Phys 07:079
24. Frederix R, Tsinikos I (2020) Subleading EW corrections and spin-correlation effects in $t\bar{t}W$ multi-lepton signatures. arXiv:2004.09552
25. CMS Collaboration (2018) Performance of the DeepJet b tagging algorithm using 41.9 /fb of data from proton-proton collisions at 13 TeV with Phase 1 CMS detector. Technical Report CMS-DP-2018-058, CERN, Geneva
26. CMS Collaboration (2020) Performance of the CMS level-1 trigger in proton-proton collisions at $\sqrt{s} = 13$TeV. J Instrum 15(10):P10017
27. CMS Collaboration (2013) CMS technical design report for the level-1 trigger upgrade. Technical Report CERN-LHCC-2013-011. CMS-TDR-12, CERN, Geneva
28. Frixione S, Nason P, Oleari C (2007) Matching NLO QCD computations with parton shower simulations: the POWHEG method. J High Energy Phys 2007(11):070
29. Czakon M, Heymes D, Mitov A, Pagani D, Tsinikos I, Zaro M (2017) Top-pair production at the LHC through NNLO QCD and NLO EW. J High Energy Phys 2017(10):186
30. CMS Collaboration (2017) CMS luminosity measurements for the 2016 data taking period. Technical Report CMS-PAS-LUM-17-001, CERN, Geneva
31. CMS Collaboration (2019) CMS luminosity measurement for the 2018 data-taking period at $\sqrt{s} = 13$ TeV. Technical Report CMS-PAS-LUM-18-002, CERN, Geneva
32. CMS Collaboration (2018) CMS luminosity measurement for the 2017 data-taking period at $\sqrt{s} = 13$ TeV. Technical Report CMS-PAS-LUM-17-004, CERN, Geneva
33. ATLAS Collaboration and CMS Collaboration (2011) Procedure for the LHC Higgs boson search combination in Summer 2011. Technical Report CMS-NOTE-2011-005, ATL-PHYS-PUB-2011-11, CERN, Geneva
34. Barlow R, Beeston C (1993) Fitting using finite Monte Carlo samples. Comput Phys Commun 77(2):219–228
35. Cowan G, Cranmer K, Gross E, Vitells O (2011) Asymptotic formulae for likelihood-based tests of new physics. Eur Phys J C 71:1554. [Erratum: Eur Phys J C 73:2501 (2013)]

Chapter 7
Extraction of the $t\bar{t}H$ and tH Signals with Machine Learning Techniques

The search for the $t\bar{t}H$ and tH processes in multileptonic final states follows the general strategy described in Chap. 6. Once the signal regions are defined, the rate of the signals is determined by means of the maximum likelihood (ML) fit to the data in distributions of certain observables in the signal and control regions, following the statistical tools presented in Sect. 6.7. The observables employed for this purpose are chosen so that the maximal separation power between the signals and the backgrounds is achieved. Once the discriminating variable is identified, one could choose to select events that pass an optimized cut on this variable to reduce the background contributions in the signal regions, followed by a straight-forward statistical analysis of the data. This approach, known as *cut-and-count*, is suboptimal when the variable in question does not allow for a significant background rejection while retaining most of the signal. The alternative approach consists in fitting the shape of the distribution itself, allowing for events in kinematic regions which suffer from a low signal-to-background ratio to be retained, hence enhancing the sensitivity of the analysis. The distributions used can be derived from single observables (*single-variable* analysis), or by combining several observables into a single optimized variable (*multivariate* analysis). The former is the method used for signal extraction in the multileptonic categories of the control analysis. Machine learning algorithms and the Matrix Element Method, used in the main and control analysis, respectively, are examples of the latter.

This chapter provides an overview of the multivariate methods used in the main analysis, which are based on machine learning techniques, namely Boosted Decision Trees (BDT) and Deep Neural Networks (DNN). Machine learning methods have become increasingly powerful tools for binary classification problems in high energy physics, where the goal is to separate the signal from the backgrounds. These algorithms are capable of learning the characteristics of the signal and the background in a multidimensional phase space, taking advantage of complex features and cor-

C. Martin Perez, *Search for the Higgs Boson Produced in Association with Top Quarks with the CMS Detector at the LHC*, Springer Theses,
https://doi.org/10.1007/978-3-030-90206-3_7

relations among many variables, in order to best estimate the physical process from which a given event originates. This is instrumental in analysis such as the one presented here, where rare signals are targeted in the presence of multiple backgrounds. Machine learning algorithms can also solve multiclassification problems: a given event can be categorized as tH, $t\bar{t}H$ or one specific background, meaning the signals can be separated between themselves, which is especially beneficial when both are measured simultaneously, as is the case in the main analysis.

This chapter is structured as follows. An overview of the signal extraction strategy in the main analysis is given in Sect. 7.1. A series of BDT-based taggers used to identify specific signal topologies are described in Sect. 7.2. The optimization and performance of the DNNs and the BDTs are shown in Sects. 7.3 and 7.4, respectively. The results of the measurements of the $t\bar{t}H$ and tH cross sections are presented in Sect. 7.5; they include both SM and BSM interpretations in the context of the κ framework. Prospects for future measurements of these signals at the HL-LHC are evaluated in Sect. 7.7. The alternative measurement performed in the control analysis is described in Chap. 8.

7.1 Signal Extraction Strategy

The main analysis is designed to extract the $t\bar{t}H$ and tH signals simultaneously through a maximum likelihood fit across the 10 categories presented in Sect. 6.2. The discriminating observables employed are the output of machine learning algorithms: DNNs in 3 categories and BDTs in the remaining 7, as outlined in Table 7.1. They have been optimized independently in each channel, each presenting very different statistical power and relative signal and background contributions, illustrated in Fig. 7.1. To enhance the sensitivity, the 3ℓ and 4ℓ control regions (see Sect. 6.5), enriched in the $t\bar{t}Z$, WZ and ZZ backgrounds, are simultaneously fitted. The $t\bar{t}W(W)$ and $t\bar{t}Z$ normalizations are allowed to float around the initial value estimated from the simulation; this way, they can be estimated using the data instead of relying solely on the simulation. A factor to correct for their normalization is derived this way and applied to the rate of the processes in the signal regions.

The $2\ell ss + 0\tau_h$, $3\ell + 0\tau_h$ and $2\ell ss + 1\tau_h$ channels are the most sensitive of the analysis. As illustrated in Fig. 7.1, these categories profit from a high signal purity and a significant statistical power. The main backgrounds in these regions arise from the irreducible $t\bar{t}W$ and $t\bar{t}Z$ processes, while the reducible backgrounds -whose estimation is subject to higher uncertainties- present minor contributions. Additionally, these channels are constructed to accept a significant number of tH events. These features make these categories ideal use cases for DNN-based signal extraction. The available statistics allow for multiclassification to be conducted: dedicated nodes are designed to target individual processes. Despite the high degree of overlap in their experimental signatures, the $t\bar{t}H$ and tH processes are separated into dedicated nodes for the purpose of enhancing the sensitivity of the simultaneous measurement. Given the high yields of the $t\bar{t}Z$ and $t\bar{t}W$ background processes, they can likewise

Table 7.1 List of signal and control regions and their respective observables used in the maximum likelihood fit in the main analysis. The control regions are described in Sect. 6.5

Main analysis		
Region	Category	Observable
Signal regions	$2\ell ss + 0\tau_h$	Deep neural networks
	$3\ell + 0\tau_h$	
	$2\ell ss + 1\tau_h$	
	$4\ell + 0\tau_h$	Boosted decision trees
	$1\ell + 1\tau_h$	
	$2\ell os + 1\tau_h$	
	$3\ell + 1\tau_h$	
	$1\ell + 2\tau_h$	
	$2\ell + 2\tau_h$	
	$0\ell + 2\tau_h$	
Control regions	3ℓ-CR	Object multiplicity
	4ℓ-CR	

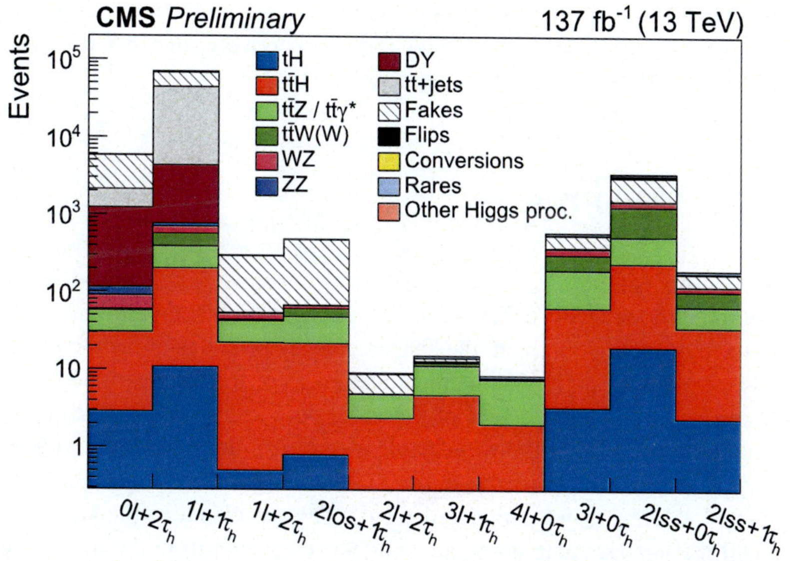

Fig. 7.1 Event yields of the signal and background processes considered in the 10 categories of the main analysis, obtained from the simulation. The yields are shown before any fit is performed to the data

be classified into high-purity nodes, bringing additional control over the dominant irreducible backgrounds.

The $0\ell + 2\tau_h$ and $1\ell + 1\tau_h$ categories profit from a comparable presence of tt̄H and tH yields with respect to the previous categories. The sensitivity of these categories is nonetheless degraded by the large irreducible Drell-Yan and tt̄+jets backgrounds, which are difficult to model in the simulation. The $1\ell + 2\tau_h$ and $2\ell os + 1\tau_h$ categories have similar statistical power and purity to the DNN-based categories, but in this case the overwhelming fake background contribution constitutes more than 10 times the yield of the signal processes. The $2\ell + 2\tau_h$, $3\ell + 1\tau_h$ and $4\ell + 0\tau_h$ channels present an adequate signal-to-background ratio and suffer only marginally from the presence of fake background. Owing to the high lepton and τ_h multiplicities, these categories have nonetheless very poor statistical power and a negligible tH yield. Hence, little sensitivity gain is expected from the multiclassification approach in these 7 categories; BDTs are used instead to separate the sum of the signals from the sum of the backgrounds.

The observables used by the DNNs and the BDTs learn the features of the signal and the background processes based on a series of input variables. These are chosen independently in each category: not only do the topologies differ from one another, but also the discriminating power of a given variable changes. Typical observables used are the multiplicity and kinematic properties of the reconstructed objects, together with global event quantities such as $E_{\mathrm{T}}^{\mathrm{miss}}$ or the angular separations between the objects. In order to further exploit the signature of the signal, dedicated BDTs are used to identify top quark decaying hadronically and the Higgs boson decaying to a WW pair. These topological taggers are denoted as *hadronic top tagger* (HTT) and *Higgs boson jet tagger* (Hj), respectively; they are used as input to the BDTs and DNNs in the signal extraction.

7.2 Topological Taggers

Hadronic Top Tagger

The hadronic top tagger is a BDT-based algorithm which computes the likelihood of three jets to be compatible with the hadronic decay products of the top quark via the process $t \rightarrow Wb \rightarrow q\bar{q}'b$, which constitutes ~2/3 of the top quark decays. The jets resulting from the hadronization of these quarks must be reconstructed separately and must not overlap with any fakeable electron, muon or τ_h within a cone size $\Delta R < 0.4$. The BDT is trained with the XGBoost algorithm [1] using a combination of $t\bar{t}$H, $t\bar{t}$W, $t\bar{t}$Z and $t\bar{t}$+jets simulated events. The signal is identified as the jet triplet where all the jets are matched to generator-level quarks from the top decay; if one of the jets fails to satisfy the matching criteria, the event is considered background. The BDT is trained once for all categories in which hadronic top quarks are expected ($2\ell ss + 0\tau_h$, $3\ell + 0\tau_h$, $1\ell + 1\tau_h$, $2\ell ss + 1\tau_h$, $2\ell os + 1\tau_h$, $1\ell + 2\tau_h$ and $0\ell + 2\tau_h$).

The HTT algorithm iterates over all jet triplets present in the event. In each iteration, a b-jet candidate is assigned (b-jet), along with two light jet candidates from the W boson decay (W_1-jet and W_2-jet). For each triplet, the following variables are evaluated:

- the b-tagging discriminator of the b-jet, W_1-jet and W_2-jet,
- the quark-gluon discriminator [2] of the W_1-jet and W_2-jet,

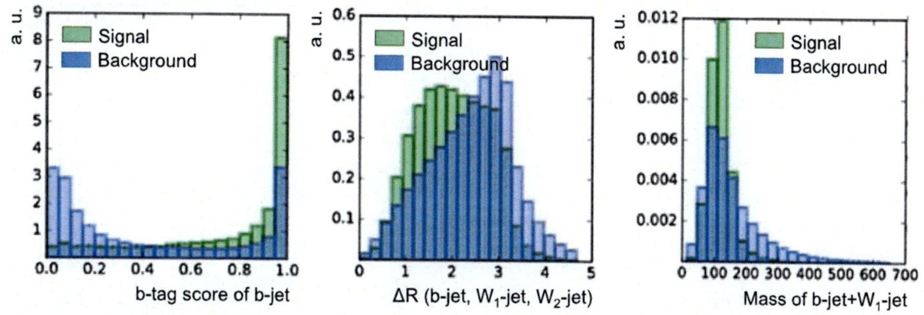

Fig. 7.2 Distributions of various input variables to the hadronic top tagger BDT used tp reconstruct hadronic top decays. The signal corresponds to jet triplets matched to generator-level quarks from the hadronic top decay and the background corresponds to the case where one or more jets fail this matching criterion

- the p_{T} of the b-jet and the W_2-jet,
- the p_{T} of the b+W_1+W_2-jets system,
- the mass of the b-jet, W_1-jet and W_2-jet,
- the masses of the b+W_1-jets, the b+W_2-jets and the b+W_1+W_2-jets systems,
- the ratio between the masses of the W_1+W_2-jets and the b+W_1-jets systems,
- the ΔR between the W_1-jet and W_2-jet,
- the ΔR between the b-jet and the W_1+W_2-jets system.

The distributions of the signal and the background of some of these variables are shown in Fig. 7.2. For each triplet, a probability to originate from a hadronic top quark is assigned. The HTT scores for the $\mathrm{t\bar{t}H}$ signal and the main backgrounds are illustrated in Fig. 7.3. After evaluating all permutations, the jet triplet with the highest probability is considered the hadronic top candidate; its kinematic properties are used as input to the signal extraction BDTs and the DNNs.

Higgs Boson Jet Tagger

The Higgs boson jet tagger is a BDT-based discriminant that targets the identification of jets originating from the $\mathrm{H} \rightarrow \mathrm{WW}^*$ decay in the $2\ell ss + 0\tau_h$ category, which is the most sensitive in the analysis. The selection criteria requires the presence of two leptons with the same sign and at least four jets, meaning one of the two W bosons decays hadronically. The Hj algorithm is trained on simulated samples of $\mathrm{t\bar{t}H}$ and $\mathrm{t\bar{t}W}$ events with the XGBoost algorithm [1]. The events with a pair of jets originating from the $\mathrm{H} \rightarrow \mathrm{W}^*\mathrm{W} \rightarrow (\mathrm{b}\nu\ell)(\mathrm{b}\nu\mathrm{qq}')$ are considered as signal, while the rest are considered as background. The input variables to the Hj tagger are

- the p_{T} of the jets,
- the b-tagging score of the jets,
- the quark-gluon discriminator [2] of the jets,
- the minimum and maximum ΔR between jets and any fakeable lepton.

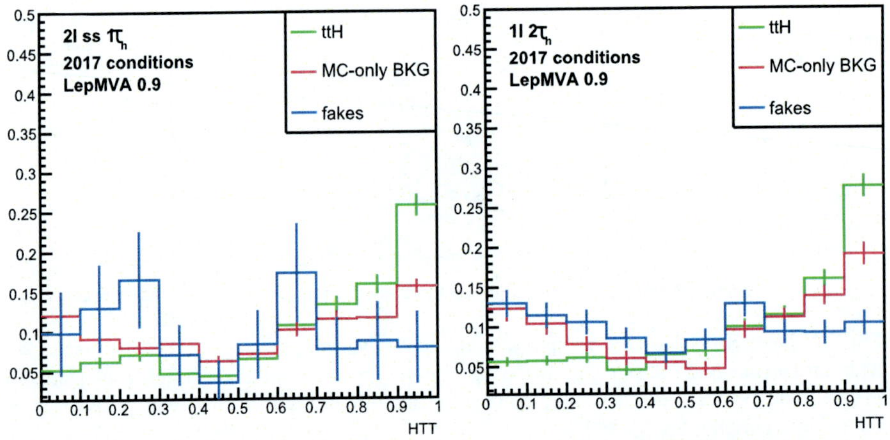

Fig. 7.3 Distributions of the hadronic top tagger (HTT) output score obtained in 2017 data-taking conditions, for the $2\ell ss + 1\tau_h$ (left) and $1\ell + 2\tau_h$ (right) categories. The label "tt̄H" corresponds to hadronic top candidates matched to generator-level top quarks in tt̄H events. The label "MC-only BKG" corresponds to hadronic top candidates matched to generator-level leptons in tt̄Z, tt̄W, WW+jets and WZ+jets events. The label "fakes" corresponds to jet triplets in the reducible fake background

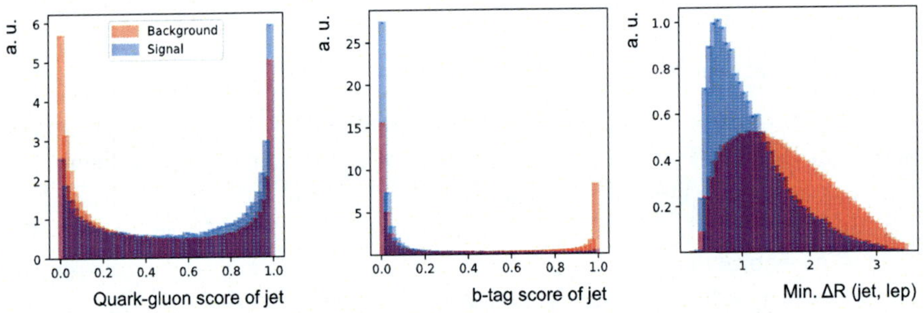

Fig. 7.4 Distributions of various input variables to the Higgs boson jet tagger BDT in the $2\ell ss + 0\tau_h$ category. The signal corresponds to the jets matched to generator-level $H \to W^*W \to (b\nu\ell)(b\nu qq')$ decays and the background corresponds to the case where one or more jets fail this matching criterion

The distributions of some of these observables are given in Fig. 7.4. Their separation power results in a significant discrimination of the $H \to W^*W$ decay, illustrated in Fig. 7.5; it is exploited in the DNN used for signal extraction in the $2\ell ss + 0\tau_h$ category.

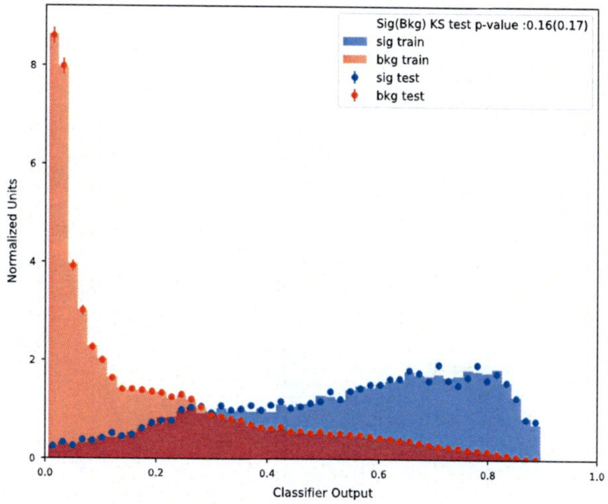

Fig. 7.5 Distributions of the Higgs boson jet tagger output score in the $2\ell ss + 0\tau_h$ category. The signal corresponds to the jets matched to generator-level $H \rightarrow W^*W \rightarrow (b\nu\ell)(b\nu qq')$ decays and the background corresponds to the case where one or more jets fail this matching criterion. The training and test samples are superimposed, showing no overtraining

7.3 Deep Neural Networks

In the categories that profit from higher statistics and a relatively large yield of the tH signal process ($2\ell ss + 0\tau_h$, $3\ell + 0\tau_h$ and $2\ell ss + 1\tau_h$), the observables chosen for the signal extraction are the outputs of the multiclassifier DNNs developed separately for each category. A simplified sketch of the multiclassifier DNN architecture can be found in Fig. 7.6. For each event, several input variables are provided to the input layer, which contains one node per variable holding the magnitude (or activation value) of the observable. At the other end of the network, the output layer presents one node associated to each of the targeted processes ($t\bar{t}$H, tH and the individual or collective backgrounds). After an event runs through the network, the activation values in the output nodes will be a number between 0 and 1, encoding the probability that it belongs to the associated process. Between the input and output layers stand the hidden layers. The activation values in their nodes are determined from weighted averages of the activation values of the previous nodes to which they are connected, meaning each layer picks up increasingly abstract features. The optimal weight configuration is obtained by *training* the network by providing it with a high statistics dataset which comprises events from all classes. After the training phase, the performance of the predictive model is tested with an independent *testing* sample to make sure that the model achieves the same performance, i.e. that it is not *overtrained*. An overtraining means that the model has learned features specific to the training sample and it is not able to generalize to other events.

The output classes used for each channel can be found in Table 7.2. Separate classes are allocated for the $t\bar{t}$H and tH signals in the $2\ell ss + 0\tau_h$, $3\ell + 0\tau_h$ and $2\ell ss + 1\tau_h$ categories. To achieve a better separation of the tH signal, only the tHq component is targeted: it has a higher cross section and it typically contains a higher multiplicity of forward jets, making its topology more distinct and adequate for the training. Besides, the tHW process presents similar kinematic properties as the $t\bar{t}$H

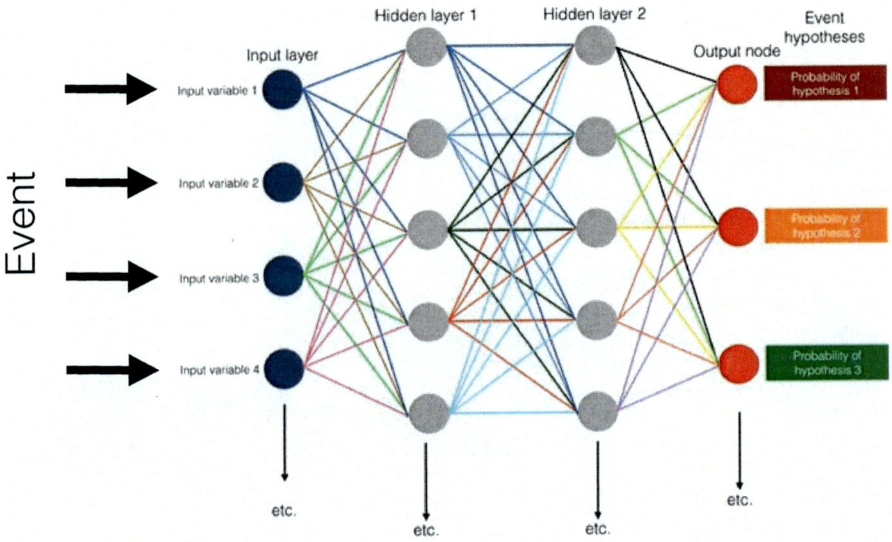

Fig. 7.6 Sketch of the multiclassifier DNN architecture. The input nodes (in dark blue) hold the values of the input variables of an event, while the output nodes (in red) hold a probability that such event corresponds to a certain physics processes. In between, the hidden layers (in grey) pick up the specific patterns that relate the former to the latter in order to make a prediction

Table 7.2 Hyperparameters of the multiclassifier DNNs used for signal extraction in the $2\ell ss + 0\tau_h$, $3\ell + 0\tau_h$ and $2\ell ss + 1\tau_h$ categories (see text and Ref. [3] for the definition of the hyperparameters)

	$2\ell ss + 0\tau_h$	$3\ell + 0\tau_h$	$2\ell ss + 1\tau_h$
Output nodes (classes)	4 ($t\bar{t}H$, tH, $t\bar{t}W$, rest)	3 ($t\bar{t}H$, tH, rest)	3 ($t\bar{t}H$, tH, rest)
Hidden layers (nodes)	1 (32) + 10 (16) + 5 (8)	2 (16) + 1 (8)	3 (32) + 2 (16)
Activation functions	ReLu	PReLu	PReLu
Dropout rate	0.01	0.1	0.1
Epochs	300	45	45

process, as they interfere at leading order; hence, targeting such process would result in a loss of discriminating power against the tHq process. The classes allocated to the backgrounds differ in each category according to the most relevant contributions. Given the high statistics of the $2\ell ss + 0\tau_h$ channel, the DNN can classify the backgrounds as either $t\bar{t}W$ or $t\bar{t}$+jets, the latter being labelled as "rest", encompassing the fakes, the flips and the conversions backgrounds. The main backgrounds in the $3\ell + 0\tau_h$ category result from the $t\bar{t}W$, $t\bar{t}Z$ and $t\bar{t}$+jets processes, which are grouped in one class named "rest". As the $2\ell ss + 1\tau_h$ category presents comparable background contributions from multiple processes, all the backgrounds are grouped in one class under the tag "rest".

Each networks takes ~40 input variables, listed in Table 7.3; they are chosen according to their separation power and the expected sensitivity in each category. Typical observables are the kinematic properties of leptons, τ_h and jets, the angular separations amongst them, the sum of charges of different combinations of leptons and τ_h, the b-tagging properties of the jets and the linear discriminant L_D. The outputs of the HTT and Hj taggers are also included. To improve the separation between the $t\bar{t}H$ and the tH signals, the multiplicity of forward jets and the kinematic properties of the leading forward jet are considered, along with its distance $\Delta\eta$ to the closest jet. The normalized distributions of some of the input variables are shown in Fig. 7.7 for the $2\ell ss + 0\tau_h$, $2\ell ss + 1\tau_h$ and $3\ell + 0\tau_h$ categories; the separation in shape between the signals and the backgrounds is visible, as well as between the signals themselves.

The training of the DNNs is performed with simulated samples of the signal processes ($t\bar{t}H$, tHq, tHW) and the main background processes ($t\bar{t}W$, $t\bar{t}Z$), allocating 80% of the statistics to the training and the remaining 20% for testing purposes. The DNN is trained with the TensorFlow algorithm [4] interfaced with Keras [3]. The objective of the training is to minimize the loss function, which represents the inaccuracy of the predictions. The weights of the DNN during the training are updated with the batch gradient descent algorithm, which looks for the minimum of such function by averaging over all the training data. The hyperparameters of the network are chosen so that the area under the ROC curve and the expected sensitivity are maximized. The main DNN parameters are shown in Table 7.2; their definition can be found in Ref. [3].

The DNN provides a normalized output score associated to each node reflecting the probability to belong to the corresponding class. Each event is then categorized under the class presenting the highest probability. This defines four categories in $2\ell ss + 0\tau_h$ and three in the other two. Events in the $2\ell ss + 0\tau_h$ category are subcategorized depending on the flavour of the leptons (ee, $e\mu$, $\mu\mu$); it is justified by the fact that the electrons present lower identification efficiency than the muons, as well as higher charge misidentification rates. In the $3\ell + 0\tau_h$ category, events are subcategorized according to the flavour of the leptons (eee, $ee\mu$, $e\mu\mu$, $\mu\mu\mu$) and also according to the number of jets that pass the medium working point of the b-tagging discriminant ("bl" with < 2, or "bt" with ≥ 2); the splitting allows to separate the fake background. No subcategorization is applied in the $2\ell ss + 1\tau_h$ category due to the limited statistics. For each node (and subcategory), the events are binned according to the DNN output score so that the sensitivity is maximized.

The distributions of the DNN outputs for the $2\ell ss + 0\tau_h$ category are found in Fig. 7.8. The $t\bar{t}H$ and the tH signals are well separated into their respective nodes, where the ratio of the signal over the background gradually increases towards the uppermost bins within each subcategory. Likewise, a noteworthy control of the background is attained: the $t\bar{t}W$ node presents high purity and a significant fraction of the total fake background is found in the background node. The distributions in the $3\ell + 0\tau_h$ and $2\ell ss + 1\tau_h$ categories are presented in Fig. 7.9; the $t\bar{t}H$ and tH signals are visibly separated into their respective nodes and an increasing signal-over-background ratio is found along the bins. These distributions are the ones included in the maximum likelihood fit to extract the $t\bar{t}H$ and tH signals.

Table 7.3 List of input variables to the multiclassifier DNNs used for signal extraction in the $2\ell ss + 0\tau_h$, $2\ell ss + 1\tau_h$ and $3\ell + 0\tau_h$ categories. The check mark (✓) indicates that the variable is used

DNN input variable	$2\ell ss + 0\tau_h$	$2\ell ss + 1\tau_h$	$3\ell + 0\tau_h$
Number of electrons	✓	✓	✓
Leptons p_T	✓	✓	✓
Leptons η	✓	✓	✓
Leptons ϕ	✓	✓	✓
Leptons transverse mass	✓	✓	✓
Leading lepton charge	✓		
Hadronic τ p_T		✓	
Hadronic τ η		✓	
Hadronic τ ϕ		✓	
Transverse mass of leptons (+τ_h) system		(✓)	✓
Invariant mass of leptons+τ_h system		✓	
Charge sum of leptons (+τ_h)		(✓)	✓
Maximum lepton η	✓		
Minimum ΔR between leptons (+τ_h)		(✓)	✓
Minimum ΔR between leptons+jets	✓	✓	✓
Minimum ΔR between τ_h+jets		✓	
Number of jets	✓	✓	✓
Jets p_T	✓	✓	✓
Jets η	✓	✓	✓
Jets ϕ	✓	✓	✓
Average ΔR between jets	✓	✓	✓
Number of forward jets	✓	✓	
Leading forward jet p_T	✓	✓	✓
Leading forward jet η	✓	✓	
Min. $\Delta\eta$ between leading forward jet + jets	✓	✓	✓
Number of loose b-jets	✓	✓	✓
Number of medium b-jets	✓	✓	✓
Invariant mass of loose b-jets		✓	✓
Invariant mass of medium b-jets	✓		
Linear discriminator L_D	✓	✓	✓
Hadronic top tagger score	✓	✓	✓
Hadronic top p_T	✓	✓	✓
Higgs boson jet tagger score	✓		
Number of variables	36	41	37

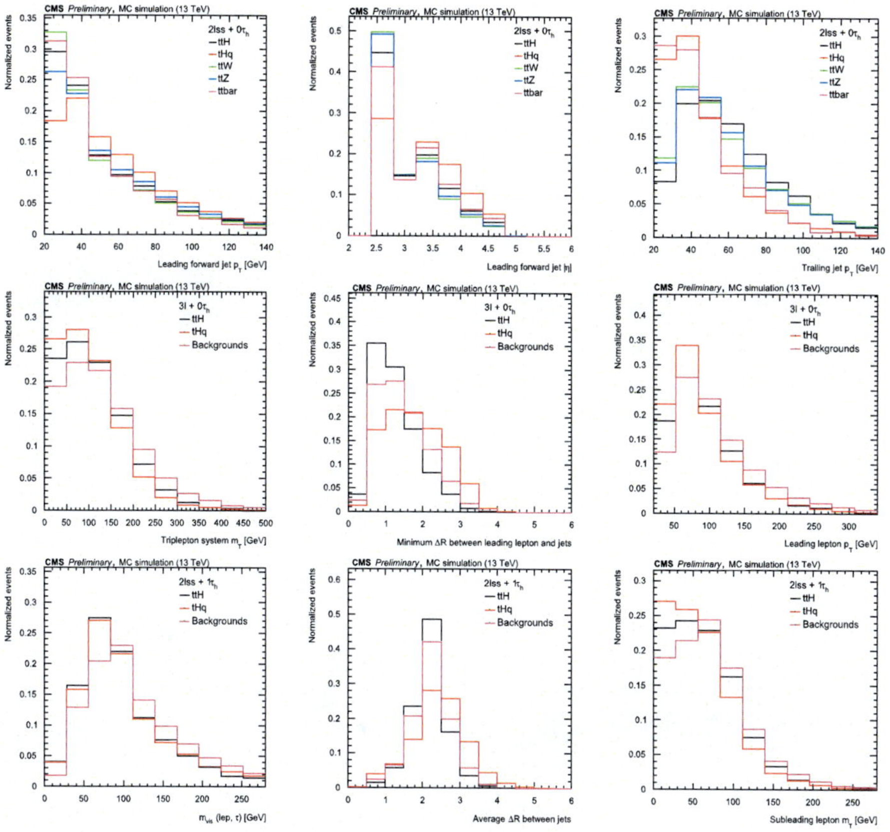

Fig. 7.7 Normalized distributions of the DNN input variables providing the largest separation between the $\mathrm{t\bar{t}H}$ signal, the tH signal and the backgrounds in the $2\ell ss + 0\tau_h$ (top row), $3\ell + 0\tau_h$ (middle row) and $2\ell ss + 1\tau_h$ (bottom row) categories

7.4 Boosted Decision Trees

The remaining signal regions either suffer from a large fake background or have too small event yield to be adequate for a multiclassification approach. The signal extraction is performed with BDTs for these regions. These are aimed to separate the $\mathrm{t\bar{t}H}$ signal from the sum of the backgrounds by providing a probability for each event to originate from the signal process. The prediction is built upon iterative partition of the data: the samples are broken down into subsequently smaller subsets in the form of a tree structure, sketched in Fig. 7.10. The algorithm starts at the root node with the complete dataset comprising events of the signal and the background. A division of the sample is done based on a cut on one of the input variables providing the best separation. The data are thus split into two separate nodes; the procedure is repeated iteratively until a stopping criterion is met, either when the maximum depth is reached or when the signal purity is maximized. The terminal nodes (*leaves*) hold a

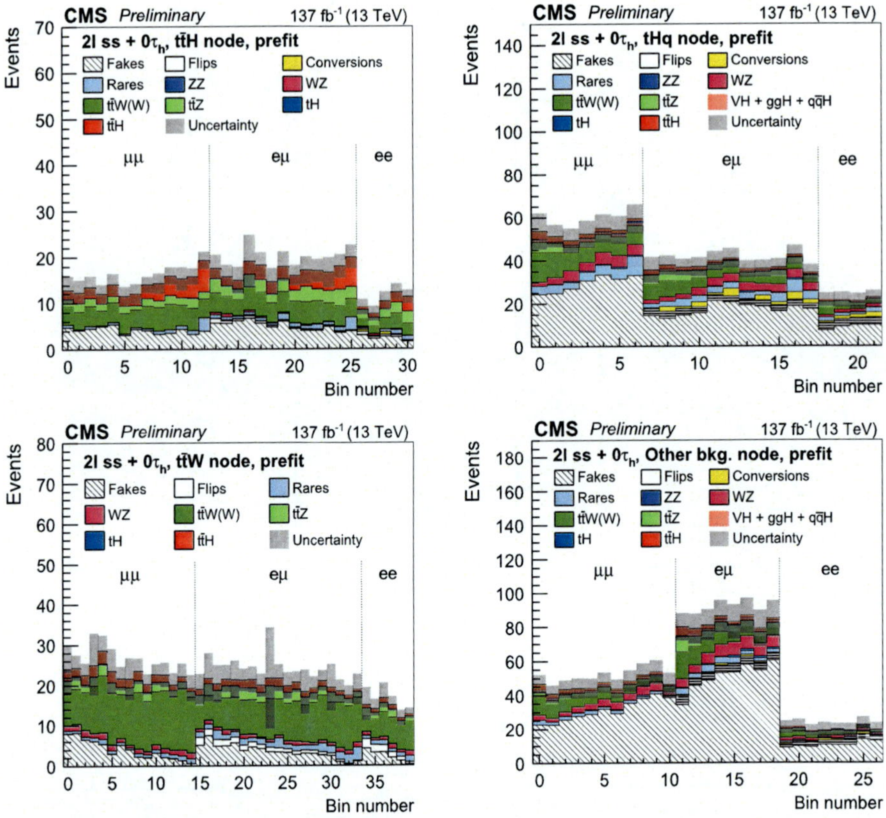

Fig. 7.8 Distributions of the DNN output scores in the $t\bar{t}H$ (top left), tH (top right), $t\bar{t}W$ (bottom left) and other background (bottom right) nodes of the $2\ell ss + 0\tau_h$ category, obtained from the simulation. The uncertainty bands correspond to systematic uncertainties. The distributions are presented before any fit is performed to the data

probability to belong to the signal according to the purity of the dataset they contain. The global BDT algorithm employs an ensemble (*forest*) of such trees, where a predefined number of trees are trained sequentially. The events entering each tree are weighted according to the degree of misclassification in the previous, so that the new tree can correct for the previous errors (Table 7.4).

The BDTs employ from 7 to 18 input variables, listed in Table 7.5. Distributions of some of these variables for the seven categories are found in Fig. 7.11. The variables are chosen according to the area under the ROC curve and the sensitivity they bring. Similarly to the DNN, such variables concern the kinematic properties of the objects, the b-tagging properties of the jets, the L_D discriminator and the HTT tagger. In the $1\ell + 2\tau_h$ and $0\ell + 2\tau_h$ categories, the jets which are not tagged in the first hadronic tagger are used to reconstruct a second hadronic top whose properties are also used in the BDT. In these same categories, the invariant mass of the $\tau\tau$ system is used; it is reconstructed with the SVFit algorithm [6]. The algorithm is based on a likelihood

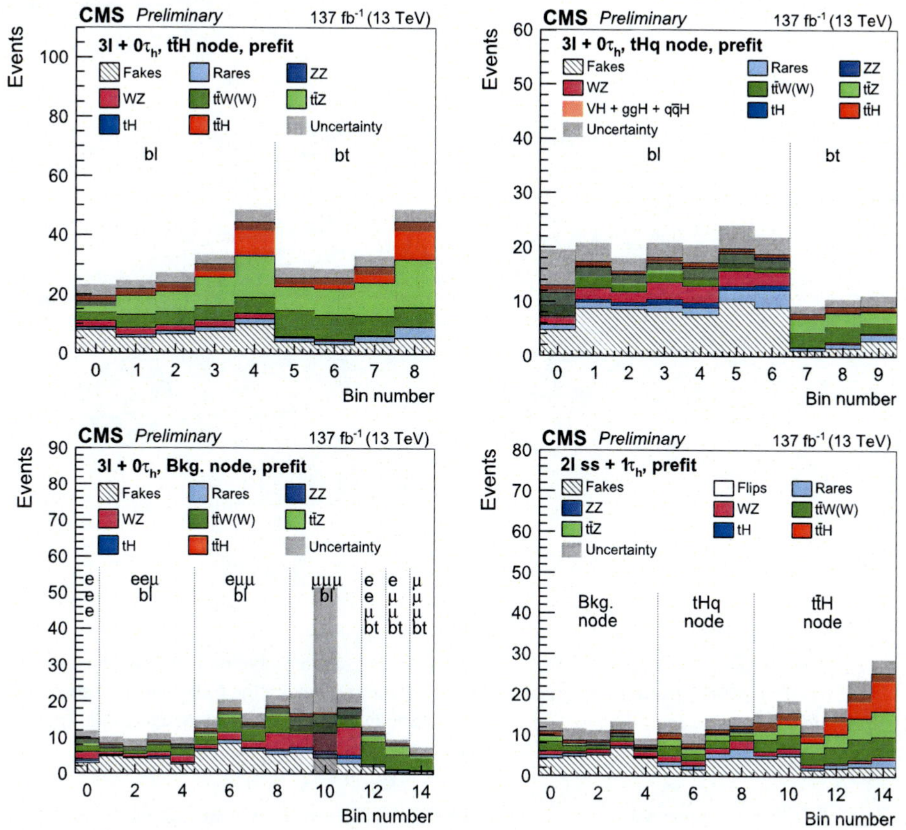

Fig. 7.9 Distributions of the DNN output scores in the $t\bar{t}H$ (top left), tH (top right) and rest (bottom left) nodes of the $3\ell + 0\tau_h$ category and in the $t\bar{t}H$, tH and other background nodes of the $2\ell ss + 1\tau_h$ category (bottom right), obtained from the simulation. The uncertainty bands correspond to systematic uncertainties. The distributions are presented before any fit is performed to the data

function that quantifies the compatibility of the Higgs boson mass with the reconstructed visible momenta of the τ decays and the E_T^{miss} of the event. In the $1\ell + 1\tau_h$, $1\ell + 2\tau_h$, $2\ell + 2\tau_h$ and $0\ell + 2\tau_h$ categories, the decay angle of the two τ leptons in the Higgs boson rest frame ($\cos\theta^*$) is considered. The training of BDT is performed with simulated samples of the $t\bar{t}H$ signal process and the main background processes pertinent to each category ($t\bar{t}W$, $t\bar{t}Z$, $t\bar{t}$+jets and Z/γ^*+jets). The XGBoost algorithm [1] is used with the hyperparameters shown in Table 7.4; their meaning is detailed in Ref. [1]. They are chosen with the objective of maximizing the area under the ROC curve.

The distributions of BDT scores for each category are shown in Figs. 7.12 and 7.13; they are used as discriminating observables in the signal extraction, with no subcategorization applied. The small $t\bar{t}H$ signal yield accumulates at the rightmost bins in the $1\ell + 1\tau_h$ and $0\ell + 2\tau_h$ categories, under an overwhelming fake, $t\bar{t}$+jets and Drell-Yan background. The $1\ell + 2\tau_h$ and $2\ell os + 1\tau_h$ categories are dominated

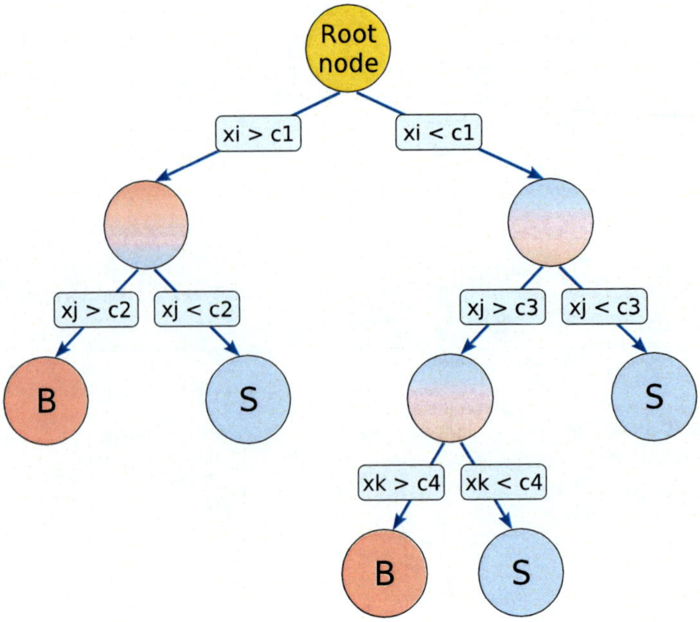

Fig. 7.10 Sketch of the structure of a decision tree to derive the probability of an event to originate from the signal (S) or from the background (B). The training data in the root node are subsequently split into subsamples according to cuts in the predefined variables ($x_{i,j,k}$) until a terminal leave with enough signal (blue) or background (red) purity is reached [5]

Table 7.4 Hyperparameters of the BDTs used for signal extraction in the $4\ell + 0\tau_h$, $2\ell os + 1\tau_h$, $3\ell + 1\tau_h$, $1\ell + 2\tau_h$, $2\ell + 2\tau_h$, $1\ell + 1\tau_h$, and $0\ell + 2\tau_h$ categories (see text and Ref. [1] for the definition of the hyperparameters)

Category	Training signals/backgrounds	Trees	Depth	Learning rate
$4\ell + 0\tau_h$	$t\bar{t}$H, tH/$t\bar{t}$W, $t\bar{t}$Z, $t\bar{t}$+jets	1000	2	0.07
$2\ell os + 1\tau_h$		1400	3	0.03
$3\ell + 1\tau_h$		300	4	0.02
$1\ell + 2\tau_h$		300	4	0.1
$2\ell + 2\tau_h$		600	3	0.1
$1\ell + 1\tau_h$	$t\bar{t}$H, tH/$t\bar{t}$W, $t\bar{t}$Z, $t\bar{t}$+jets, DY	300	3	0.1
$0\ell + 2\tau_h$		1500	3	0.02

Table 7.5 List of input variables to the BDTs used for signal extraction in the $0\ell + 2\tau_h$, $1\ell + 1\tau_h$, $1\ell + 2\tau_h$, $2\ell os + 1\tau_h$, $2\ell + 2\tau_h$, $3\ell + 1\tau_h$ and $4\ell + 0\tau_h$ categories. The check mark (✓) indicates that the variable is used

BDT input variable	$0\ell +$ $2\tau_h$	$1\ell +$ $1\tau_h$	$1\ell +$ $2\tau_h$	$2\ell os +$ $1\tau_h$	$2\ell +$ $2\tau_h$	$3\ell +$ $1\tau_h$	$4\ell +$ $0\tau_h$
Lepton(s) p_T		✓	✓	(✓)		(✓)	(✓)
Lepton(s) transv. mass		✓	✓	(✓)			
τ_h(s) p_T	✓	✓	(✓)	✓	(✓)	✓	
τ_h(s) transverse mass	(✓)	✓					
ℓ+τ_h transverse mass			✓				
$\tau_h\tau_h$ ($\tau_h\ell$) visible mass	✓	(✓)	✓	(✓)	✓	(✓)	
$\tau_h\tau_h$ ($\tau_h\ell$) SVFit mass	✓	(✓)					
Charge sum lepton+τ_h		✓					
Lepton or τ_h max. $\lvert\eta\rvert$	✓	✓	✓	✓			
ΔR leptons				✓			
ΔR τ_h	✓		✓				
ΔR lepton and τ_h		✓					
ΔR OS lepton and τ_h			✓	✓			
ΔR SS lepton and τ_h			✓	✓			
Min. ΔR leptons + τ_h	✓			✓			
Min. ΔR lepton(s) + jets		✓	✓	(✓)	(✓)		
Min. ΔR τ_h(s) + jets	(✓)			✓			
cos*θ of $\tau_h\tau_h$ ($\tau_h\ell$)	✓	(✓)	✓		✓		
Number of jets		✓					
Average ΔR jets	✓	✓	✓	✓	✓		
Number of loose b-jets		✓					
Number of medium b-jets	✓						
Loose b-jets inv. mass	✓	✓	✓	✓	✓		
Linear discriminator L_D	✓	✓	✓	✓	✓	✓	✓
Has SFOS loose leptons						✓	✓
Loose ℓ min. inv. mass						✓	✓
Hadronic top tagger	✓	✓	✓	✓			
2nd hadronic top tagger		✓					
Hadronic top p_T	✓		✓	✓			
2nd hadronic top p_T	✓						
Number of variables	15	16	17	18	9	9	7

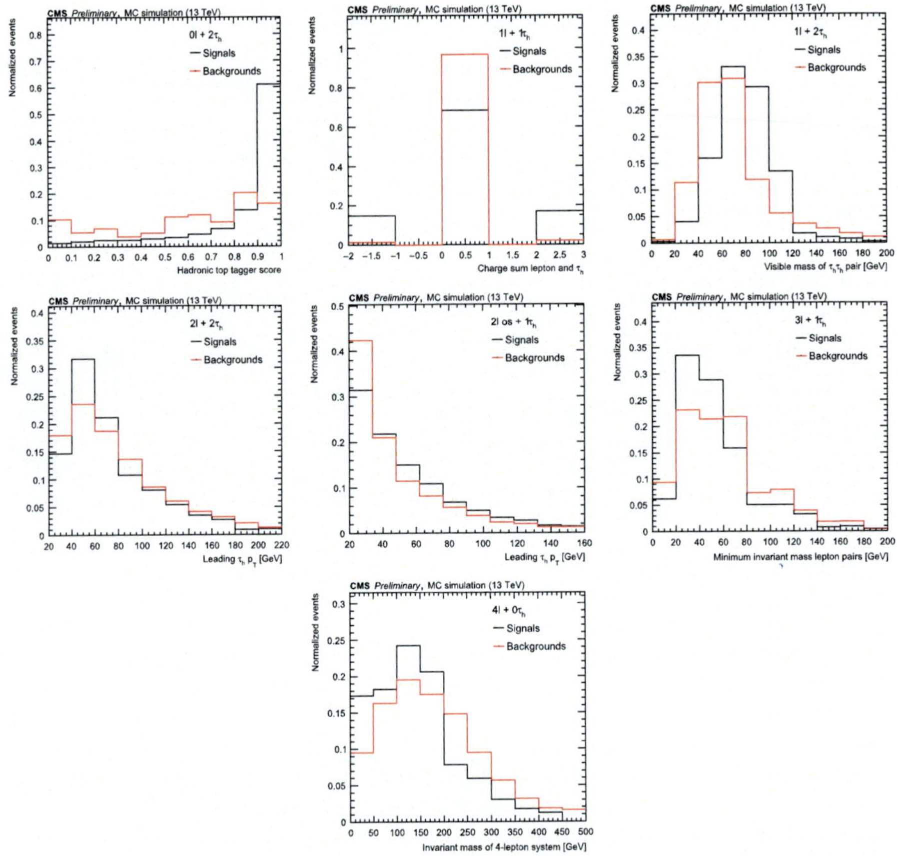

Fig. 7.11 Normalized distributions of the BDT input variables providing the largest separation between the $t\bar{t}H$ signal and the sum of the backgrounds in the $0\ell + 2\tau_h$, $1\ell + 1\tau_h$, $1\ell + 2\tau_h$, $2\ell os + 1\tau_h$, $2\ell + 2\tau_h$, $3\ell + 1\tau_h$ and $4\ell + 0\tau_h$ categories

by the fake background; however, the BDT manages to achieve a relatively pure last bin. Given the low statistical power of the $2\ell + 2\tau_h$, $3\ell + 1\tau_h$ and $4\ell + 0\tau_h$ categories, the BDT scores are rebinned to achieve a maximal separation between the signal and the backgrounds.

7.5 Results

The results presented in this section are derived on the full dataset collected with the CMS detector at a centre-of-mass energy of $\sqrt{s} = 13$ TeV during Run 2 (2016–2018), corresponding to a total integrated luminosity of 137 fb^{-1}. The signal strength modifiers $\mu_{t\bar{t}H}$ and μ_{tH} are extracted by performing a maximum likelihood fit of the observed yields to the expected signal and background yields in the discriminat-

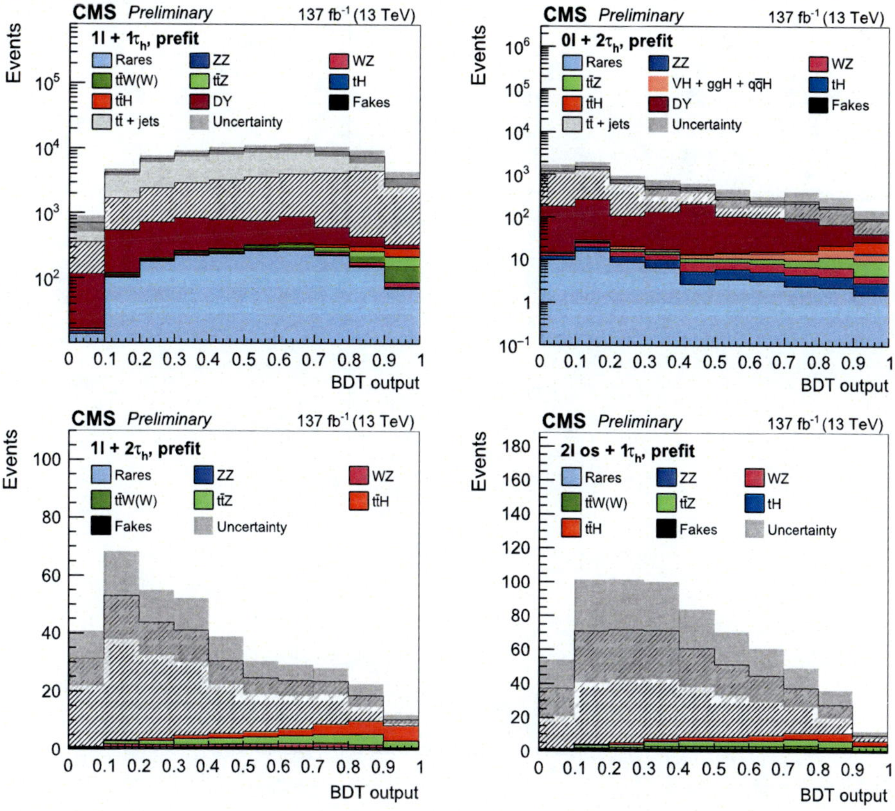

Fig. 7.12 Distributions of the BDT output scores in the $1\ell + 1\tau_h$, $0\ell + 2\tau_h$, $1\ell + 2\tau_h$ and $2\ell os + 1\tau_h$ categories, obtained from the simulation. The uncertainty bands correspond to systematic uncertainties. The distributions are presented before any fit is performed to the data

ing observables chosen for each signal and control region, following the statistical methods described in Sect. 6.7. These include the outputs of the DNNs in the different nodes and subcategories of the $2\ell ss + 0\tau_h$, $3\ell + 0\tau_h$ and $2\ell ss + 1\tau_h$ categories (Figs. 7.8 and 7.9), the BDT outputs in the $4\ell + 0\tau_h$, $1\ell + 1\tau_h$, $2\ell os + 1\tau_h$, $3\ell + 1\tau_h$, $1\ell + 2\tau_h$, $2\ell + 2\tau_h$ and $0\ell + 2\tau_h$ categories (Figs. 7.12 and 7.13) and the observables of the 3ℓ-CR and the 4ℓ-CR (Fig. 6.17). The rate of the $\mathrm{t\bar{t}W(W)}$ and $\mathrm{t\bar{t}Z}$ processes is left unconstrained in the fit. Two separate sets of results are derived: the first interpretation assumes the top Yukawa coupling to be SM-like, while the second one contemplates variations in the magnitude and sign of this coupling in the context of the κ_t framework.

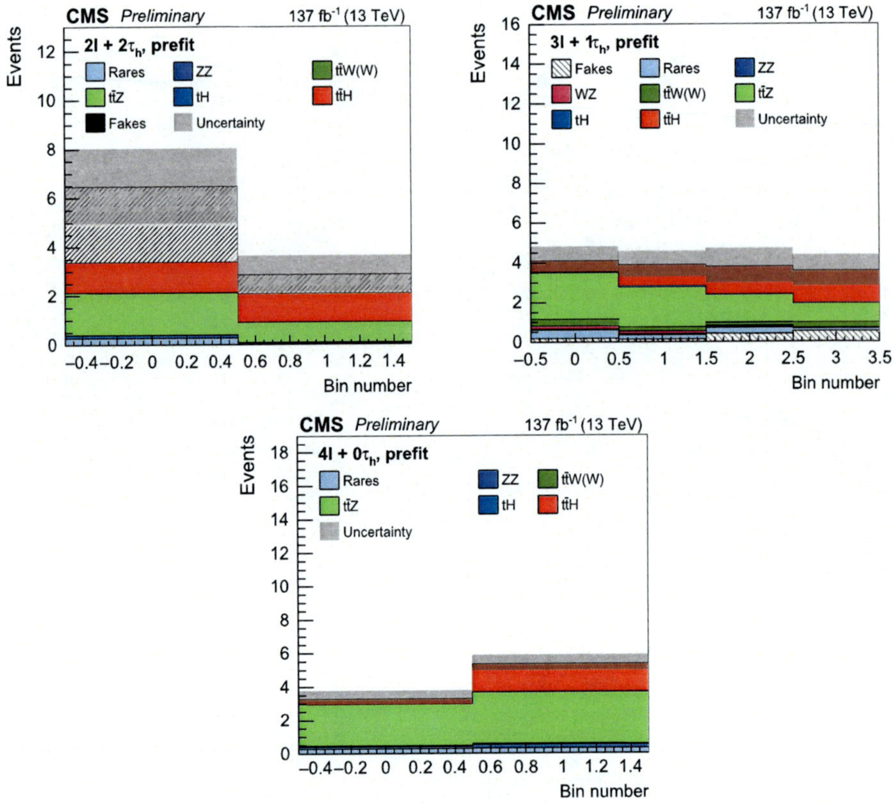

Fig. 7.13 Distributions of the BDT output scores in the $2\ell + 2\tau_h$, $3\ell + 1\tau_h$ and $4\ell + 0\tau_h$ categories, obtained from the simulation. The uncertainty bands correspond to systematic uncertainties. The distributions are presented before any fit is performed to the data

7.5.1 Standard Model Interpretation

The set of results presented in this section are derived assuming that the Higgs coupling modifiers are consistent with the SM ($\kappa_V = 1$ and $\kappa_t = 1$). As a first step, the statistical model and the nuisance parameters are validated by testing the impact of the nuisance parameters on the measurement of the signal strengths, and by evaluating the χ^2 of the global fit. As no anomalous behaviours are seen, the observed data events are compared to the yields predicted by the signal and the background simulated processes in the SM scenario, separately for each of the 10 categories and their corresponding subcategories. A global fit performed to all these distributions yields the measurement of the signal strength of the $t\bar{t}$H and the tH processes, and the subsequent estimation of the statistical and systematic uncertainties on this value. As the normalization of the $t\bar{t}$W and the $t\bar{t}$Z backgrounds is left as a free parameter in the fit, the correlation between their corresponding measured scale factors and the $t\bar{t}$H and tH signal strengths is presented.

Validation of the Statistical Model

The sources of systematic uncertainties described in Sect. 6.6 yield independent nuisance parameters which can affect the shape and the normalization of the distributions of the discriminating observables. Their impact on the measurement of the $t\bar{t}H$ and tH signal strengths was assessed by evaluating the so-called *impacts* and *pulls*. The impact of a nuisance parameter θ is defined as the shift $\Delta\mu$ induced on the value of signal strength by fixing θ to its postfit value $\hat{\theta}$ plus or minus 1σ, while the other nuisance parameters are estimated as the minimum of a profiled likelihood; the method provides an estimate of the correlation between θ and μ. The pulls are defined as the difference between the best fit estimation $\hat{\theta}$ from the likelihood maximization and the initial value of the nuisance parameter, normalized to the value of the uncertainty. The impacts and pulls of the leading 30 nuisances are shown in Fig. 7.14 for the $t\bar{t}H$ and tH signals. No parameter is significantly pulled away from its nominal value or overconstrained by the fit procedure, indicating the quality of the statistical model. The theoretical uncertainties and the uncertainties associated to the fake background estimation impact the most the $t\bar{t}H$ and tH measurements, as they affect all the final states and categories simultaneously. The experimental uncertainties associated to the τ_h and b-tagging selection efficiencies and the jet energy corrections have a smaller but still sizeable impact. The bin-by-bin statistical uncertainties in the measurement of the tH signal have a non-negligible impact, as expected from the limited statistics.

Further verification of the correct modelling of the data is provided by the *goodness-of-fit* test with the saturated method [7]. It is defined as a generalization of the χ^2 method for data that are not normally distributed, as in the case of the number of events in the binned distributions of the analysis. The test is built as a likelihood ratio, but the alternative hypothesis is considered to be the one that matches the data exactly. The distribution of this test statistics is obtained from pseudo-experiments generated from the modelled processes and is compared to the one computed with the observed data. The comparison is shown in Fig. 7.15, where the observed value of the χ^2 goodness of fit, with a p-value of 0.11, lies at the tail but within the distribution of the 1000 generated pseudo-experiments, concluding that the statistical model describes well the observations.

Data and Simulated Event Yields

The expected and observed event yields for the signals and backgrounds in the signal regions of the analysis are shown in Table 7.6. For the sake of visualization, no subcategorization is applied. The signal and background yields in the table are obtained after the maximum likelihood fit (they are the *post-fit* yields), meaning they are computed for the best fit values of nuisance parameters and after applying the scale factors on the $t\bar{t}W(W)$ and $t\bar{t}Z$ backgrounds obtained in the fit. The table shows that the number of data events and the expected signal and background yields are in good agreement within the associated uncertainties in all categories. The post-fit distributions of the discriminating observables in each channel are shown in Figs. 7.16, 7.17, 7.18 and 7.19. The plots confirm a good agreement in shape between the distributions of observed and expected events in all categories.

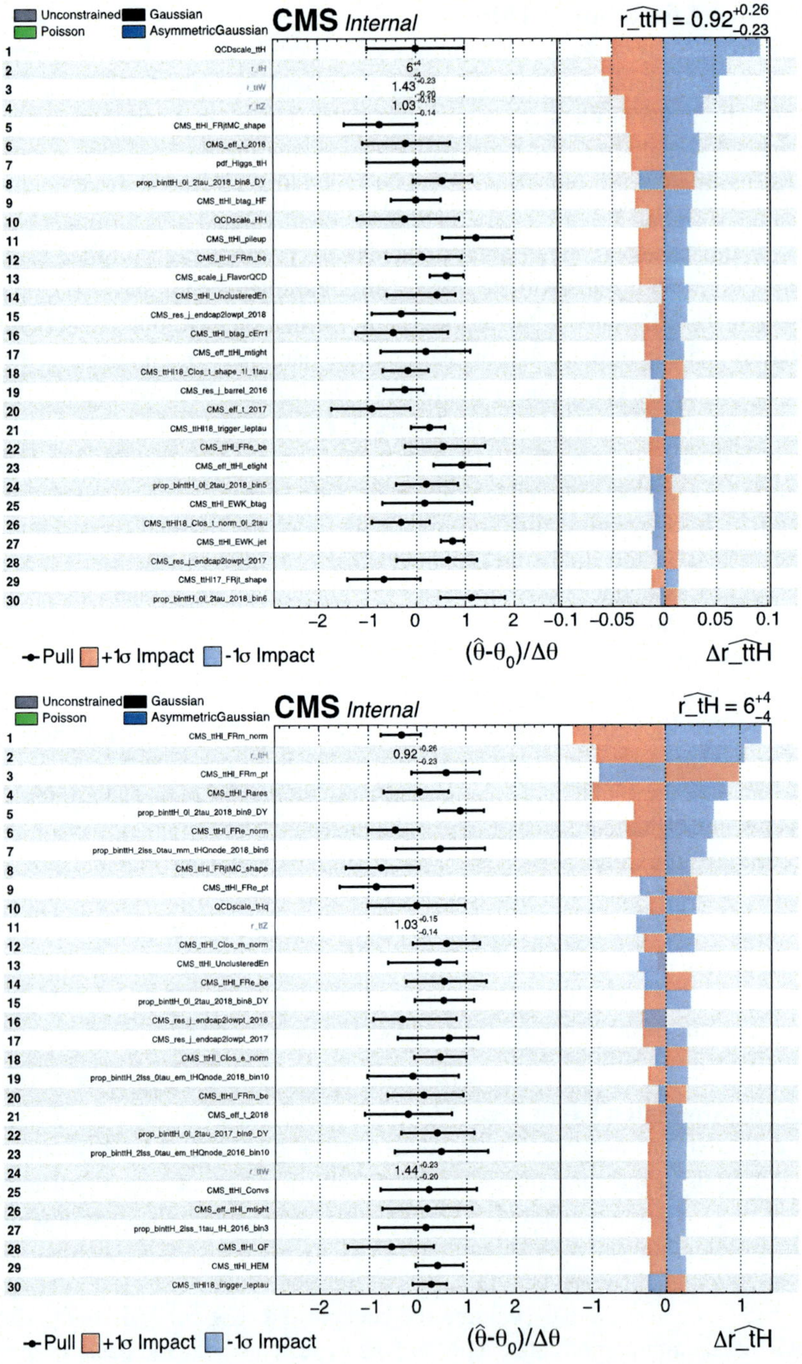

Fig. 7.14 List of the 30 nuisances parameters with the highest impact on the measurement of the $t\bar{t}H$ (top) and the tH (bottom) signal strengths. The first column corresponds to the pulls of each nuisance parameter and the second column corresponds to the +1σ and -1σ impacts on the measurement of the signal strength

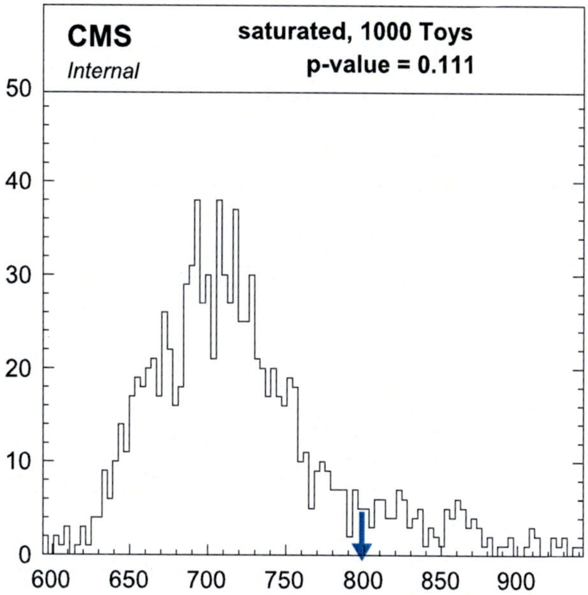

Fig. 7.15 Distribution of the χ^2 goodness of fit variable computed with the saturated model on 1000 pseudo-experiments (black line) compared to the observed value (blue arrow)

Measurement of the Signal Strengths

The measured signal rate modifiers under the hypothesis that the $\mathrm{t\bar{t}H}$ and tH processes are SM-like for the individual signal regions, and for the combination of all of them, are found in Fig. 7.20. The $\mathrm{t\bar{t}H}$ and tH signals are measured simultaneously in the fit. The measurement of the $\mathrm{t\bar{t}H}$ production rate is performed in all 10 channels, while the measurement of the tH production rate is limited to the $2\ell ss + 0\tau_h$, $3\ell + 0\tau_h$ and $2\ell ss + 1\tau_h$ channels, which employ a multiclassifier DNN to separate tH from the $\mathrm{t\bar{t}H}$ signal. The measured signal strength modifiers amount to

$$\begin{aligned} \mu_{\mathrm{t\bar{t}H}} &= 0.92\ ^{+0.19}_{-0.19}\ (\mathrm{stat})\ ^{+0.17}_{-0.13}\ (\mathrm{syst})\ , \\ \mu_{\mathrm{tH}} &= 5.67\ ^{+2.75}_{-2.68}\ (\mathrm{stat})\ ^{+2.98}_{-2.95}\ (\mathrm{syst})\ . \end{aligned} \tag{7.1}$$

The uncertainties on the measurement of the signal strengths are compatible with the expected values obtained from an Asimov dataset, $\mu^{\mathrm{exp.}}_{\mathrm{t\bar{t}H}} = 1.00^{+0.19}_{-0.18}(\mathrm{stat})^{+0.18}_{-0.13}(\mathrm{syst})$ and $\mu^{\mathrm{exp.}}_{\mathrm{tH}} = 1.00^{+2.66}_{-2.61}(\mathrm{stat})^{+2.67}_{-2.62}(\mathrm{syst})$.

Figure 7.20 shows the measured signal strength modifiers of the $\mathrm{t\bar{t}H}$ and tH processes separately for each signal regions and for the combination of all of them with the control regions. The sensitivity of each channel can be inferred from the size of the corresponding uncertainty band on the measured signal strength. The channel providing the highest sensitivity is $2\ell ss + 0\tau_h$, followed by $3\ell + 0\tau_h$ and $2\ell ss + 1\tau_h$. The measured signal strengths are consistent with the SM prediction, except in the $2\ell + 2\tau_h$ category, where the best-fit value of the $\mathrm{t\bar{t}H}$ signal rate was found to be negative due to the deficit of observed events compared to the background expectation. The value of the signal strength for this category shown in Fig. 7.20 is obtained after bounding the $\mathrm{t\bar{t}H}$ production rate to positive values. The significance of this

Table 7.6 Number of observed events in the 10 categories of the main analysis compared to the event yields expected from the $t\bar{t}H$ and tH signals and from the main background processes, computed for the values of nuisance parameters and of the parameters of interest obtained from the maximum likelihood fit. The quoted uncertainties include the statistical and systematic components

Process	$2\ell ss + 0\tau_h$	$1\ell + 1\tau_h$	$0\ell + 2\tau_h$
$t\bar{t}H$	222.1 ± 51.1	183.1 ± 40.8	24.4 ± 6.0
tH	119.3 ± 85.4	64.6 ± 45.6	16.1 ± 11.7
$t\bar{t}Z + t\bar{t}\gamma^*$	321.6 ± 24.8	203.0 ± 23.6	27.1 ± 3.8
$t\bar{t}W + t\bar{t}WW$	1153.2 ± 64.2	254.5 ± 34.3	3.8 ± 0.5
WZ	296.2 ± 31.0	198.1 ± 37.3	42.5 ± 8.7
ZZ	31.2 ± 3.3	98.5 ± 12.8	34.2 ± 4.8
Drell-Yan	–	4477.2 ± 465.0	1433.0 ± 217.5
$t\bar{t}$+jets	–	40951.2 ± 1904.0	860.5 ± 98.1
Fakes	1217.3 ± 90.7	25336.7 ± 1885.4	3788.0 ± 221.7
Flips	121.1 ± 19.4	–	–
Conversions	42.3 ± 11.9	–	–
Rares	222.0 ± 47.8	1928.1 ± 424.0	60.3 ± 13.9
Other Higgs boson processes	35.3 ± 4.0	38.5 ± 3.6	26.7 ± 3.6
Total expected background	3517.4 ± 85.1	73550.4 ± 611.6	6292.3 ± 135.9
Data	3738	73736	6310

Process	$3\ell + 0\tau_h$	$2\ell ss + 1\tau_h$	$2\ell os + 1\tau_h$	$1\ell + 2\tau_h$
$t\bar{t}H$	60.9 ± 14.7	28.9 ± 6.4	19.1 ± 4.3	19.3 ± 4.2
tH	20.1 ± 14.2	12.7 ± 9.0	4.8 ± 3.4	2.6 ± 1.9
$t\bar{t}Z + t\bar{t}\gamma^*$	145.5 ± 10.9	29.6 ± 3.3	25.5 ± 2.9	20.3 ± 2.1
$t\bar{t}W + t\bar{t}WW$	171.1 ± 9.5	47.4 ± 6.5	17.4 ± 2.4	2.6 ± 0.4
WZ	89.7 ± 9.7	19.4 ± 2.9	8.4 ± 1.6	11.8 ± 2.2
ZZ	16.2 ± 1.6	1.6 ± 0.3	1.9 ± 0.3	1.8 ± 0.3
Fakes	140.3 ± 11.3	52.0 ± 9.6	518.9 ± 28.0	250.1 ± 15.9
Conversions	5.6 ± 1.6	–	0.5 ± 0.2	–
Rares	41.0 ± 8.9	13.3 ± 3.1	5.9 ± 1.3	5.6 ± 1.3
Other Higgs boson processes	3.4 ± 0.3	1.8 ± 0.3	0.8 ± 0.1	–
Total expected background	627.3 ± 19.6	178.7 ± 12.6	583.6 ± 27.4	295.0 ± 15.9
Data	744	201	603	307

Process	$4\ell + 0\tau_h$	$3\ell + 1\tau_h$	$2\ell + 2\tau_h$
$t\bar{t}H$	2.0 ± 0.5	4.0 ± 0.9	2.2 ± 0.5
tH	0.2 ± 0.2	0.8 ± 0.6	0.3 ± 0.2
$t\bar{t}Z + t\bar{t}\gamma^*$	5.9 ± 0.4	6.6 ± 0.7	2.5 ± 0.3
$t\bar{t}W + t\bar{t}WW$	0.2 ± 0.0	1.1 ± 0.2	–
ZZ	0.6 ± 0.2	0.3 ± 0.1	0.2 ± 0.0
Fakes	–	1.5 ± 0.9	3.4 ± 0.9
Conversions	–	–	–
Rares	0.6 ± 0.1	1.0 ± 0.3	0.3 ± 0.1
Total expected background	7.4 ± 0.5	11.5 ± 1.3	6.8 ± 1.0
Data	12	18	3

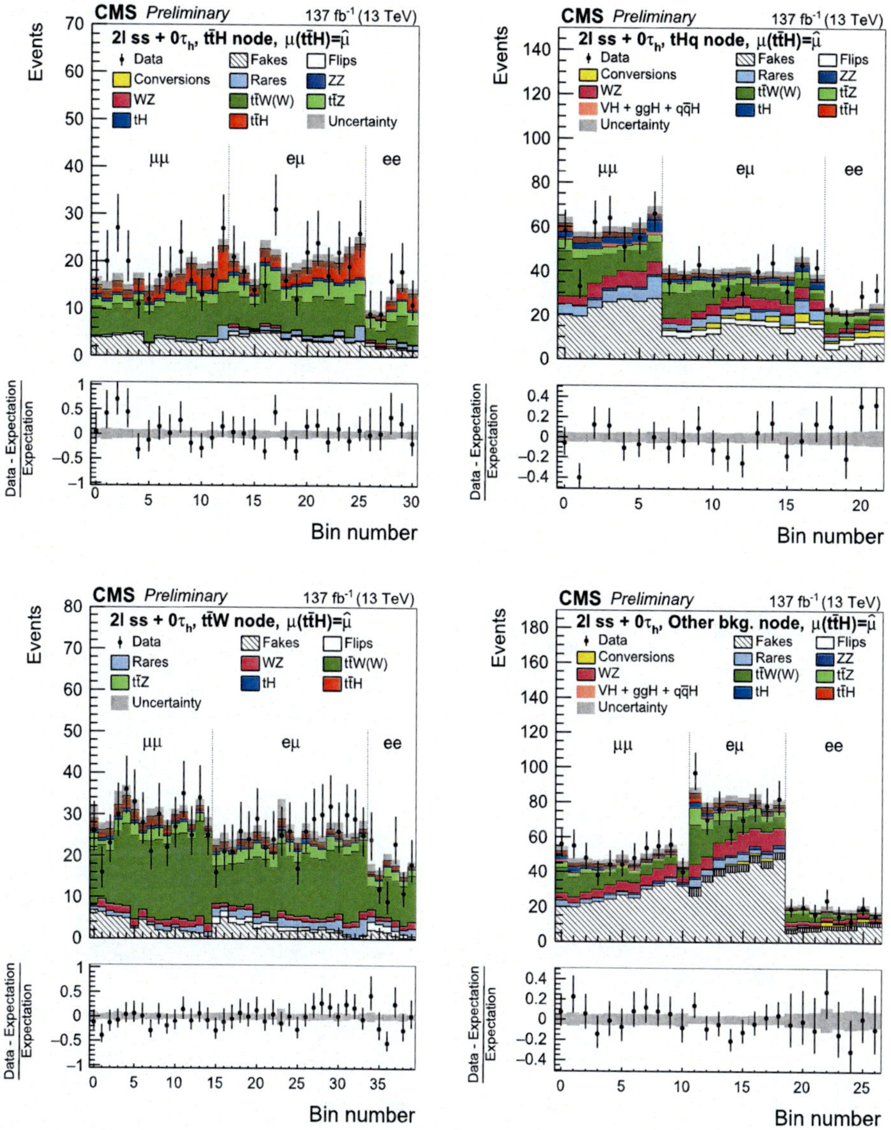

Fig. 7.16 Post-fit distributions of the discriminating observables used in the four DNN output nodes of the $2\ell ss + 0\tau_h$ category (t$\bar{\text{t}}$H, tH, t$\bar{\text{t}}$W and rest) for data and simulated events. The uncertainty bands include the statistical and systematic components

category under the signal plus background hypothesis amounts to 1.9σ; hence, the measured value is compatible with the SM expectation.

Assuming the tH production rate equivalent to that of the SM, the results correspond to an observed (expected) significance of 4.7σ (5.2σ) over the background-only hypothesis. The achieved sensitivity allows to the observation of the t$\bar{\text{t}}$H process in multilepton final states to be claimed. The observed (expected) significance for the

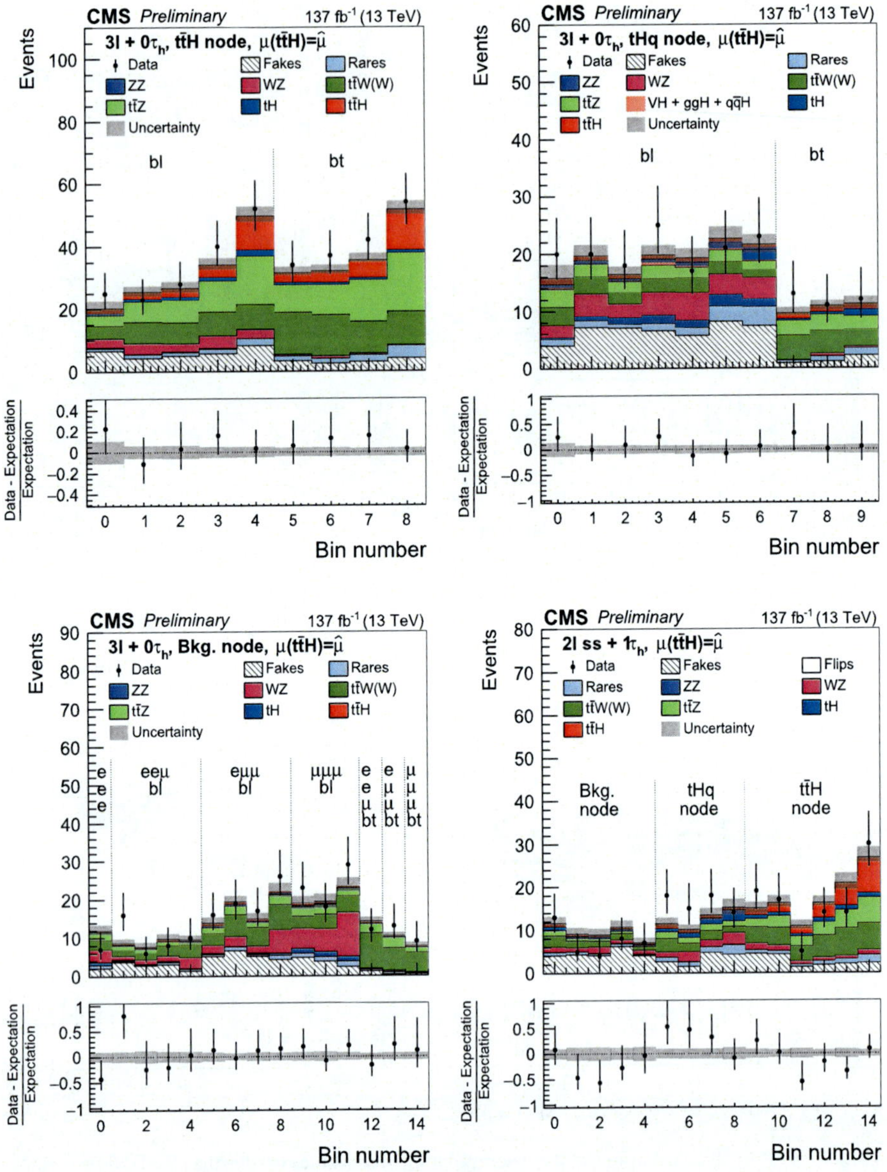

Fig. 7.17 Post-fit distributions of the discriminating observables used in the three DNN output nodes of the $3\ell + 0\tau_h$ category (tt̄H, tH and rest) and the $2\ell ss + 1\tau_h$ category (tt̄H, tH and rest) for data and simulated events. The uncertainty bands include the statistical and systematic components

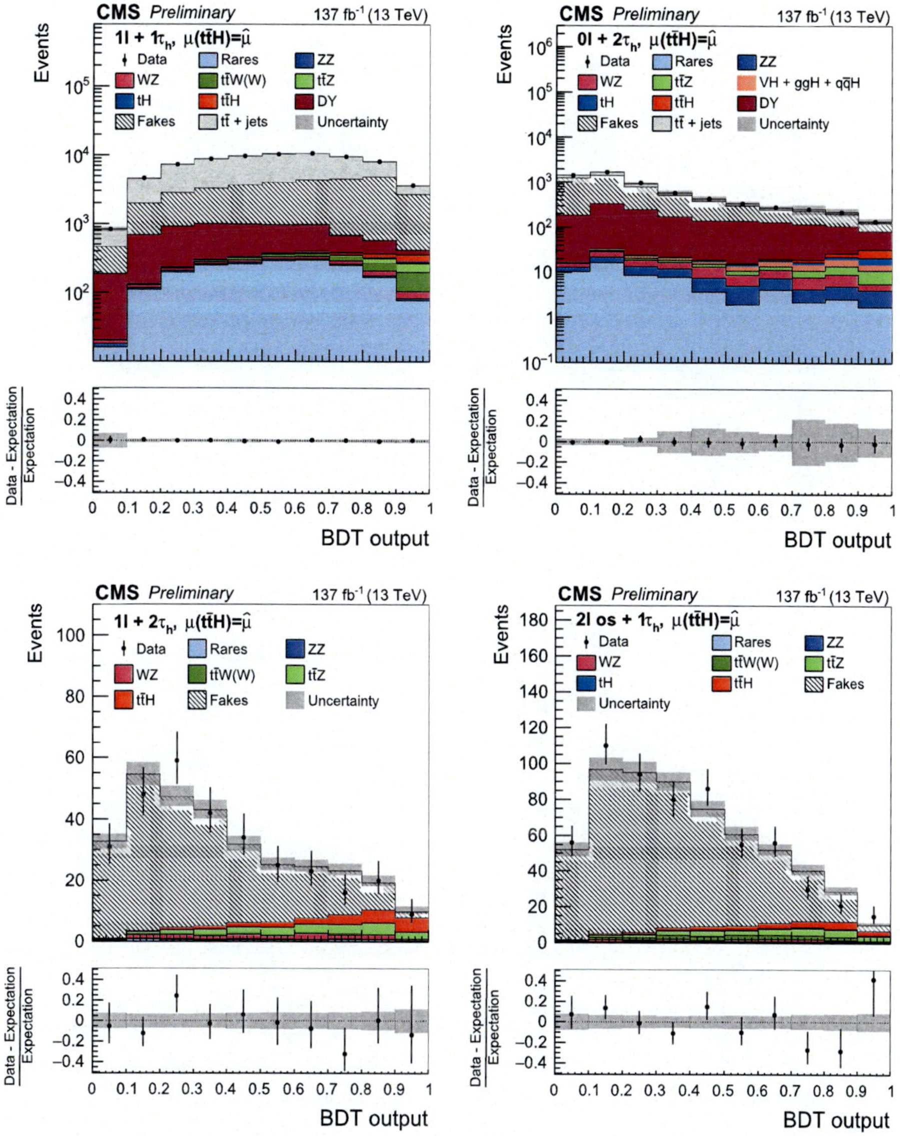

Fig. 7.18 Post-fit distributions of the discriminating observables used in the $1\ell + 1\tau_h$, $0\ell + 2\tau_h$, $1\ell + 2\tau_h$ and $2\ell os + 1\tau_h$ categories for data and simulated events. The uncertainty bands include the statistical and systematic components

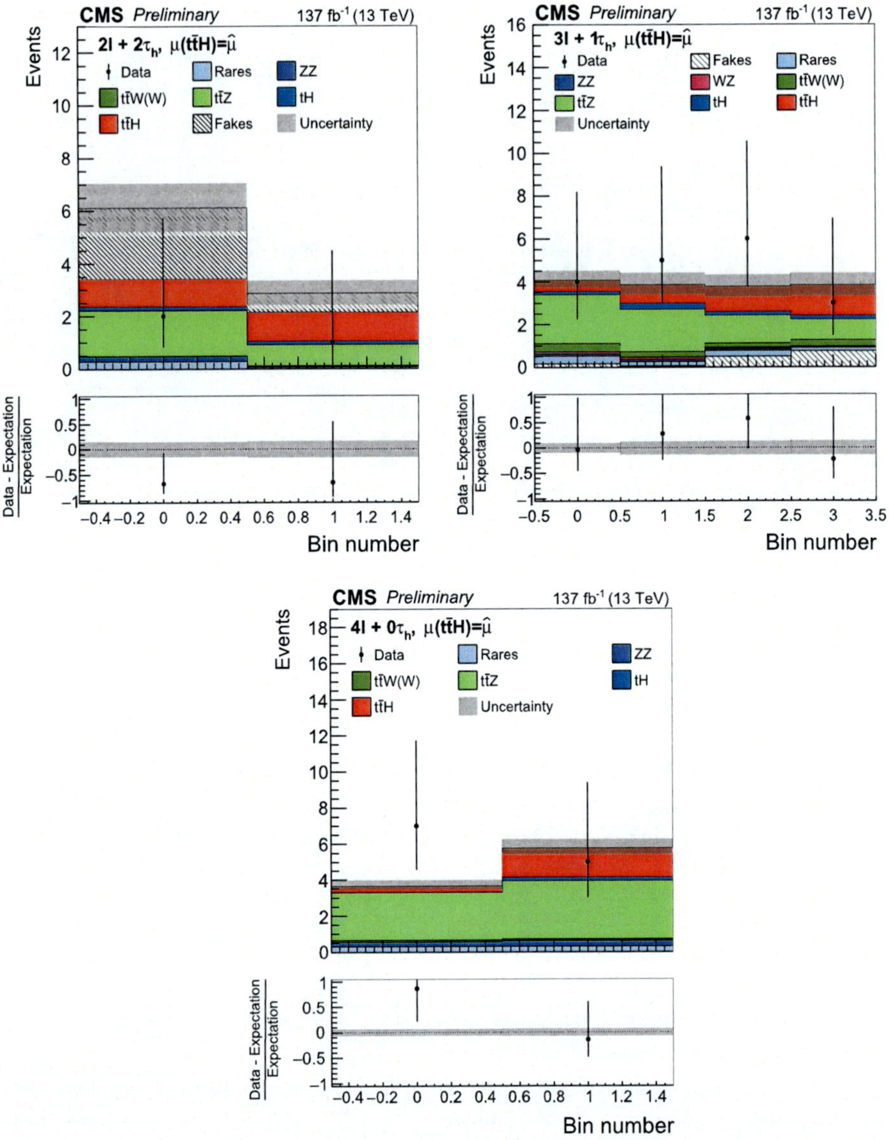

Fig. 7.19 Post-fit distributions of the discriminating observables used in the $2\ell + 2\tau_h$, $3\ell + 1\tau_h$ and $4\ell + 0\tau_h$ categories for data and simulated events. The uncertainty bands include the statistical and systematic components

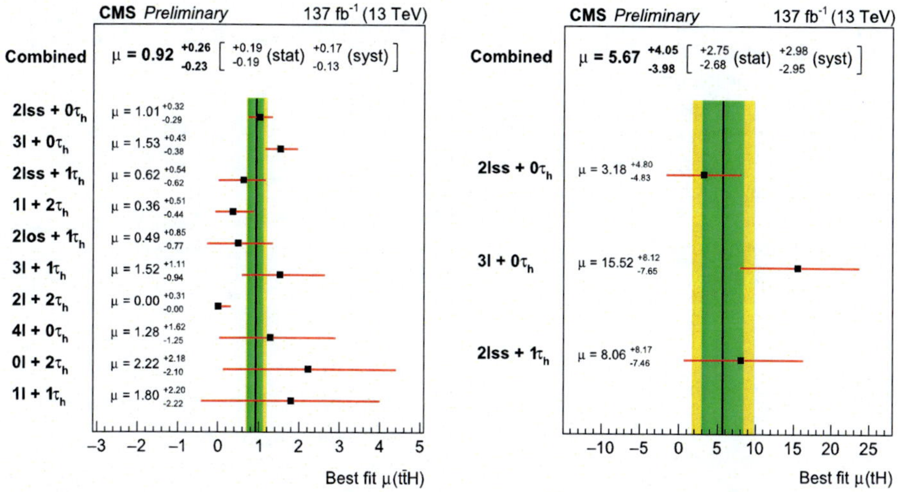

Fig. 7.20 Measured signal strengths ($\mu = \sigma/\sigma_{\mathrm{SM}}$) of the $\mathrm{t\bar{t}H}$ process (left) and the tH process (right) under the hypothesis that they are SM-like, separately for each signal region considered and for the combination of all including the control regions

tH process amounts to 1.4σ (0.3σ), also assuming the $\mathrm{t\bar{t}H}$ process to have the SM production rate. The evidence for the presence of the $\mathrm{t\bar{t}H}$ and tH signals in the data is illustrated in Fig. 7.21, which shows the logarithm of the ratio of signal events over background events including all regions. A broad excess of events with respect to the background hypothesis is visible in the rightmost bins; it is consistent with the expectation for $\mathrm{t\bar{t}H}$ and tH production in the SM.

The impact of the systematic uncertainties on the measurement of the $\mathrm{t\bar{t}H}$ and tH signal strengths is presented in Table 7.7, where the uncertainties have been grouped by source. The main impacts on the measurement of the $\mathrm{t\bar{t}H}$ and tH signal strengths are caused by the uncertainty on the reducible background estimate, the normalization of the processes estimated by the simulation, the theoretical uncertainties and the statistical uncertainties on the simulation, the data and the control regions.

Signal and Background Correlations

The normalizations of the $\mathrm{t\bar{t}W}$ and the $\mathrm{t\bar{t}Z}$ backgrounds are left unconstrained in the fit, so that they can be estimated making use of the data. The corresponding scale factors are extracted from the fit, amounting to

$$\begin{aligned} \theta_{\mathrm{t\bar{t}W(W)}} &= 1.44\,^{+0.23}_{-0.20}\,, \\ \theta_{\mathrm{t\bar{t}Z}} &= 1.03\,^{+0.15}_{-0.14}\,. \end{aligned} \tag{7.2}$$

These scale factors are applied to the normalization of the $\mathrm{t\bar{t}W}$ and $\mathrm{t\bar{t}Z}$ backgrounds in the signal regions. The measured $\mathrm{t\bar{t}Z}$ production rate is in agreement with the SM prediction, whereas an excess of the data over the SM prediction is observed in the $\mathrm{t\bar{t}W(W)}$ process. This tension between the data and the simulation has been

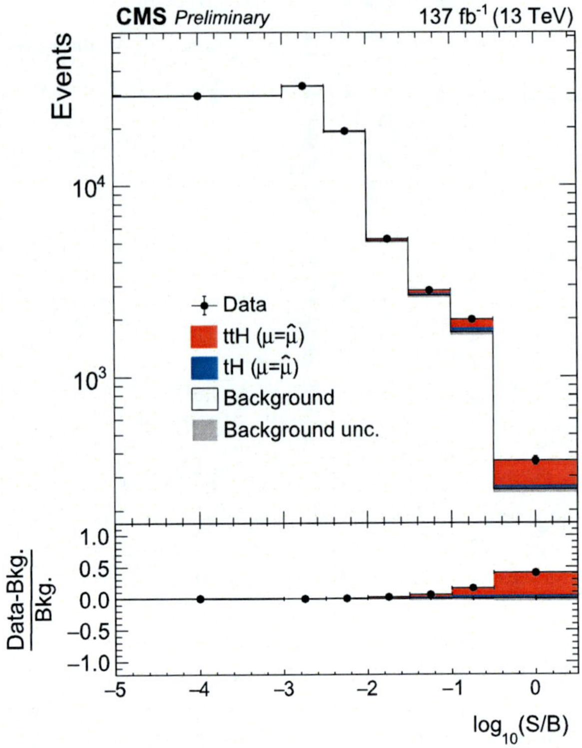

Fig. 7.21 Distribution in the decimal logarithm of the ratio between the expected $t\bar{t}H$ and tH signals and the expected sum of background contributions, compared to the observed data. A broad excess of events with respect to the background-only expectation is visible

Table 7.7 Main sources of systematic uncertainty and their impact on the measurement of the $t\bar{t}H$ and tH signal strengths. The quantity $\Delta\mu_x/\mu_x$ corresponds to the change in uncertainty when fixing the nuisances associated to it in the fit

Uncertainty source	$\Delta\mu_{t\bar{t}H}/\mu_{t\bar{t}H}$ (%)	$\Delta\mu_{tH}/\mu_{tH}$ (%)
Trigger efficiency	2.3	7.3
Electron and muon selection efficiency	2.9	7.1
τ_h selection efficiency	4.6	9.1
b-tagging efficiency	3.6	13.6
Jet energy scale and resolution	3.4	8.3
Reducible background estimate	6.0	36.8
Normalization of MC estimation processes	13.3	12.3
Theoretical sources	4.6	18.2
Prefiring	0.4	3.5
Integrated luminosity	2.2	4.6
MC and sideband statistical uncertainty	7.1	27.2
Data statistical uncertainty	20.9	48.0

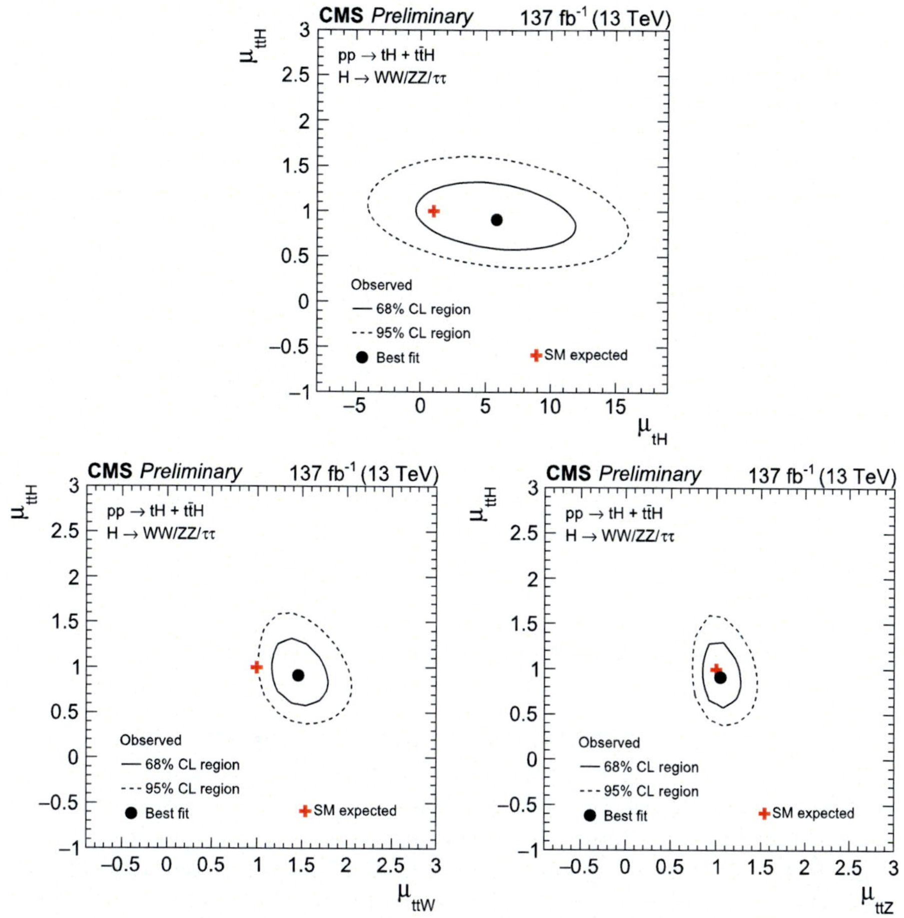

Fig. 7.22 Two-dimensional contours of $-2\Delta\ln\mathcal{L}$ as a function of the $t\bar{t}H$ and tH (top), $t\bar{t}H$ and $t\bar{t}W$ (bottom left) and $t\bar{t}H$ and $t\bar{t}Z$ (bottom right) production rates. For each case, the two production rates not shown on either the x or y axis are profiled

previously reported by both the CMS and ATLAS collaborations, either in dedicated searches [8, 9] or in the context of the $t\bar{t}H$ multilepton searches [10, 11]. The excess is observed despite the latest electroweak corrections in the modelling of the $t\bar{t}W$ process [12] being used.

The correlations between the measured $t\bar{t}H$ and tH signal rates and the $t\bar{t}W$ and $t\bar{t}Z$ production rates are illustrated in Fig. 7.22, depicting the two-dimensional contours of the likelihood function corresponding to the 68% and 98% confidence levels. The correlations are shown for the pairs of processes ($t\bar{t}H$, tH), ($t\bar{t}H$, $t\bar{t}W$) and ($t\bar{t}H$, $t\bar{t}Z$). The figures show the level of agreement with the SM expectation and the moderate correlation amongst the processes, demonstrating the performance of the DNN in disambiguating the signals and the backgrounds.

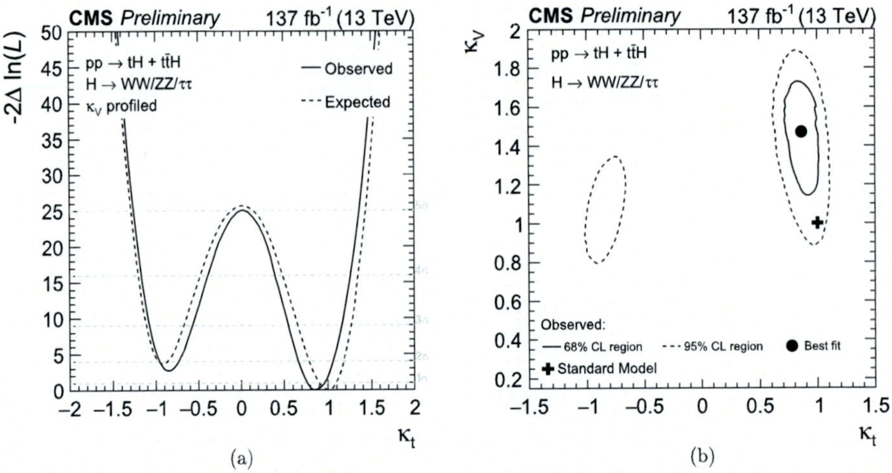

Fig. 7.23 **a** Scan of $-2\Delta\ln\mathcal{L}$ as a function of the κ_t coupling modifier, while profiling over the κ_V coupling modifier. **b** Two-dimensional contours of $-2\Delta\ln\mathcal{L}$ as a function of the κ_t and κ_V coupling modifiers

7.5.2 *Beyond the Standard Model Interpretation*

The results presented in this section no longer assume the Higgs boson couplings to the top quark and the W boson to be consistent with the SM. The simultaneous measurement of the $t\bar{t}H$ and tH processes allows modified coupling scenarios to be probed (see Sect. 2.2.2); they are explored in terms of κ_t and κ_V modifiers, which encode the ratio between the measured value and the expectation. In the SM, $\kappa_t = \kappa_V = 1$; the opposite configuration corresponds to the ITC scenario, with $\kappa_t = -1$ and $\kappa_V = +1$. Variations in κ_t and κ_V affect the cross section of the $t\bar{t}H$ and tH processes, the Higgs boson branching ratios and the kinematics of the tH process. To perform the interpretation, the statistical model used in the SM scenario is used, fixing the signal strengths to the SM values. The likelihood function is evaluated profiling all the nuisances for several hypothesis of κ_t and κ_V; the point with the maximum likelihood corresponds to the best fit for κ_t and κ_V.

Figure 7.23a shows the function $-2\Delta\log\mathcal{L}$ for different values of κ_t; it is compared to the expected result obtained from an Asimov dataset. In the scan, the value of κ_V is profiled. The value of κ_t is constrained to the intervals $-0.9 < \kappa_t < -0.7$ or $0.7 < \kappa_t < 1.1$ at 95% confidence level. The two-dimensional 68% and 98% confidence regions of $-2\Delta\log\mathcal{L}$ as a function of κ_t and κ_V are shown in Fig. 7.23b, fixing the rest of the coupling modifiers to their SM value. The minimum of the likelihood ratio is found at $\kappa_t = 0.9$ and $\kappa_V = 1.5$, meaning the data favour the SM.

7.6 Conclusions

The production of the Higgs boson in association with either one (tH) or two ($t\bar{t}H$) top quarks has been measured in events containing electrons, muons and hadronically decaying τ leptons in the final state. The measurement targets the WW, ZZ and $\tau\tau$ decay modes of the Higgs boson, and both hadronic and leptonic decays of the top quark. Signal regions are built upon different combinations of lepton and τ_h multiplicity, and selections are applied according to the targeted topology. The results presented are derived with an unprecedented dataset of 137 fb^{-1}, the largest integrated luminosity collected up to date, recorded during the complete Run 2 of the LHC. It constitutes the first analysis in which the cross sections of the $t\bar{t}H$ and the tH processes are measured simultaneously. The study of these processes is well motivated, as they provide unique insights on important SM parameters and are also sensitive to potential deviations introduced by new physics manifestations. The complementarity between both processes sets stringent constraints on the magnitude of the top Yukawa coupling and its relative sign with respect to the coupling of the Higgs boson to the massive electroweak bosons. To overcome the challenges related to the complexity of the final states and the rarity of the signals, regions enriched in the signal and the backgrounds are defined with advance machine learning techniques (DNNs and BDTs); they are able learn the features of the processes of interest in a complex high-dimensional phase space and separate the signals from the backgrounds and also the signals themselves.

A first set of results has been presented assuming a SM-like coupling of the Higgs boson to the top quark and to the vector bosons. The measured cross sections of the $t\bar{t}H$ and tH processes remain consistent with the SM expectation, with signal strengths of $\mu_{t\bar{t}H} = 0.92^{+0.26}_{-0.23}$ and $\mu_{tH} = 5.67^{+4.05}_{-3.98}$. The observed (expected) significance of the $t\bar{t}H$ process amounts to 4.7σ (5.2σ) over the background-only hypothesis; the measurement is performed with enough sensitivity to claim the observation of the $t\bar{t}H$ process in multilepton final states. The sensitivity is significantly increased with respect to the previous $t\bar{t}H$ multileptons search by CMS with 77.4 fb^{-1} of data, where a significance of 3.2σ was observed [11]. Likewise, the tH cross section is measured with the highest sensitivity up to date; the previous upper limit set by CMS with an analyzed dataset of 35.9 fb^{-1} amounted to 25 times the SM expectation [13] upon combination with the $b\bar{b}$ and $\gamma\gamma$ final states. To further validate these results, a complementary measurement of the $t\bar{t}H$ signal rate is performed using a set of alternative observables in the maximum likelihood fit, based on the Matrix Element Method and simple kinematic variables, presented in Chap. 8. A more precise determination of the $t\bar{t}H$ and tH cross sections will be achieved upon combination with other Higgs boson decay modes ($H \rightarrow b\bar{b}$ and $H \rightarrow \gamma\gamma$) and with the 300 fb^{-1} of integrated luminosity foreseen to be collected during Run 3.

The exploration of BSM scenarios is conducted in a second set of results. They allow for deviations of the top Yukawa coupling from the value predicted by the SM to be probed, to which both the $t\bar{t}H$ and the tH processes are sensitive. The value of the coupling modifier κ_t has been constrained to the observed intervals of $-0.9 < \kappa_t <$

-0.7 or $0.7 < \kappa_t < 1.1$ at 95% confidence level. The SM scenario is favoured by the data, with the maximum likelihood point located at $\kappa_t = 0.9$. Such intervals are significantly more stringent than the previous measurement performed by CMS with a dataset of 35.9 fb^{-1}, with constrained values of κ_t within the intervals $-0.9 < \kappa_t < -0.5$ or $0.9 < \kappa_t < 2.1$ at 95% confidence level; the data were favouring the SM scenario as well [13]. Other explorations of modified top-Higgs coupling scenarios include the measurement of CP-violation in the top Yukawa sector. The presence of a CP-odd component in the coupling is highly disfavoured but still allowed with the latest interpretation of the $t\bar{t}H$ analysis in the $H \rightarrow \gamma\gamma$ final state [14, 15]. Significant sensitivity to CP-violation can be gained upon combination with the multileptons final states.

7.7 Future Prospects

Exhaustive projection studies, taking into account the upcoming detector upgrades and state-of-the-art machine learning analysis techniques, have been conducted by the CMS and ATLAS collaborations to estimate the $t\bar{t}H$ and tH physics reach during the HL-LHC era [16, 17]. The projections are based on the analyses with 36 fb^{-1} of data collected in 2016 at $\sqrt{s} = 13$ TeV; they assume a total integrated luminosity of 3000 fb^{-1} at $\sqrt{s} = 14$ TeV in the HL-LHC. They are evaluated for different scenarios of the evolution of the systematic uncertainties with increased data samples and improved theoretical predictions. In both cases, it is assumed that the CMS and ATLAS upgrades will provide the same level of detector and trigger performance as in Run 2. The first scenario (S1) assumes that the systematic uncertainties are kept constant with the integrated luminosity, with the performance of the CMS and ATLAS detector unchanged with respect to the reference analyses used for the projection. The second scenario (S2) scales the theoretical uncertainties down by a factor 2, while the experimental uncertainties are scaled down with $\sqrt{\mathcal{L}}$ until they reach a defined minimum value based on the estimated achievable accuracy with the upgraded detector.

Table 7.8 shows the breakdown of the contributions to the expected uncertainties on the measurement of the $t\bar{t}H$ cross section in the multileptons final state with the ATLAS experiment. Uncertainties of 11% and 7% are foreseen for the signal strength $\mu_{t\bar{t}H}$ with the S1 and S2 scenarios, respectively; they represent a ~60% and ~70%

Table 7.8 Contributions to the expected uncertainties on the measurement of the $t\bar{t}H$ signal strength in multileptons final state for the two scenarios considered in the HL-LHC with the ATLAS detector (see text) [16]

Scenario	Δ_{tot}/σ_{SM}	$\Delta_{stat}/\sigma_{SM}$	Δ_{exp}/σ_{SM}	Δ_{sig}/σ_{SM}	Δ_{bkg}/σ_{SM}	$\Delta\mu_{sig}$
S1	$^{+0.27}_{-0.28}$	$^{+0.07}_{-0.07}$	$^{+0.23}_{-0.23}$	$^{+0.09}_{-0.08}$	$^{+0.12}_{-0.12}$	$^{+0.11}_{-0.11}$
S2	$^{+0.25}_{-0.25}$	$^{+0.07}_{-0.07}$	$^{+0.22}_{-0.22}$	$^{+0.05}_{-0.05}$	$^{+0.07}_{-0.07}$	$^{+0.07}_{-0.07}$

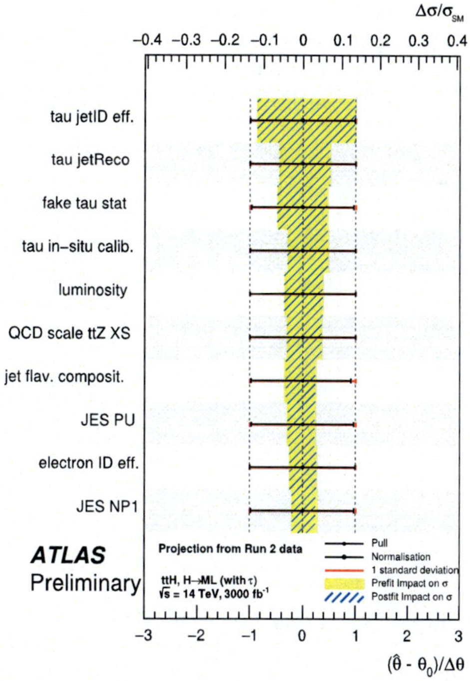

Fig. 7.24 Ranking of the 10 most significant systematic uncertainties on the measurement of the $t\bar{t}H$ signal strength in multileptons final state under the S2 scenario in the HL-LHC at the ATLAS experiment (see text). The uncertainties are listed according to their post-fit impact [16]

improvement with respect to the measured value in this thesis. Figure 7.24 shows the ranking of the 10 most significant systematic uncertainties under the S2 scenario, listed according to their post-fit impact on the measurement of the $t\bar{t}H$ cross section. The dominant experimental uncertainty originates from the τ_h identification, while the main theoretical uncertainty is related to the modelling of the $t\bar{t}Z$ background. The fake lepton uncertainties are moderately constrained as there is no reduction factor contemplated for such uncertainties under any of the considered scenarios. For what concerns the tH process, the evolution of the expected uncertainty on the measurement of μ_{tH} at CMS assuming SM rate is given in Table 7.9 for the case where the $t\bar{t}H$ process is left freely floating in the fit. Values of ±1.2 and ±0.9 are obtained for the S1 and S2 scenarios, respectively; the improvement in precision with respect to the results included in this thesis scales similarly to the $t\bar{t}H$ measurement. The projected constraints on the value of the coupling modifier κ_t obtained from the combination of the tH and $t\bar{t}H$ processes in CMS at the HL-LHC is shown in

Table 7.9 Expected $\pm1\sigma$ uncertainties on the measurement of the tH signal strength in multileptons final state for the two scenarios considered in the HL-LHC with the CMS detector (see text) [16]

Scenario	$\pm1\sigma$ ($\mu_{tH} = 1$)
S1	$^{+1.2}_{-1.2}$
S2	$^{+0.9}_{-0.9}$

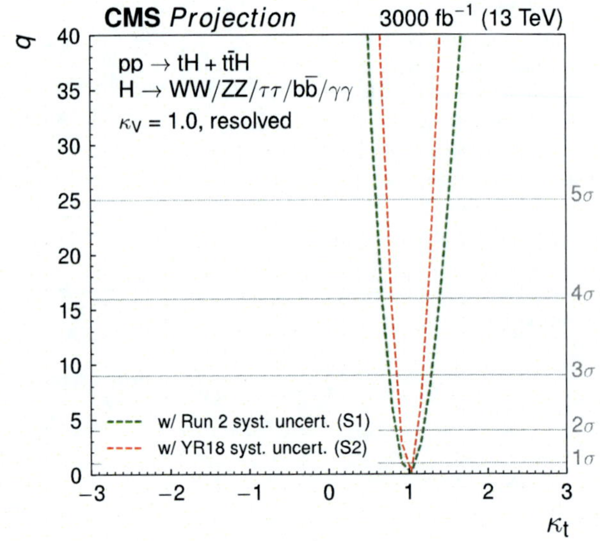

Fig. 7.25 Likelihood scan as a function of κ_t extracted from the measurement of the tH and $t\bar{t}H$ signal strengths under the S1 and S2 scenarios in the HL-LHC at the CMS detector for the combination of the multileptons, $b\bar{b}$ and $\gamma\gamma$ final states (see text) [17]

Fig. 7.25. In this case, the $b\bar{b}$ and $\gamma\gamma$ final states of the Higgs boson are included in the fit along with the multileptons final states. A negative value of κ_t disfavoured in both scenarios with a significance larger than 5σ.

References

1. Chen T, Guestrin C (2016) XGBoost. In: Proceedings of the 22nd ACM SIGKDD international conference on knowledge discovery and data mining
2. CMS Collaboration (2018) Identification of heavy-flavour jets with the CMS detector in pp collisions at 13 TeV. JINST 13(05):P05011–P05011
3. Keras documentation. https://keras.io/ (last visited 29 June 2020)
4. Martín A et al (2016) TensorFlow: large-scale machine learning on heterogeneous distributed systems. arXiv:1603.04467
5. Hoecker A et al (2007) TMVA - toolkit for multivariate data analysis. arXiv:0703039
6. Bianchini L, Conway J, Friis EK, Veelken C (2014) Reconstruction of the Higgs mass in $H \to \tau\tau$ Events by Dynamical Likelihood techniques. J Phys Conf Ser 513:022035
7. Cousins RD (2013) Generalization of chisquare goodness-of-fit test for binned data using saturated models with application to histograms. http://www.physics.ucla.edu/~cousins/stats/cousins_saturated.pdf
8. CMS Collaboration (2018) Measurement of the cross section for top quark pair production in association with a W or Z boson in proton-proton collisions at $\sqrt{s} = 13$TeV. J High Energy Phys 2018(8):11
9. ATLAS Collaboration (2019) Measurement of the $t\bar{t}Z$ and $t\bar{t}W$ cross sections in proton-proton collisions at $\sqrt{s} = 13$TeV with the ATLAS detector. Phys Rev D 99:072009
10. ATLAS Collaboration (2019) Analysis of $t\bar{t}H$ and $t\bar{t}W$ production in multilepton final states with the ATLAS detector. Technical Report ATLAS-CONF-2019-045, CERN, Geneva
11. CMS Collaboration (2018) Measurement of the associated production of a Higgs boson with a top quark pair in final states with electrons, muons and hadronically decaying τ leptons in data recorded in 2017 at $\sqrt{s} = 13$ TeV. Technical Report CMS-PAS-HIG-18-019, CERN, Geneva

12. Frederix R, Tsinikos I (2020) Subleading EW corrections and spin-correlation effects in $t\bar{t}W$ multi-lepton signatures. arXiv:2004.09552
13. CMS Collaboration (2019) Search for associated production of a Higgs boson and a single top quark in proton-proton collisions at $\sqrt{s} = 13$TeV. Phys Rev D 99:092005
14. CMS Collaboration (2020) Measurements of $t\bar{t}H$ production and the CP structure of the Yukawa interaction between the Higgs boson and top quark in the diphoton decay channel. https://arxiv.org/abs/2003.10866
15. ATLAS Collaboration (2020) CP properties of Higgs Boson interactions with top quarks in the $t\bar{t}H$ and tH processes using $H \rightarrow \gamma\gamma$ with the ATLAS detector. Phys Rev Lett 125:061802
16. Cepeda M et al (2018) Report from working group 2: Higgs physics at the HL-LHC and HE-LHC. Technical Report CERN-LPCC-2018-04, CERN, Geneva
17. CMS Collaboration (2018) Sensitivity projections for Higgs boson properties measurements at the HL-LHC. Technical Report CMS-PAS-FTR-18-011, CERN, Geneva

Chapter 8
Extraction of the $t\bar{t}H$ Signal with the Matrix Element Method

The measurement of the $t\bar{t}H$ and tH cross sections based on machine learning techniques (DNNs, BDTs) has been described in Chap. 7. The measured $t\bar{t}H$ signal strength following such approach is $\mu_{t\bar{t}H} = 0.92^{+0.26}_{-0.23}$, with an observed significance of 4.7σ over the background-only hypothesis; the measured tH signal strength modifier is $\mu_{tH} = 5.67^{+4.05}_{-3.98}$, with a significance of 1.4σ. The control analysis presented in this chapter constitutes a complementary measurement of the $t\bar{t}H$ cross section using a set of alternative observables in the maximum likelihood fit. The signal extraction relies on the separation power of the Matrix Element Method (MEM) in the $2\ell ss + 1\tau_h$ category and optimized single variables in the purely leptonic categories. As no machine learning techniques are used, the goal of this alternative version of the analysis is to quantify the gain achieved by the supervised methods of the main analysis, in particular concerning the separation power of the signal and the backgrounds. It is designed to be a robust cross check of the main analysis that serves to consolidate the measurement, validate the estimation of the signal and backgrounds yields and probe the correctness of the statistical model.

This chapter is mostly devoted to the description and performance of the MEM in the $2\ell ss + 1\tau_h$ category of the control analysis, which constitutes my personal contribution to the analysis. The MEM was used for signal extraction in the $t\bar{t}H$ multilepton analysis with the data collected in 2016 [1, 2]; the usage for Run 2 essentially follows the implementation performed then [2], with the necessary updates in terms of object reconstruction and event selection. The MEM presents many advantages and is highly suited for physics analysis, as it can be applied to a wide range of processes as long as the matrix element can be computed. It profits from a long history in high energy physics, leading to stringent measurements such as the top quark mass at the DØ collaboration [3]. The precision of the method is the result of the collaboration between theorists and experimentalists: it inherently connects the calculation of a theoretical differential cross section of a process to the associated

C. Martin Perez, *Search for the Higgs Boson Produced in Association with Top Quarks with the CMS Detector at the LHC*, Springer Theses,
https://doi.org/10.1007/978-3-030-90206-3_8

experimental resolution and detector effects. In contrast to supervised machine learning algorithms, the ability of the MEM in separating the signal from the backgrounds is purely physics-driven, making the physics underlying sensitivity transparent. In addition, it does not require any training, which is beneficial when no large simulated samples are available for the signals or backgrounds considered. However, the numerical integration required for its computation is highly CPU time-consuming; the latest technological advances in parallel processing have nonetheless made the application of the method significantly more practical.

This chapter is structured as follows. The signal extraction strategy of the control analysis is described in Sect. 8.1. An overview of the MEM and its performance is given in Sect. 8.2. The optimization of the single variable analysis in the purely leptonic categories is summarized in Sect. 8.3. The measured results upon combination of the MEM and the single variables together with the control regions are presented in Sect. 8.4 and compared to the results of the main analysis.

8.1 Signal Extraction Strategy

The control analysis was developed in parallel to the main analysis and essentially follows the same analysis strategy described in Chap. 6; the main differences are summarized in this section. It is restricted to 4 of the 10 signal regions included in the main analysis: the purely leptonic categories ($2\ell ss + 0\tau_h, 3\ell + 0\tau_h$ and $4\ell + 0\tau_h$) and the $2\ell ss + 1\tau_h$ category. The choice is motivated by the high sensitivity of these categories: they constitute a simplified but robust scrutiny of the main result for an easier interpretation of the results. The discriminating observables included in the maximum likelihood fit of the control analysis are outlined in Table 8.1. The multivariate MEM is used for signal extraction in the $2\ell ss + 1\tau_h$ category, while single variable distributions corresponding to the invariant of the leptonic system are used in the leptonic categories. The discriminating power of these variables is enhanced with a tuned subcategorization and binning of the distributions, inheriting the strategies of the previous tt̄H searches in leptonic final states [4].

No dedicated measurement of the tH production rate is performed in the control analysis; only the tt̄H signal strength is measured, while the tH process is considered as background. Correspondingly, the jet multiplicity selections in the signal regions are targeted to the tt̄H process (see Sect. 6.2) and no BSM interpretations of the results are derived. The reasons to restrict the measurement to the tt̄H signal are technical, as the extraction techniques are not optimized to measure the tH process, but also instrumental, as the sensitivity of the SM-like analysis is largely driven by the tt̄H process. Being the categories of the control analysis dominated by the irreducible tt̄V and electroweak backgrounds, the 3ℓ-CR and 4ℓ-CR, enriched in the tt̄Z, WZ and ZZ backgrounds, are included in the fit. Since the control analysis does not profit from a dedicated tt̄W as in the main analysis, a tt̄W-enriched control region with

Table 8.1 List of signal and control regions and their respective observables used in the maximum likelihood fit in the control analysis. The control regions are described in Sect. 6.5

Control analysis		
Region	Category	Observable
Signal regions	$2\ell ss + 1\tau_h$	Matrix element method
	$2\ell ss + 0\tau_h$	Invariant mass of leptonic system
	$3\ell + 0\tau_h$	
	$4\ell + 0\tau_h$	
Control regions	2ℓss+3j-CR	
	3ℓ-CR	Object multiplicity
	4ℓ-CR	

two same-sign leptons and three jets (see Sect. 6.5) is added to the fit to constrain this background. Like in the main analysis, the $t\bar{t}W(W)$ and $t\bar{t}Z$ normalizations are left floating freely in the fit so that they can be constrained from the data.

8.2 Overview of the Matrix Element Method

The Matrix Element Method is a powerful technique universally used in many analyses, such as the measurement of the top quark mass at the DØ collaboration [3] and many searches in the top quark and the Higgs boson sector both at the Tevatron and at the LHC [5–9]. The method consists in computing a probability that an event arises from a given physical process (or hypothesis), modelled by a Feynman diagram. The process can be a signal or a background, in which case the MEM is used for signal-background discrimination, like is done in the search presented here. The MEM can also be interpreted as a differential probability as a function of a varying parameter, like in the case of the top quark mass, where the maximum probability provides the best estimate of the parameter. The compatibility of an event with a given hypothesis is based on the calculation of the cross section at a given phase space point corresponding to the reconstructed kinematic configuration of the event. For this purpose, a multidimensional integration is computed based on numerical Monte Carlo techniques, taking into account the theoretical description of the physics process, via their matrix element, together with the detector resolution effects, encoded in the so-called transfer functions. These functions describe the reconstruction of the physics objects in the CMS detector; they also account for the neutrinos produced in the decays of the particles involved and the overall boost of the system.

The MEM is used in the $2\ell ss + 1\tau_h$ final state, illustrated in the event display in Fig. 8.1. The choice of this particular channel is driven by its several advantages. First, the requirement of two leptons of the same sign significantly reduces the backgrounds arising from various SM processes, yielding a higher signal purity. Second, as one

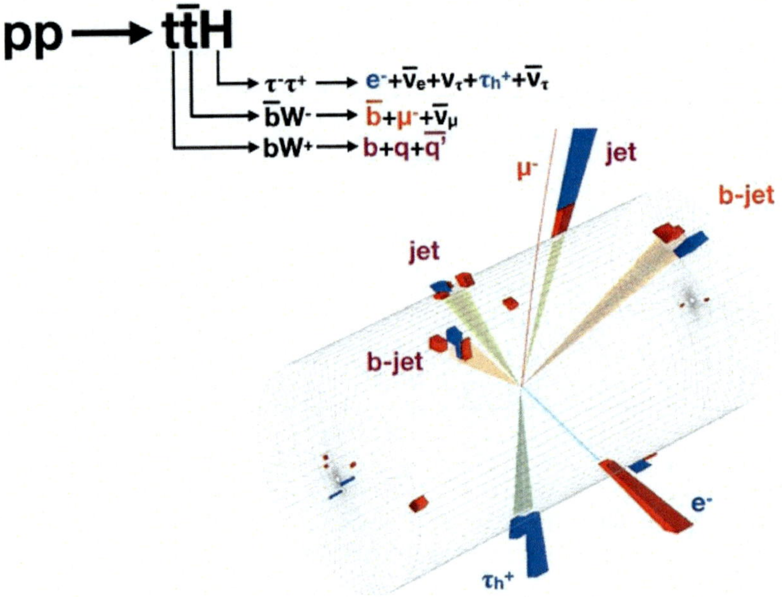

Fig. 8.1 An event candidate for the $t\bar{t}$H production in the $2\ell ss + 1\tau_h$ final state. The Higgs boson decays into a $\tau^+\tau^-$ pair. The former decays hadronically into a τ_h (in blue) and a neutrino, while the latter decays into an electron (in blue) plus neutrinos (undetected). The top quark decays into two light jets and one b-jet (in purple). The top antiquark decays into a muon of the same sign as the electron and into an additional b-jet (in red) [10]

hadronic top is expected in this final state, this category has a larger branching ratio than those which require higher lepton multiplicities, and therefore profits from a larger event yield. Third, the main background contribution to this channel arises from the well-modelled irreducible $t\bar{t}$Z process, while the reducible backgrounds are not as sizeable as in other categories. Finally, among all channels of the analysis, the $2\ell ss + 1\tau_h$ category is the most sensitive to the $t\bar{t}$H process with H $\rightarrow \tau^+\tau^-$.

The discriminating power of the MEM is based on the Neyman-Pearson lemma [11], which states that the optimal test statistics to compare two hypothesis H_0 and H_1 is given by the ratio of their likelihood functions,

$$\Lambda(\mathbf{y}) = \frac{L(H_0|\mathbf{y})}{L(H_1|\mathbf{y})}\,, \tag{8.1}$$

where $L(H_0|\mathbf{y})$ and $L(H_1|\mathbf{y})$ are the likelihoods for the hypothesis H_0 and H_1, respectively, as a function of the data $\mathbf{y}$. In the case of the MEM, the hypothesis H_1 would be that an event arises from a signal process, while the hypothesis H_0 would be that it arises from a background process.

The aim of the MEM is to provide an estimate of the likelihood function associated to the signal or background hypotheses with the so-called MEM weight $w_\Omega(\mathbf{y})$. It encodes the probability that an event reconstructed with the observables $\mathbf{y}$ arises from a physics process Ω. It is expressed as

$$
\begin{aligned}
w_\Omega(\mathbf{y}) = \frac{1}{\sigma_\Omega} \sum_p \int d\mathbf{x}\, dx_a\, dx_b \cdot \frac{f_i(x_a, Q)\, f_j(x_b, Q)}{x_a x_b s} \\
\cdot\, \delta^4(x_a P_a + x_b P_b - \sum p_k) \cdot |\mathcal{M}_\Omega(\mathbf{x})|^2 \cdot W(\mathbf{y}||\mathbf{x})\,.
\end{aligned}
\tag{8.2}
$$

The MEM weight represents the differential cross section of the process Ω with respect to the observables $\mathbf{y}$, while integrating over the unmeasured or poorly measured phase space of the final-state particles ($\mathbf{x}$) and the momentum fractions of the incoming partons (x_a, x_b). The normalization factor σ_Ω in Eq. 8.2 corresponds to the product of the cross section of the process Ω times the detector acceptance and the efficiency of the analysis, derived by imposing

$$
\int d\mathbf{y}\; w_\Omega(\mathbf{y}) = 1\,. \tag{8.3}
$$

The sum $\sum_p$ extends over all the potential associations between the reconstructed objects and the final-state particles (see Sect. 8.2.3). The functions $f_i(x, Q)$ are the parton distribution functions (PDF) associated to the incoming partons of flavour i carrying a fraction x of the proton energy in a process of scale Q; they are derived with LHAPDF [12]. The variable s corresponds to the usual Mandelstam variable related to the centre-of-mass energy, while $P_{a,b}$ are the four-vectors of the colliding protons and p_k are the four-vectors of the final-state particles. The energy-momentum conservation is imposed with the delta function δ. The symbol $\mathcal{M}_\Omega(\mathbf{x})$ corresponds to the matrix element of the process Ω at leading order (see Sect. 8.2.1), which depends on a set of variables describing the initial and final-state particles; it is derived using the MadGraph5 Monte Carlo generator [12]. The squared matrix element is convolved with the transfer function $W(\mathbf{y}||\mathbf{x})$ (see Sect. 8.2.2), which reflects the probability of measuring the observables $\mathbf{y}$ given a point $\mathbf{x}$ in the phase space of the final-state particles. It describes the decays of the final-state particles and takes into account the resolution of the reconstruction techniques; it also encodes the possible boost of the system in the transverse plane. Given the non-analytic parametrization of the function to integrate, the MEM weight is computed numerically for each event using the VEGAS algorithm [13] (see Sect. 8.2.4).

The weights obtained with Eq. 8.2 are derived for the signal hypothesis (w_S) and a relevant set of background hypotheses B (w_B). They are combined into the MEM *likelihood ratio* (LR), expressed as

$$
\mathrm{LR}(\mathbf{y}) = \frac{w_S(\mathbf{y})}{w_S(\mathbf{y}) + \sum_B w_B(\mathbf{y})}\,. \tag{8.4}
$$

With values ranging from 0 to 1, the LR is interpreted as the probability that a given event with observables $\mathbf{y}$ originates from the signal process rather than from one of the background processes.

8.2.1 Scattering Amplitude

The scattering amplitude $|\mathcal{M}_\Omega(\mathbf{x})|^2$ in Eq. 8.2 corresponds to that of the process of interest for which the MEM weight is obtained (signal or backgrounds). In QFT, the scattering amplitude is most generally expressed as

$$|\mathcal{M}_\Omega(\mathbf{x})|^2 = \sum_{\text{flavour}} f_1(x_1, Q)\, f_2(x_2, Q) \sum_{\text{spin}} |\mathcal{M}(p_1 p_2 \to X)|^2 , \tag{8.5}$$

where p_1 and p_2 denote the incoming interacting partons (gluons, quarks) and X denotes the final-state particles (leptons, neutrinos, quarks). The first sum extends over the flavours of the incoming partons, weighted by their associated PDF. For the signal and the background processes, both the gluon-gluon and the quark-antiquark configurations are considered in the computation. The second sum extends over the possible spins of the incoming and outgoing particles. In the current implementation, the hard scattering amplitude $|\mathcal{M}_\Omega(\mathbf{x})|^2$ is evaluated in the centre-of-mass of the incoming and outgoing particles at leading order using the MADGRAPH5 Monte-Carlo generator [12].

The MEM weight w_S associated to the signal hypothesis is derived from the scattering amplitude of the t$\bar{\text{t}}$H process. As shown in Fig. 6.6, with the event selection applied in the $2\ell ss + 1\tau_h$ signal region, ~53% of the t$\bar{\text{t}}$H signal yield arises from the $\text{H} \to \tau^+\tau^-$ decay, while the remaining ~45% and ~2% originate from the $\text{H} \to \text{WW}^*$ and $\text{H} \to \text{ZZ}^*$ decays, respectively. Being the dominant Higgs boson decay, the matrix element of the signal hypothesis encodes the $\text{H} \to \tau^+\tau^-$ decay configuration; the corresponding Feynman diagram is shown in Fig. 8.3, with decay chain

$$\begin{aligned} \text{gg} / \text{q}\bar{\text{q}} &\to \text{t}\bar{\text{t}}\text{H} \\ &\to (\text{b}\, \ell^+\nu_\ell)\, (\bar{\text{b}}\, \text{q}\bar{\text{q}})\, \tau^+\tau^- \\ &\to (\text{b}\, \ell^+\nu_\ell)\, (\bar{\text{b}}\, \text{q}\bar{\text{q}})\, (\ell^+\nu_\ell\, \nu_\tau)\, (\tau_h\nu_\tau) . \end{aligned} \tag{8.6}$$

The addition of the dedicated $\text{H} \to \text{WW}^*$ matrix element is in principle possible, but faces two main challenges. First, the contribution of this process to the signal region originates equally from events where the reconstructed τ_h is genuine and events where the reconstructed τ_h is a misidentified light jet, as seen in Fig. 8.2. As both cases result in comparable yields, the inclusion of a matrix element for the WW* decay requires the design of separate weights reflecting each configuration. Second, the $\text{H} \to \text{WW}^*$ decay is less constrained kinematically than the $\text{H} \to \tau^+\tau^-$ decay, as one of the W bosons from the Higgs boson decay is off-shell. It means that

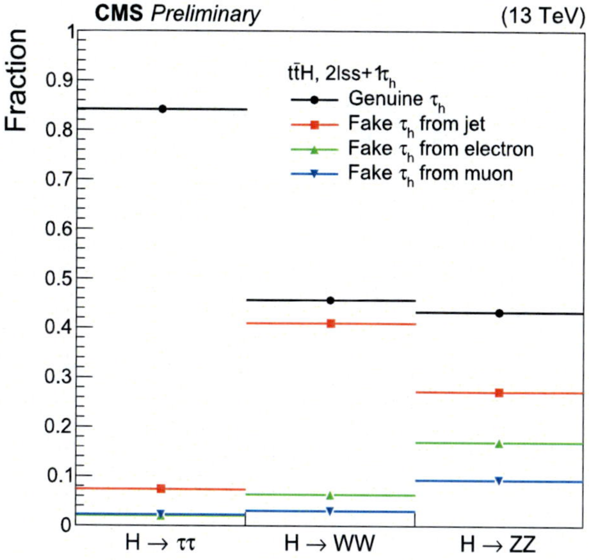

Fig. 8.2 Fraction of genuine τ_h and fake τ_h from jets, electrons and muons in the $2\ell ss + 1\tau_h$ category, obtained for $\mathrm{t\bar{t}H}$ simulated events, split according to the Higgs boson decay mode

two additional weights would have to be derived for the on-shell and the off-shell configurations; this amounts to a total of 4 independent $\mathrm{t\bar{t}H}$ weights to be derived. The addition of the $\mathrm{H} \rightarrow \mathrm{ZZ}^*$ decay would bring little benefit, since its yield is negligible in the $2\ell ss + 1\tau_h$ category.

Three background weights are developed and included in the LR of Eq. 8.4. They originate from the $\mathrm{t\bar{t}Z}$ and $\mathrm{t\bar{t}}$ processes; the corresponding Feynman diagrams are shown in Fig. 8.3. Two configurations are considered for the $\mathrm{t\bar{t}Z}$ process, leading to two independent weights. The first one is the irreducible component, where the Z boson decays into two τ leptons via the decay chain

$$\begin{aligned} \mathrm{gg} \,/\, \mathrm{q\bar{q}} &\rightarrow \mathrm{t\bar{t}Z} \\ &\rightarrow (\mathrm{b}\,\ell^+\nu_\ell)\,(\mathrm{\bar{b}}\,\mathrm{q\bar{q}})\,\tau^+\tau^- \\ &\rightarrow (\mathrm{b}\,\ell^+\nu_\ell)\,(\mathrm{\bar{b}}\,\mathrm{q\bar{q}})\,(\ell^+\nu_\ell\,\nu_\tau)\,(\tau_h\nu_\tau)\,. \end{aligned} \tag{8.7}$$

The second $\mathrm{t\bar{t}Z}$ contribution is the reducible component, where the Z boson decays into two electrons or muons, one of them being misidentified as a τ_h, namely

$$\begin{aligned} \mathrm{gg} \,/\, \mathrm{q\bar{q}} &\rightarrow \mathrm{t\bar{t}Z} \\ &\rightarrow (\mathrm{b}\,\ell^+\nu_\ell)\,(\mathrm{\bar{b}}\,\mathrm{q\bar{q}})\,\ell^+(\ell^-\,[\rightsquigarrow \tau_h])\,, \end{aligned} \tag{8.8}$$

where the symbol $\rightsquigarrow$ represents the misidentification. Both the irreducible and the reducible components have a comparable contribution in the $2\ell ss + 1\tau_h$ region. The last source of background considered is the $\mathrm{t\bar{t}}$ process, in which a non-prompt lepton from a semileptonic B hadron decay is misidentified as a prompt lepton via the decays

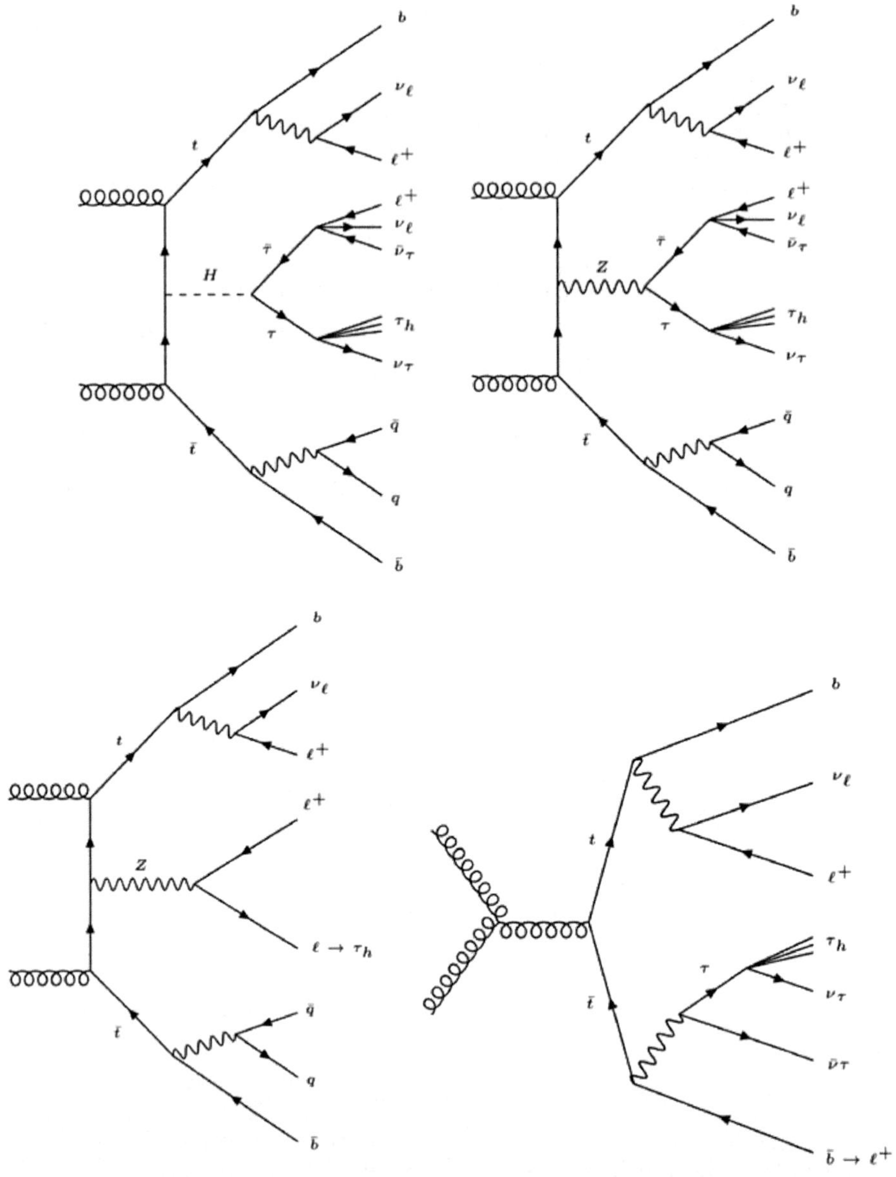

Fig. 8.3 Leading-order Feynman diagrams of the processes for which a dedicated MEM weight is developed in the $2\ell ss + 1\tau_h$ category: the t$\bar{\rm t}$H signal process (top left), the t$\bar{\rm t}$Z irreducible background process (top right), the t$\bar{\rm t}$Z reducible background process (bottom left) and the t$\bar{\rm t}$ reducible background process (bottom right). Both the gg and q$\bar{\rm q}$ initial state configurations are considered in the MEM computation

$$\begin{aligned} gg\ /\ q\bar{q} &\rightarrow t\bar{t} \\ &\rightarrow (b\,\ell^{+}\nu_{\ell})\ (\bar{b}\,[\rightarrow \ell^{+}]\ \tau\ \nu_{\tau}) \\ &\rightarrow (b\,\ell^{+}\nu_{\ell})\ (\bar{b}\,[\rightarrow \ell^{+}]\ \tau_h\nu_{\tau}\ \nu_{\tau})\ . \end{aligned} \tag{8.9}$$

For the reducible $t\bar{t}Z$ and $t\bar{t}$ backgrounds, besides the matrix element, the transfer functions and the integration strategies are also modified as described later.

In the computation of the signal and background weights, the hard processes are modelled with the LO matrix element, meaning a perfect energy balance in the transverse plane is assumed. This is not the case in data, where there is energy imbalance in the transverse plane due to the presence of initial- and final-state radiation, as well as particles that might fall out the detector acceptance or are possibly misreconstructed. The 4-dimensional δ function of the MEM, expressing the total energy-momentum conservation, is thus replaced by a 2-dimensional δ function enforcing the conservation of energy and of the longitudinal momentum component. This way, the total transverse momentum of the system is not constrained in the MEM computation. It is however not entirely free: the boost of the system translates into a recoil that can be compared to the measured recoil through a dedicated transfer function, as explained in Sect. 8.2.2. This recoil transfer function is crucial in the computation of the MEM integral, as it regularizes the very boosted configurations. Hence, most of the effects of the higher-order corrections not considered in the LO matrix element are taken into account in the evaluation of the weights.

8.2.2 *Transfer Functions*

The transfer function $W(\mathbf{y}||\mathbf{x})$ in Eq. 8.2 encodes the probability of reconstructing the set of observables $\mathbf{y}$ at detector level given the set of four-momenta $\mathbf{x}$ of the outgoing partons, as illustrated in Fig. 8.4. These functions take into account the showering and hadronization effects, as well as the experimental resolution and reconstruction. The set of observables $\mathbf{y}$ considered in the MEM are

- the three-momenta of the charged leptons ($\hat{\vec{\ell}}$[1]),
- the three-momenta of the visible decay products from the hadronic τ decay ($\hat{\vec{\pi}}$),
- the energy ($\hat{E}_{\mathrm{j}}$) and the direction ($\hat{\vec{e}}_{\mathrm{j}}$) of the jets,
- the missing transverse energy ($\hat{E}_{\mathrm{T}}^{\mathrm{miss}}$).

Given the excellent performance of the CMS object reconstruction (see Sect. 3.4), the momenta of the charged leptons and the visible decay products of the τ_h are considered perfectly measured, both in direction and in magnitude. Likewise, the direction of the jets is assumed to be perfectly measured and aligned to the direction of the quark from which it originates. Being the object reconstruction mostly uncorrelated, the overall transfer function can be decomposed into the product of

[1] Quantities with a hat represent measured quantities.

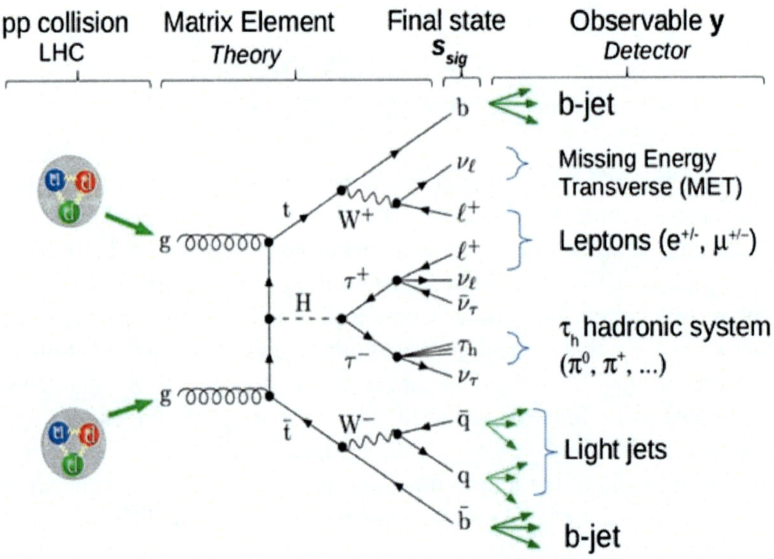

Fig. 8.4 Leading-order Feynman diagram of the t$\bar{\rm t}$H production in the $2\ell ss + 1\tau_h$ final state. The MEM weight w_S is computed as the differential cross section of the t$\bar{\rm t}$H process in a phase-space point given by the observables **y**, which are related to the final-state particles via dedicated transfer functions

the transfer functions associated to each individual object. Based on the assumptions made, the function is written as

$$W(\mathbf{y}||\mathbf{x}) = \prod_{\ell} \delta(\hat{\vec{\ell}} - \vec{\ell})\; T_\ell(\hat{\ell}|\tau_\ell) \cdot \prod_{\pi} \delta(\hat{\vec{\pi}} - \vec{\pi})\; T_h(\hat{\pi}|\tau_h)$$
$$\cdot \prod_{\mathrm{q}\in\mathcal{A}} \delta(\hat{\vec{e}}_{\mathrm{j}} - \vec{e}_{\mathrm{q}})\; T_{\mathrm{j}}(\hat{p}_{\mathrm{Tj}}|p_{\mathrm{Tq}}, \eta_{\mathrm{q}}) \cdot \prod_{\mathrm{q}\notin\mathcal{A}} A_{\mathrm{q}}(p_{\mathrm{Tq}}, \eta_{\mathrm{q}}) \cdot T_{E_{\mathrm{T}}}(\hat{\vec{\rho}}_{\mathrm{T}}|\vec{P}_{\mathrm{T}})\,. \tag{8.10}$$

Since the transfer function is interpreted as a conditional probability to measure the observables **y** given the final-state particle phase space **x**, the normalization condition

$$\int d\mathbf{y}\; W(\mathbf{y}||\mathbf{x}) = 1, \forall \mathbf{x} \tag{8.11}$$

is imposed.

The functions $T_\ell(\hat{\ell}|\tau_\ell)$ and $T_h(\hat{\pi}|\tau_h)$ in Eq. 8.10 relate the momentum of the leptonically and hadronically decaying τ leptons, respectively, to the momentum of their visible decay products. The notation $\mathrm{q} \in \mathcal{A}$ refers to quarks which are associated to at least one of the jets, while the notation $\mathrm{q} \notin \mathcal{A}$ refers to those which are not associated to any jet. In the first case, the function $T_{\mathrm{j}}(\hat{p}_{\mathrm{Tj}}|p_{\mathrm{Tq}}, \eta_{\mathrm{q}})$ relates the transverse momentum of the reconstructed jet to the one of the associated quark. In the second case, $A_{\mathrm{q}}(p_{\mathrm{Tq}}, \eta_{\mathrm{q}})$ expresses the probability that no jet is reconstructed given the kinematics of the quark. Finally, $T_{E_T}(\hat{\vec{\rho}}_T|\vec{P}_T)$ relates the opposite of the measured recoil summed over the reconstructed objects ($\hat{\vec{\rho}}_T = \sum_{\ell,\tau_h,j} \vec{p}_T + \vec{E}_T^{miss}$) to the recoil summed over the mother particles ($\vec{P}_T = \sum_{t,H,Z} \vec{p}_T$).

In the computation of the reducible $t\bar{t}Z$ background weight, an additional transfer function $T_{\ell \to \tau_h}(\hat{\tau}_h|\ell)$ is used to relate the transverse momentum of the misreconstructed τ_h to the one of the original lepton. Likewise, in the computation of the $t\bar{t}$ background weight, the transfer function $T_{b\to\ell}(\hat{\ell}|b)$ relating the transverse momentum of the reconstructed non-prompt lepton to that of the original b-quark is included.

Jet Transfer Function

The jet transfer function $T_j(\hat{p}_{Tj}|p_{Tq}, \eta_q)$ accounts for the jet energy response and resolution. It encodes the conditional probability to measure a jet with transverse momentum $\hat{p}_{Tj}$ assuming a quark with transverse momentum p_{Tq} and pseudorapidity η_q. The dependence with η_q is introduced to account for the different jet energy resolution in the barrel and the endcaps. The probability is modelled with the weighted sum of two Gaussian distributions,

$$\begin{aligned} T_j(\hat{p}_{Tj}|p_{Tq}, \eta_q) = f \cdot \mathcal{G}\left(\hat{p}_{Tj}|\mu_1(p_{Tq}, \eta_q), \sigma_1(p_{Tq}, \eta_q)\right) \\ + (1-f) \cdot \mathcal{G}\left(\hat{p}_{Tj}|\mu_2(p_{Tq}, \eta_q), \sigma_2(p_{Tq}, \eta_q)\right) , \end{aligned} \tag{8.12}$$

with

$$\mathcal{G}(\hat{p}_{Tj}|\mu_i, \sigma_i) = \frac{1}{\sigma_i\sqrt{2\pi}} \exp\left(-\frac{1}{2}\left(\frac{\hat{p}_{Tj}-\mu_i}{\sigma_i}\right)^2\right) . \tag{8.13}$$

The procedure to derive the parameters of the transfer function consists in obtaining, for different bins of generator-quark p_T, the corresponding reconstructed jet p_T distributions, separately for the barrel and for the endcap; they are then fitted with the double Gaussian to obtain the parameters. It is done in simulated $t\bar{t}$+jets events, separately for the jets matched to light quarks and the jets matched to b-quarks. The distributions in the barrel region are shown in Fig. 8.5. The p_T of the reconstructed jets matched to b-quarks presents larger tails in the low-p_T due to the neutrinos produced in the semileptonic B hadron decays. The response (μ_i) and the resolution (σ_i) across all bins are fitted with the functions

$$\begin{aligned} \mu_i(p_{Tq}, \eta_q) &= m_i(\eta_q) \cdot p_{Tq} + n_i(\eta_q) , \\ \sigma_i(p_{Tq}, \eta_q) &= a_i(\eta_q) \oplus b_i(\eta_q) \cdot \sqrt{p_{Tq}} \oplus c_i(\eta_q) \cdot p_{Tq} . \end{aligned} \tag{8.14}$$

The numerical values of the coefficients and the weight applied to the Gaussians are shown in Table 8.2.

Quark Acceptance Transfer Function

The function $A_q(p_{Tq}, \eta_q)$ expresses the probability that there is no reconstructed jet given the kinematics of the final-state quark. The function depends on the kinematics of the quark but also of the nearest lepton found due to the overlap removal applied in the analysis. It encodes four separate scenarios, expressed as

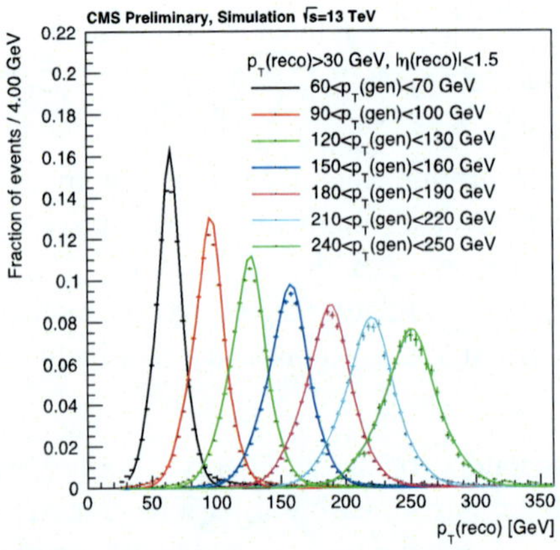

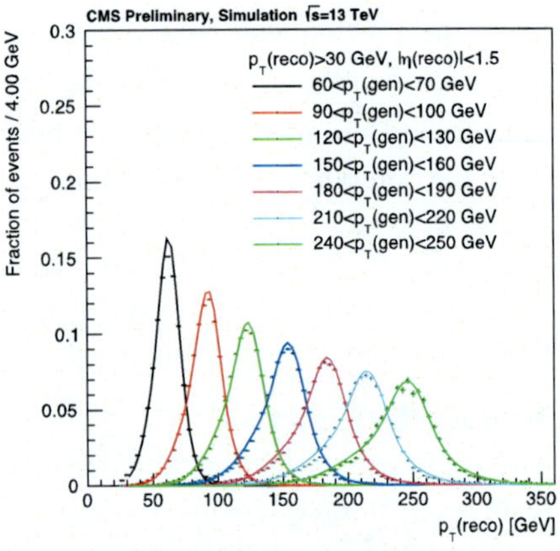

Fig. 8.5 Distributions of p_T of the reconstructed jets matched to light quarks (left) and to b-quarks (right), obtained from simulated $t\bar{t}$+jets events, in different bins of quark p_T found in $|\eta| < 1.5$. The solid line represents the double Gaussian fit used in the jet transfer function [2]

Table 8.2 Numerical values of the coefficients used to parametrize the jet transfer functions, obtained in different regions of the η position of the quark, separately for jets matched to light quarks and to b-quarks [2]

Flavour	$\|\eta_q\|$	(m_1, m_2)	(n_1, n_2)	(a_1, a_2)	(b_1, b_2)	(c_1, c_2)	f
Light quarks	[0.0, 1.5]	(0.99, 0.94)	(−2.33, 9.39)	(0.00, 0.00)	(1.53, 0.78)	(0.13, 0.05)	0.81
	[1.5, 2.4]	(0.94, 0.82)	(10.79, 51.65)	(4.26, 2.56)	(0.58, 1.99)	(0.00, 0.00)	0.80
b-quarks	[0.0, 1.5]	(0.99, 0.97)	(−6.24, −8.12)	(0.00, 0.00)	(1.07, 0.00)	(0.05, 0.23)	0.66
	[1.5, 2.4]	(0.98, 0.94)	(−6.24, −10.31)	(0.00, 3.86)	(1.14, 0.47)	(0.00, 0.21)	0.52

$$A_q(p_{Tq}, \eta_q) = \begin{cases} 1 \text{ if } |\eta_q| > \eta_q^c \\ 0 \text{ else if } \Delta R(q, \ell) < R^c_{q,\ell} \text{ and } \frac{E_q}{E_\ell} > f^c \\ 1 \text{ else if } \min_{j'} \Delta R(q, j') < R^c_j \\ \int_0^{p^c_{Tj}} d\hat{p}'_{Tj}\, T_j(\hat{p}'_{Tj}|p_{Tq}, \eta_q) \text{ else} \end{cases}. \tag{8.15}$$

A uniform prior $A_q = 1$ is assigned to the quarks that fall out of the detector acceptance, which extends up to $\eta_q^c = 2.4$. On the contrary, if the quark overlaps with a charged lepton within $R^c_{q,\ell} = 0.4$ and the quark-to-lepton energy ratio is larger than the relative isolation value of $f^c = 0.4$, such event would not pass the event selection (see Sect. 6.1) and is therefore not considered in the MEM integration ($A_q = 0$). Instead, if the quark direction is close to one of the jets by less than the jet algorithm radius $R^c_j = 0.4$, it is assumed to be merged with another quark q′ into the same jet. In this case, the acceptance function of the quark q is set to $A_q = 1$; the jet transfer function $T_{\mathrm{j}}(\hat{p}_{\mathrm{Tj}}|p_{\mathrm{Tq}'}, \eta_{\mathrm{q}'})$ is modified to use the p_{T} of the qq$'$ pair instead of the p_{T} of the single quark q$'$. When none of the cases above hold, the acceptance function is the cumulative distribution of the jet transfer function evaluated at the momentum threshold $p^{c}_{\mathrm{Tj}} = 25$ GeV used in the baseline jet selection.

Recoil Transfer Function

The recoil transfer function does not have a direct correspondence with any integration variable of the MEM. Nonetheless, it plays a crucial role, as it accounts for the presence of initial or final-state radiation, not described by the matrix element at LO, as well as the resolution of the reconstructed missing transverse energy. The function expresses the compatibility between the predicted $E_{\mathrm{T}}^{\mathrm{miss}}$ and the measured one via the two-dimensional Gaussian [14]

$$T_{E_{\mathrm{T}}}(\hat{\vec{\rho}}_{\mathrm{T}}|\vec{P}_{\mathrm{T}}) = \frac{1}{\sqrt{2\pi|\mathbf{V}|}} \exp\left(-\frac{1}{2}\left(\hat{\vec{\rho}}_{\mathrm{T}} - \vec{P}_{\mathrm{T}}\right)^{T} \mathbf{V}^{-1}\left(\hat{\vec{\rho}}_{\mathrm{T}} - \vec{P}_{\mathrm{T}}\right)\right). \tag{8.16}$$

The value $\hat{\vec{\rho}}_{\mathrm{T}}$ is the opposite of the recoil, defined as $\hat{\vec{\rho}}_{\mathrm{T}} = \sum_{\ell,\pi,\mathrm{j}} \hat{\vec{p}}_{\mathrm{T}} + \hat{\vec{E}}_{\mathrm{T}}^{\mathrm{miss}}$. It is the vectorial sum of the transverse momenta of the particles that are not clustered in jets or identified as leptons: it corresponds to the measured overall boost of the system. The opposite of the recoil is compared to the transverse boost of the system, namely $\vec{P}_{\mathrm{T}} = \sum_{\mathrm{t,H,Z}} \vec{p}_{\mathrm{T}}$. If the observables in the detector were measured perfectly, the vector $(\hat{\vec{\rho}}_{\mathrm{T}} - \vec{P}_{\mathrm{T}})$ would correspond to the difference between the measured $\vec{E}_{\mathrm{T}}^{\mathrm{miss}}$ and the $\vec{p}_{\mathrm{T}}$ of the neutrinos. The matrix $\mathbf{V}$ is the $\vec{E}_{\mathrm{T}}^{\mathrm{miss}}$ covariance matrix, expressed in its diagonal form as

$$\mathbf{U} = \begin{pmatrix} \sigma^2_{E_{\mathrm{T}}} & 0 \\ 0 & E^2_{\mathrm{T}}\sigma^2_{\phi} \end{pmatrix}, \tag{8.17}$$

where $\sigma_{E_{\mathrm{T}}}$ is the resolution on the magnitude of $\vec{E}_{\mathrm{T}}^{\mathrm{miss}}$ and σ_{ϕ} is its resolution in the ϕ direction. A non-diagonal (x, y) representation of this matrix is used in the MEM.

Hadronic τ Transfer Function

The τ leptons which decay into hadrons and neutrinos have an associated transfer function $T_h(\hat{\pi}|\tau_h)$ that relates the momentum of the visible decay products to the momentum of the τ lepton before decaying. The function is derived from an analytic parametrization of the τ decay width, assuming the τ to be unpolarized, which is the case for the H $\rightarrow \tau^+\tau^-$ decay. The simplest case corresponds to the 1-prong decay $(\tau \rightarrow \pi\nu_{\tau})$, where the normalized transfer function is derived from the 2-body decay width as [2, 14]

$$T_{h,1-\mathrm{prong}}(\hat{E}_{\pi}|E_{\tau}) = \frac{1}{\left(1 - \dfrac{\hat{m}^2_{\pi}}{m^2_{\tau}}\right)} \cdot \theta\left(\frac{\hat{m}^2_{\pi}}{m^2_{\tau}}\right), \tag{8.18}$$

where θ is the Heaviside function. The function translates into a flat distribution of the variable $z = \hat{E}_{\pi}/E_{\tau}$, illustrated in Fig. 8.6 (right), corresponding to the fraction of the visible reconstructed energy over the τ energy.

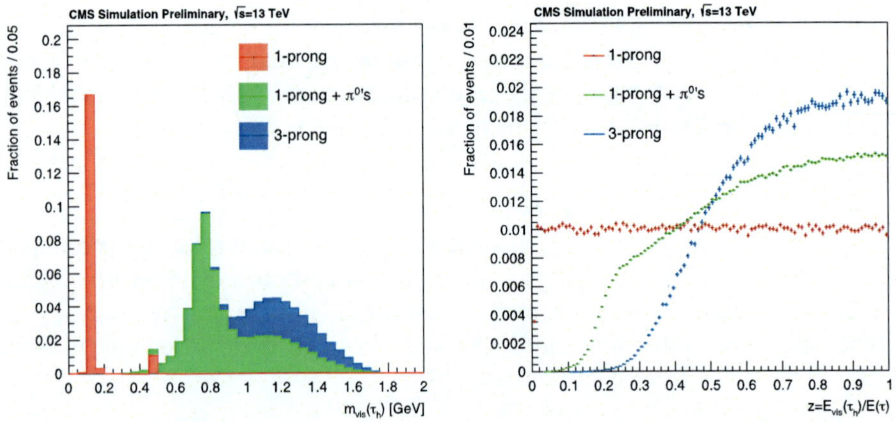

Fig. 8.6 Distributions of the invariant mass of the hadronic τ visible decay products (left) and of the variable $z = \hat{E}_\pi / E_\tau$ (right), separately for the 1-prong, 1-prong+π^0 and 3-prongs decay modes, obtained from simulated tt̄H events with $\mathrm{H} \to \tau^+\tau^-$ [2]

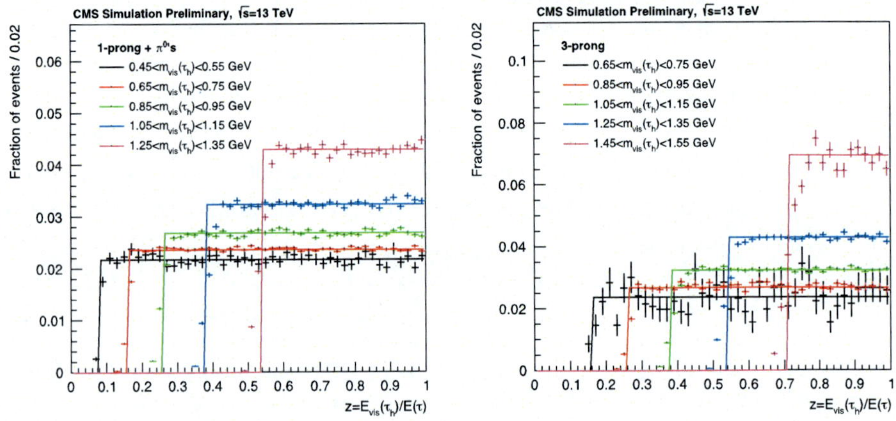

Fig. 8.7 Distributions of the variable $z = \hat{E}_\pi / E_\tau$ obtained from simulated tt̄H events with $\mathrm{H} \to \tau^+\tau^-$ (points) and the corresponding transfer function (solid line), in bins of the visible mass, for the 1-prong+π^0 (left) and 3-prongs (right) decay modes [2]

As the 1-prong+π^0's and 3-prong decay modes imply intermediate resonances, a simple analytical description is no longer possible. Unlike the 1-prong decay, the invariant mass of the visible decay products cannot be assumed to be fixed, but defined as $\hat{m}_\pi^2 = \hat{E}_\pi^2 - |\hat{\vec{\pi}}|^2$, as shown in Fig. 8.6 (left). However, for a fixed value of m_π, one can assume a 2-body decay; in this case, a flat distribution of the variable z is found between m_π^2/m_τ^2 and 1, illustrated in Fig. 8.7. Hence, a convolution between $T_{h,1-\mathrm{prong}}(\hat{E}_\pi | E_\tau)$ and the visible τ mass spectrum is performed to recover the inclusive z distribution. The predicted distribution is compared to the one obtained from the simulation in Fig. 8.8 for the 1-prong+π^0 and 3-prong decays, showing an excellent agreement.

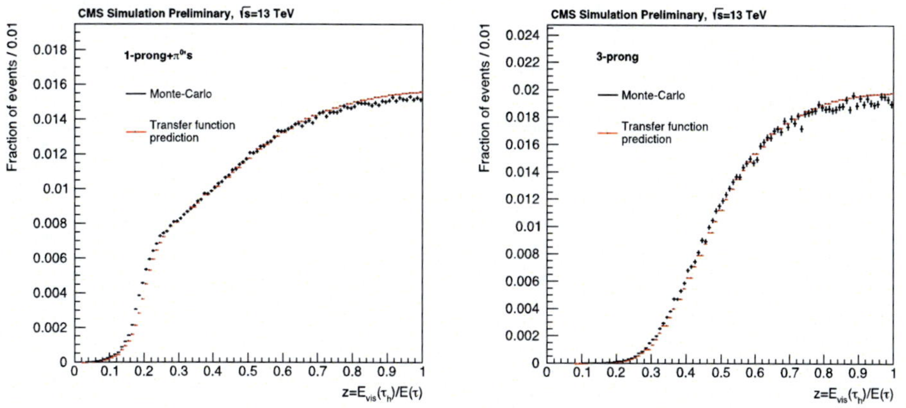

Fig. 8.8 Comparison between the variable $z = \hat{E}_\pi / E_\tau$ obtained with the convolution of the hadronic τ transfer function with the visible τ mass spectrum (red line) compared to the one obtained from simulated tt̄H events with $H \to \tau^+\tau^-$ (black line), for the 1-prong+π^0 (left) and 3-prongs (right) decays [2]

Leptonic τ Transfer Function

For a τ lepton decaying into an electron or a muon and neutrinos, the transfer function $T_\ell(\hat{\ell}|\tau_\ell)$ relates the momentum of the final-state lepton to the momentum of the τ lepton before decay. Similarly to the hadronic τ, the function is derived with an analytic computation of the decay width associated to the leptonic decay averaged over the spin configurations. It expressed as [2, 14]

$$T_\ell(\hat{\ell}|\tau) = \frac{1}{3E_\tau}\,(1-z)\,(5+5z-4z^2)\,, \tag{8.19}$$

where z is the fraction of the τ energy carried by the lepton, $z = \hat{E}_\ell / E_\tau$. An excellent agreement between the predicted and the simulated distributions is observed, as illustrated in Fig. 8.9 separately for electrons and muons.

Fake τ_h Transfer Function

The reducible background from tt̄Z with $Z \to \ell\ell$ implies the presence of a reconstructed τ_h from a misidentified electron or muon. The τ_h misidentification can happen with prompt isolated leptons originating from gauge bosons, easily misidentified as 1-prong τ_h, or electrons which produce bremsstrahlung photons in the tracker material, hence reconstructed as 1-prong+π^0's or even 3-prongs when there is photon conversion. In this analysis, an antilepton discriminator is applied in the τ_h selection (see Table 6.3), but very loose working points are used, as they are found to be more beneficial to the sensitivity. Hence, residual contributions can enter the signal regions. Since the τ_h reconstruction algorithm does not contemplate the lepton misidentification, the mismatch between the p_T of the reconstructed τ_h and that of the lepton is taken into account via a dedicated transfer function $T_{\ell\to\tau_h}(\hat{\tau}_h|\ell)$, separately for electrons and muons. In the case of electrons, the function is modelled by the

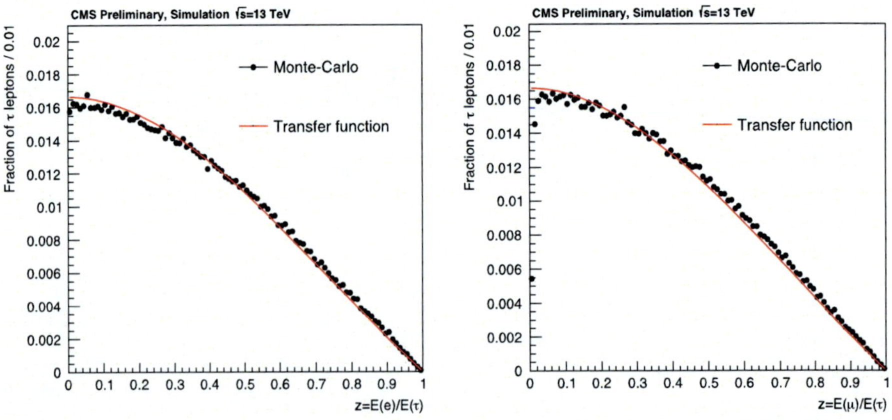

Fig. 8.9 Comparison between the variable $z = \hat{E}_\ell / E_\tau$ obtained with the leptonic τ transfer function (red line) compared to the one obtained from simulated $t\bar{t}H$ events with $H \to \tau^+\tau^-$ (black line) [2]

Table 8.3 Numerical values of the coefficients used to parametrize the fake τ_h transfer function [2]

Lepton	a	b	c	d	e	f	α
Electron	0.96	0.84	0.04	4.69	–	–	–
Muon	0.97	1.17	0.02	4.46	−0.015	2.29	1.2

Breit-Wigner distribution

$$T_{e\to\tau}(\hat{\tau}_h|e) = \frac{1}{1 + \left(\dfrac{\hat{p}_{T\tau} - \mu}{\sigma}\right)^2} . \tag{8.20}$$

In the case of misidentified muons, the p_T spectra of the τ_h presents larger tails, mostly due to accidental superimposed hadronic deposits from surrounding particles. Thus, it is better modelled with the Crystal-Ball distribution

$$T_{\mu\to\tau}(\hat{\tau}_h|\mu) = \begin{cases} 0 \text{ if } \left|\dfrac{\hat{p}_{T\tau} - \mu}{\sigma}\right| > 5 \\ \exp\left(-0.5\left(\dfrac{\hat{p}_{T\tau} - \mu}{\sigma}\right)^2\right) \text{ if } \dfrac{\hat{p}_{T\tau} - \mu}{\sigma} < \alpha \\ \left(\dfrac{n}{\alpha}\right)^n \exp(-0.5\alpha^2)\left(\dfrac{n}{\alpha} - \alpha + \dfrac{\hat{p}_{T\tau} - \mu}{\sigma}\right)^{-n} \text{ else} \end{cases} . \tag{8.21}$$

The parameters of the transfer functions are obtained from the p_T distributions of the reconstructed τ_h matched to generator-level leptons, derived in bins of lepton p_T, as

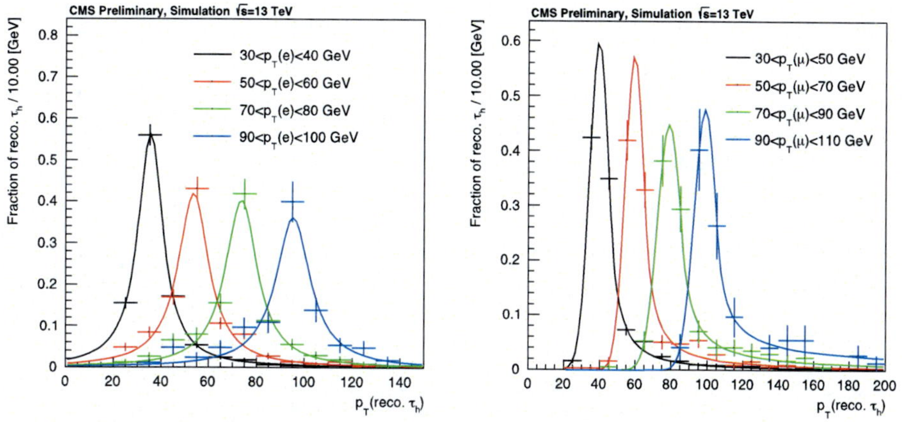

Fig. 8.10 Distribution of p_T of the reconstructed τ_h matched to generator-level electrons (left) and muons (right) in bins of lepton p_T, obtained from simulated $t\bar{t}Z$ events. The solid lines represent the Breit-Wigner (left) and Crystal-Ball (right) fits employed in the fake τ transfer function [2]

shown in Fig. 8.10. They are fitted for a generic lepton p_T, separately for electrons and muons, with the linear functions

$$\mu(p_{T,\ell}) = a \cdot p_{T,\ell} + b \,, \tag{8.22}$$

$$\sigma(p_{T,\ell}) = c \cdot p_{T,\ell} + d \,, \tag{8.23}$$

$$n(p_{T,\ell}) = e \cdot p_{T,\ell} + f \,, \tag{8.24}$$

while α is a constant. The numerical values of the coefficients in Eq. 8.24 are shown in Table 8.3.

Non-prompt Lepton Transfer Function

Non-prompt leptons originating from the decays of B hadrons can possibly pass the tight lepton selection criteria. In those cases, the overlapping b-jet is removed, but its momentum can still be evaluated based on the lepton momentum. It is done with a dedicated transfer function $T_{b\rightarrow\ell}(\hat{\ell}|b)$ that relates the momentum of the initial b-quark to that of the resulting reconstructed lepton, assuming their direction is the same, which is a good approximation given the usage of PF jets [15]. The function is modelled by a Gaussian distribution, as seen in Fig. 8.11, where the distributions of p_T of the reconstructed leptons are shown in different bins of b-quark p_T. The mean and width of the Gaussian are fitted as a function of the p_T of the b-quark as

$$\begin{aligned} \mu(p_{T,b}) &= m \cdot p_{T,b} + n \,, \\ \sigma(p_{T,b}) &= a \oplus b \cdot \sqrt{p_{T,b}} \oplus c \cdot p_{T,b} \,, \end{aligned} \tag{8.25}$$

with the coefficients taking the numerical values listed in Table 8.4.

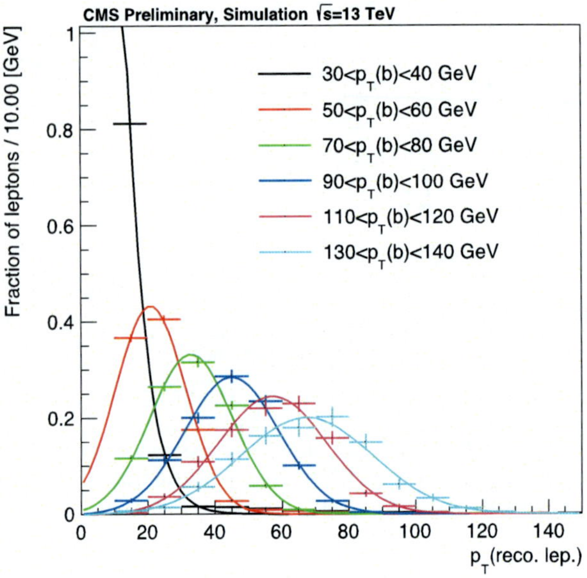

Fig. 8.11 Distribution of p_{T} of the reconstructed leptons matched to generator-level b-quarks in bins of p_{T} of the b-quark, obtained from simulated t$\bar{\text{t}}$+jets events. The solid line represents the Gaussian fit used in the non-prompt lepton transfer function [2]

Table 8.4 Numerical values of the coefficients used to parametrize the non-prompt lepton transfer function [2]

m	n	a	b	c
0.58	−10.74	6.26	0.0	0.14

8.2.3 Object Assignment and Subcategorization

The t$\bar{\text{t}}$H signal targeted in the MEM is characterized by the presence of two leptons of the same charge sign and one τ_h, along with two b-jets from and two light jets. Consequently, the event selection in the $2\ell ss + 1\tau_h$ category requires the presence of exactly two leptons of the same sign and one τ_h; however, the jet multiplicity is relaxed to require at least three jets, out of which at least two must pass the loose working point of the b-tagging discriminator or at least one must pass the medium working point. Nevertheless, the multiplicity of reconstructed jets in an event is often higher, due to the next-to-leading order QCD contributions and the presence of pileup. Evaluating the MEM weights for every possible combination of 4 jets would be very CPU consuming, and thus an assignment algorithm is used to select the jet inputs. The two jets with the highest b-tagging score are associated to the final-state b-quarks, leading to a correct assignment in ~62% of the t$\bar{\text{t}}$H events [2]. From the remaining jets, the pair with the invariant mass closest to the mass of the W boson is associated to the quark pair from the W boson. This leads to an overall correct assignment of 35% when at least 4 jets are reconstructed [2].

An additional ambiguity arises in the assignment of the reconstructed leptons and b-jets. For a t$\bar{\text{t}}$H event, the two leptons cannot be unambiguously assigned to either the leptonic decay of the τ or to the leptonic decay of the W boson. Similarly, the reconstructed b-jet can be originating from the top quark decaying leptonically or from the top quark decaying hadronically. These ambiguities in the object assignment are solved by computing the MEM weights for each of the 4 possible lepton-bottom permutations; their value is summed up to define a total MEM weight. In the case of the weight associated to the $t\bar{t}$ background, only two b-jets and no pair of W-tagged jets are expected; the permutations are computed nonetheless to decide which of the two b-jets are used.

Additionally, there are cases in which no jet is associated to one of the light quarks expected from the W boson decay. The reasons are that either the hadronization products fall out of the detector acceptance or the jet has been merged with another jet during reconstruction. To tackle this situation, the MEM weights are computed independently for the two scenarios:

- **No-missing-jet**: all 4 jets expected in the event are reconstructed,
- **Missing-jet**: one of the light jets from the W boson decay is not reconstructed.

These two exclusive subcategories of events are treated as independent in the maximum likelihood fit used in the signal extractions. The missing jet hypothesis is used when the invariant mass of the two jets associated to the W boson is not compatible with the hadronic W boson decay, namely $m_{jj} < 60$ GeV or $m_{jj} > 100$ GeV. In this category, only three jets are taken as inputs to the MEM computation, and the phase-space integration is expanded by adding the two integration variables corresponding to the direction of the missing jet. As it is not possible to know which jet comes from the W decay, the MEM weight is computed for each light-jet permutation in the event; it leads to $4 \times n_{\text{light jet}}$ MEM integrations.

Considering the lepton-bottom permutations and the light-jet permutations, the MEM integrations to be performed can reach up to $4 \times 4 \times n_{\text{light jet}}$ in missing-jet events, which is very computation-intensive. In practice, to reduce the number of integrations, certain kinematic filters are applied so that only the permutations which are physically feasible are computed and the spurious combinatorial background is removed. These filters are based on a series of compatibility checks of the Higgs boson and top quark masses; the integration is skipped in case the invariant masses of the decay products are not within the expected windows. Detailed explanations of the dimensionality reduction via the kinematic reconstruction of the Higgs boson and the top quark can be found in Ref. [2].

8.2.4 Numerical Integration

The MEM integration is performed over the phase space of the final-state particles in the matrix element after applying the assumptions mentioned before. The integration variables can be decomposed as the product of the variables related to each particle, namely $d\mathbf{x} = \prod d\mathbf{x}_k$. Being the integrand non-analytic, the integration is done numerically event-by-event making use of MC integration methods. The dimensionality of the integral requires a large number of integration points to reach the convergence; however, it is reduced with the assumptions made in the derivation of the transfer functions (see Sect. 8.2.2) and assuming that the intermediate particles (Higgs boson, top quark, W boson, Z boson and τ lepton) are on their mass shell. Such approximation is valid, as the off-shell contributions would anyway be strongly suppressed by a small value of the matrix element. In order to further speed up the integration and to allow for weight for many processes to be computed, an adaptive multidimensional MC integration method based on the VEGAS algorithm is used [13]; it is described in the following.

Given an n-dimensional integral of the type

$$I = \int_{\Omega} d^n\mathbf{x}\, f(\mathbf{x}) \,, \tag{8.26}$$

where Ω is the phase-space volume, the VEGAS algorithm performs an estimation of I by averaging the values of the function $f(x)$ obtained over a large number N of integration points $\mathbf{x}_i$. The first estimation is derived as

$$S = \frac{1}{N}\sum_{i=1}^{N}\frac{f(x_i)}{\rho(x_i)} \,, \tag{8.27}$$

where $\rho(x)$ is the density with which the random points are distributed in Ω. The uncertainty associated to this estimate is

$$\sigma_j = \frac{1}{N-1}\left(\frac{1}{N}\sum_{i=1}^{N}\frac{f(x_i)^2}{\rho(x_i)} - S_j^2\right) . \tag{8.28}$$

Successive estimates of S ($\{S_j\}_{j=0}^{m}$) are used to build a cumulative estimate $\bar{S}$,

$$I \approx \bar{S} = \bar{\sigma}^2 \sum_{j=1}^{m}\frac{S_j}{\sigma_j^2} \,, \tag{8.29}$$

with

$$\frac{1}{\bar{\sigma}^2} = \sum_{j=1}^{m}\frac{1}{\sigma_j^2} \,. \tag{8.30}$$

In order to quantify how consistent these estimations are with one another, the value of χ^2/dof is evaluated as

$$\chi^2/\text{dof} = \frac{1}{m-1}\sum_{j=1}^{m}\frac{(S_j-\bar{S})^2}{\sigma_j^2} \,. \tag{8.31}$$

It tests whether the distribution is Gaussian and expects values close to 1.

In the simplest form of the Monte Carlo integration, the N integration points are uniformly distributed, i.e. $\rho(\mathbf{x}) = const$. For the VEGAS algorithm, the density function $\rho(\mathbf{x})$ is modified in each successive estimate to minimize the associated uncertainty σ_j in the estimation. Uniformly distributed points are employed in the first iteration; in subsequent iterations the information about $f(\mathbf{x})$ is used to define a new density $\rho'(\mathbf{x})$ which reduces the value of σ_j^2; it is minimized when

$$\rho(\mathbf{x}) = \frac{|f(\mathbf{x})|}{\int_\Omega |f(\mathbf{x})|\, d^n\mathbf{x}}, \qquad (8.32)$$

i.e. when the sample points are concentrated where the integrand is largest in magnitude. Assuming the convergence of the sampling, the uncertainty $\bar{\sigma}$ decreases as $1/\sqrt{Nm}$.

Alongside the gain achieved with the VEGAS algorithm, significant improvements in the calculation speed have resulted from the developments carried out in the analysis software itself, which is now able to process multiple events in parallel using multi-core graphics processing units (GPUs). The migration to the GPU platform was performed for the $\mathrm{t\bar{t}H}$ multileptons analysis with the data collected in 2016 [2]. It was found to speed up the computational time by a factor 200 with respect to the classical CPU platforms [16], significantly increasing the computational resources available for the analysis. The results presented here were timely derived provided the impressive computing power of the GPU platform in the IN2P3 Computing Centre in Lyon, equipped with 10 nodes of 4 Nvidia Tesla K80 accelerators [16].

8.2.5 *Performance*

The MEM LR is defined as the ratio between the weight associated to the $\mathrm{t\bar{t}H}$ signal and the sum of the weights associated to the $\mathrm{t\bar{t}H}$ signal and the weights associated to the backgrounds; the latter are multiplied by individual κ coefficients. Considering the three background hypotheses used in the MEM ($\mathrm{t\bar{t}Z}\ \mathrm{Z} \to \tau\tau$, $\mathrm{t\bar{t}Z}\ \mathrm{Z} \to \ell\ell$ and $\mathrm{t\bar{t}}$ dilepton), the LR is expressed as

$$\mathrm{LR} = \frac{w_{\mathrm{t\bar{t}H}}}{w_{\mathrm{t\bar{t}H}} + \kappa_{\mathrm{t\bar{t}Z,\,Z}\to\tau\tau} \cdot w_{\mathrm{t\bar{t}Z,\,Z}\to\tau\tau} + \kappa_{\mathrm{t\bar{t}Z,\,Z}\to\ell\ell} \cdot w_{\mathrm{t\bar{t}Z,\,Z}\to\ell\ell} + \kappa_{\mathrm{t\bar{t}}} \cdot w_{\mathrm{t\bar{t}}}}. \qquad (8.33)$$

The closer the LR is to 1, the more signal-like the event is. The κ coefficients associated to each background are optimized to achieve the maximum separation between the signal and the sum of the backgrounds. The LR computed from Eq. 8.33 is denoted as the *combined likelihood ratio*, as it includes the weights associated to all the relevant backgrounds. Alternatively, one can define a *specific likelihood ratio* for a single background. In this case, the values k coefficients are equal to zero for all backgrounds except for the one of interest, which does not necessarily take the same

Table 8.5 Values of the k coefficients associated to each background used in the specific and combined MEM likelihood ratios, separately for the no-missing-jet and the missing-jet subcategories

Category	Likelihood ratio	$k_{t\bar{t}Z,Z\to\tau\tau}$	$k_{t\bar{t}Z,Z\to\ell\ell}$	$k_{t\bar{t}}$
No-missing-jet	Specific $t\bar{t}Z, Z \to \tau\tau$	0.5	–	–
	Specific $t\bar{t}Z, Z \to \ell\ell$	–	1.0	–
	Specific $t\bar{t}$	–	–	10^{-15}
	Combined	0.1	0.2	10^{-18}
Missing-jet	Specific $t\bar{t}Z, Z \to \tau\tau$	0.05	–	–
	Specific $t\bar{t}Z, Z \to \ell\ell$	–	0.1	–
	Specific $t\bar{t}$	–	–	10^{-12}
	Combined	0.05	0.5	$5 \cdot 10^{-15}$

value as in the combined likelihood ratio. The specific likelihood ratio is useful to evaluate the discrimination power against an individual background; the combined likelihood ratio is used for simultaneous discrimination against all backgrounds in the maximum likelihood fit.

The numerical values derived for these coefficients in the combined and the specific likelihood ratios are shown in Table 8.5 for the no-missing-jet and missing-jet subcategories. The discriminating power obtained in each configuration is illustrated in the ROC curves in Fig. 8.12. They are presented for the $t\bar{t}H$ signal against each individual background, obtained from the simulation. Due to the limited statistics of the $t\bar{t}$ sample, the curves for this background are obtained with events where at least one of the leptons fails the tight selection, reweighted by their fake rates obtained in the data-driven estimation of the fake background of the analysis (see Sect. 6.3.3). The figures show how, for each background, the corresponding specific likelihood ratio provides the highest discrimination power. However, using a single specific likelihood ratio results in a poor discrimination against the backgrounds for which it is not optimized. The combined likelihood, despite providing slightly less discriminating power against the individual backgrounds, shows an adequate performance in the simultaneous discrimination against all of them.

The combined MEM likelihood ratio constitutes the signal extraction observable used in the $2\ell ss + 1\tau_h$ category of the control analysis, included in the maximum likelihood fit via two separate categories (no-missing-jet and missing-jet). It represents the probability of an event to originate from the $t\bar{t}H$ process and takes values from 0 to 1. The distributions used for signal extraction are binned in 10 uniform bins; this configuration was found to give the best sensitivity. They are shown in Fig. 8.13 for the signal and all the background processes considered in the analysis. The gradual increase of the signal-over-background ratio is clearly visible, demon-

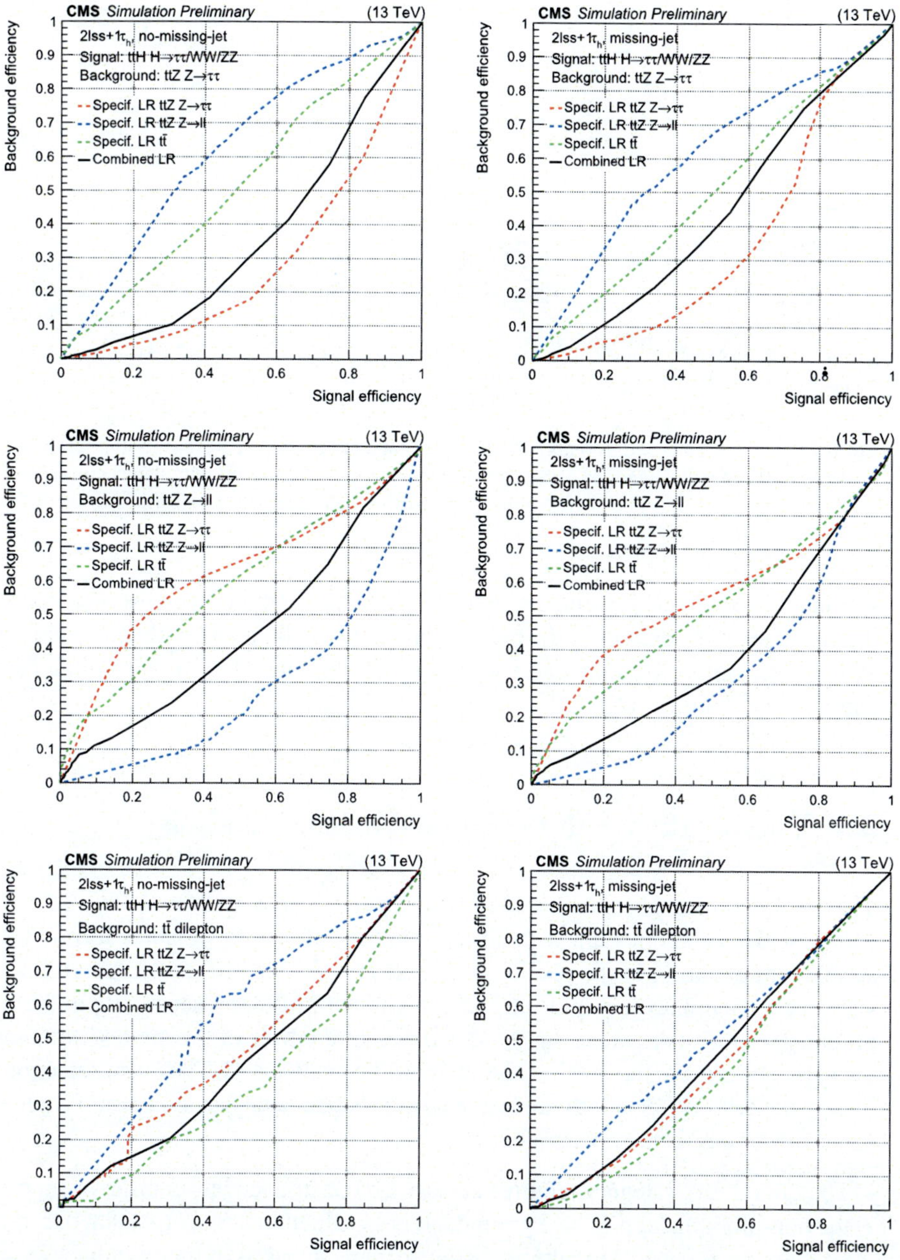

Fig. 8.12 Background efficiency as a function of the signal efficiency obtained with the specific and combined MEM likelihood ratios in the no-missing-jet (left column) and missing-jet (right column) subcategories of the $2\ell ss + 1\tau_h$ category. The discrimination is presented for the $t\bar{t}H$ signal against the $t\bar{t}Z$ $Z \rightarrow \tau\tau$ background (top row), the $t\bar{t}Z$ $Z \rightarrow \ell\ell$ background (middle row) and the $t\bar{t}$ dilepton background (bottom row), obtained from simulated events

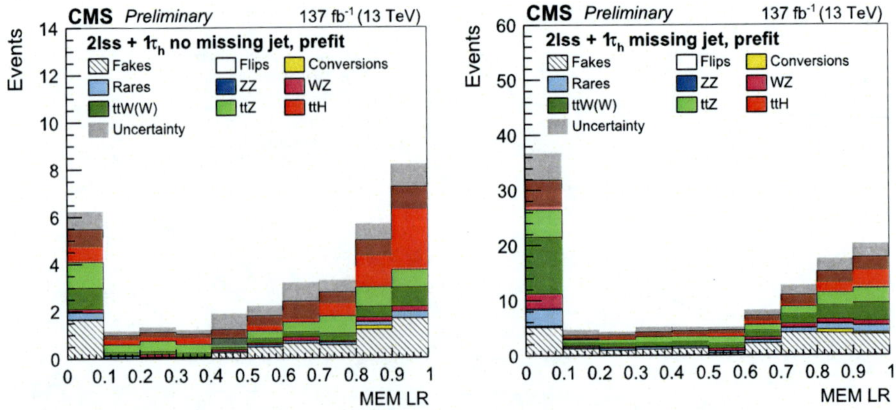

Fig. 8.13 Distributions of the combined MEM likelihood ratio used in the no-missing-jet (left) and missing-jet (right) subcategories of the $2\ell ss + 1\tau_h$ category. The uncertainty bands correspond to systematic uncertainties. The distributions are presented before any fit is performed to the data

strating the excellent performance achieved with the method. The observed data are fitted to these distributions in the maximum likelihood fit, along with the single variables of the purely leptonic categories and distributions of the control regions, to extract the $t\bar{t}H$ signal rate.

8.3 Combination with the Single-Variable Analysis

The discriminating observables used in the purely leptonic $2\ell ss + 0\tau_h$, $3\ell + 0\tau_h$ and $4\ell + 0\tau_h$ categories are derived from optimized single variables to provide a simple interpretation of the results. As the $2\ell ss + 0\tau_h$ and $3\ell + 0\tau_h$ categories are statistically rich, they are split into subcategories of different sensitivity and background composition, enhancing the discriminating power of the method following the approach used in the previous iteration of the analysis [4]. The event selection in these categories is essentially the same as in the main analysis; however, it tightens some cuts to target the $t\bar{t}H$ signal alone.

- The $2\ell ss + 0\tau_h$ category requires at least 4 reconstructed jets in order to suppress the non-prompt and $t\bar{t}W$ backgrounds. It is classified in terms of the flavour of the leptons (ee, $e\mu$ or $\mu\mu$) to take advantage of the flavour asymmetry of background processes such as charge flips and photon conversions. These subcategories are further split according to the number of reconstructed jets into a low-jet region ($N_{\text{jet}} < 6$) or a high-jet region ($N_{\text{jet}} \geq 6$); it helps separating the phase spaces dominated by the fake background from the irreducible $t\bar{t}W$ and $t\bar{t}Z$ backgrounds. As the ee and $e\mu$ subcategories have enough statistical power, they are further split according to the lepton charge ($++$ or $--$) to profit from the charge asymmetry

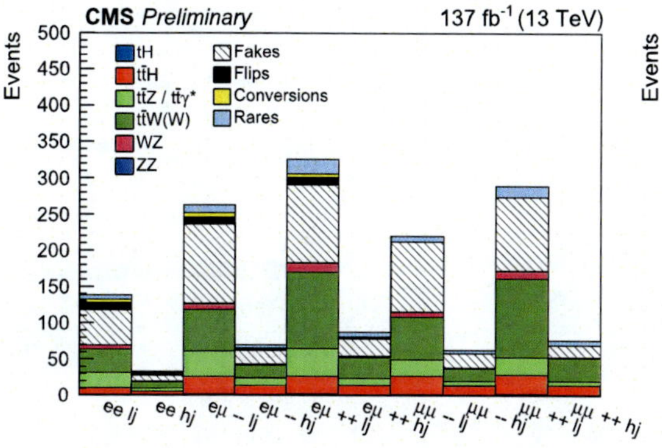

Fig. 8.14 Number of signal and backgrounds expected events in the subcategories of the $2\ell ss + 0\tau_h$ (left) and $3\ell + 0\tau_h$ (right) categories, obtained from the simulation. The label "lj" ("hj") refers to the low-jet (high-jet) region, and the labels "neg" ("pos") refer to negative (positive) total charge. The distributions are presented before any fit is performed to the data

of the $t\bar{t}W$ process. The signal and background yields for the 10 subcategories of the $2\ell ss + 0\tau_h$ channel can be found in Fig. 8.14.

- In the $3\ell + 0\tau_h$ category, at least two reconstructed jets are required. The subcategorization follows the same motivations as the previous category; it is done in terms of jet multiplicity, with a low-jet region ($N_{\text{jet}} < 4$) and a high-jet region defined ($N_{\text{jet}} \geq 4$), and in terms of the sum of the charge of the three leptons (positive or negative). The signal and background yields for the 4 subcategories of the $3\ell + 0\tau_h$ final state can be found in Fig. 8.14.
- The $4\ell + 0\tau_h$ category is included in the fit but presents no subcategorization, as it suffers from low statistics.

Several variables were studied as potential discriminating observables: the lepton kinematics (p_{T}, η, invariant mass, angular distances), the jet properties (total hadronic energy, jet and b-jet multiplicities) and global event quantities ($E_{\text{T}}^{\text{miss}}$ and angular distances between any lepton-jet pair). The variable bringing the highest sensitivity was found to be the invariant mass of the two-, three- and four-lepton system in the $2\ell ss + 0\tau_h$, $3\ell + 0\tau_h$ and $4\ell + 0\tau_h$ categories, respectively. It show good discrimination power against the $t\bar{t}W$ and $t\bar{t}Z$ background processes, thus complementing the separation against the non-prompt background achieved with the subcategorization. The binning is chosen so that the signal yield remains approximately constant across the bins. The distributions of these variables for the simulated signal and backgrounds processes can be found in Figs. 8.15, 8.16 and 8.17. For the sake of visualization, no charge subcategorization is applied in the $2\ell ss + 0\tau_h$ and $3\ell + 0\tau_h$ categories.

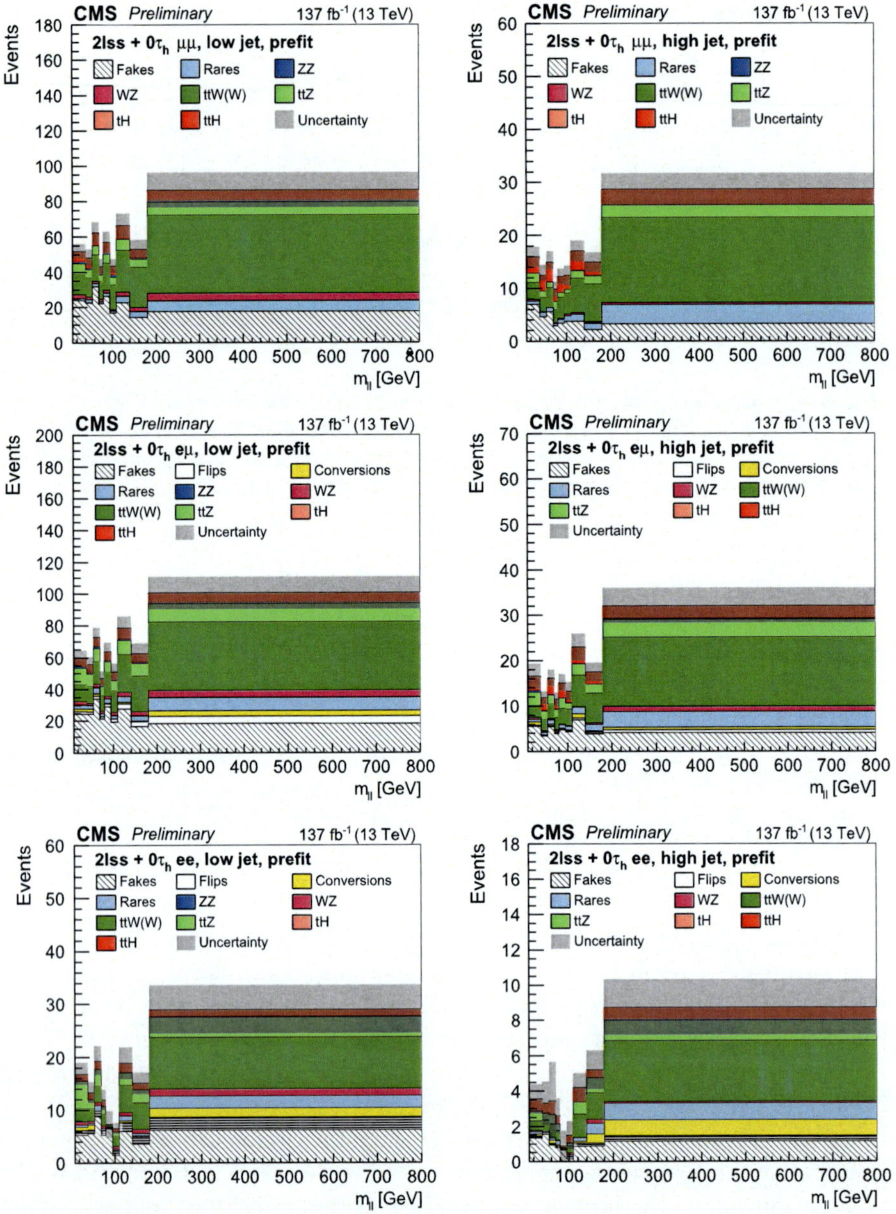

Fig. 8.15 Distributions of the invariant mass of the leptonic system used for the signal extraction in the $2\ell ss + 0\tau_h$ category of the control analysis. No subcategorization in terms of lepton charge is applied. The uncertainty bands correspond to the systematic uncertainties. The distributions are presented before any fit is performed to the data

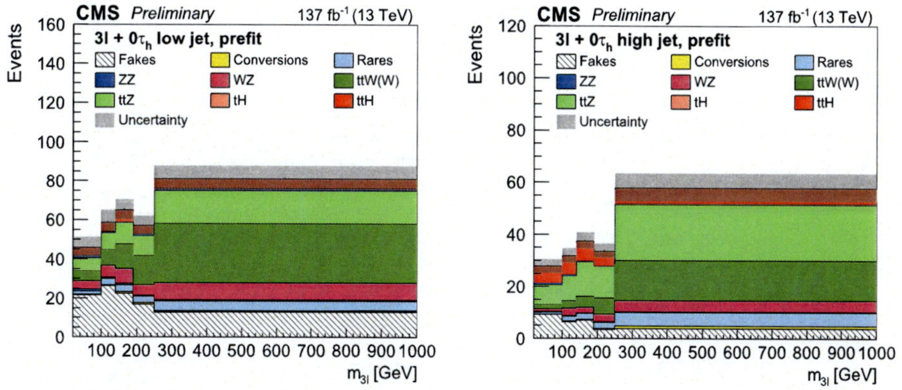

Fig. 8.16 Distributions of the invariant mass of the leptonic system used for the signal extraction in the $3\ell + 0\tau_h$ category of the control analysis. No subcategorization in terms of total charge is applied. The uncertainty bands correspond to the systematic uncertainties. The distributions are presented before any fit is performed to the data

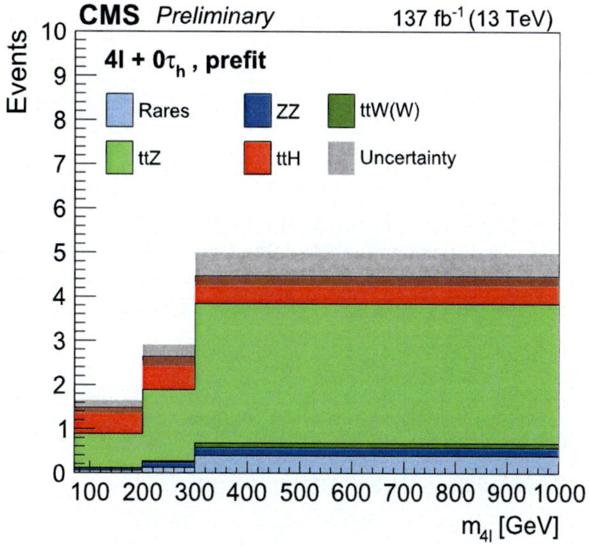

Fig. 8.17 Distributions of the invariant mass of the leptonic system used for the signal extraction in the $4\ell + 0\tau_h$ category of the control analysis. The uncertainty bands correspond to the systematic uncertainties. The distributions are presented before any fit is performed to the data

8.4 Standard Model Results

A maximum likelihood fit of the discriminating observables is performed across the four signal regions of the control analysis under the assumption that the Higgs boson coupling modifiers are consistent with the SM. The 3ℓ-CR and 4ℓ-CR, enriched in $t\bar{t}Z$, WZ and ZZ events, are included in the fit; additionally, the $2\ell ss$+3j-CR is added to constrain the $t\bar{t}W$ background (see Sect. 6.5), which is not the case in the main analysis. The results are derived following the same strategy as the main analysis: the statistical model is validated first, followed by the comparison of the observed and expected event yields and the fit of the data to the expectation to obtain the

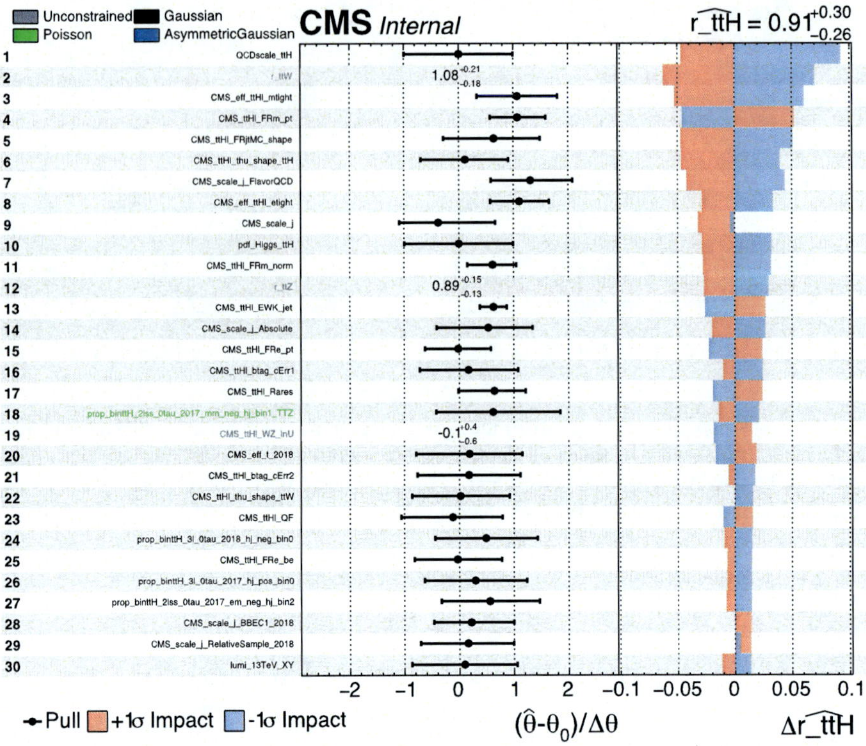

Fig. 8.18 List of the 30 nuisances parameters with the highest impact on the measurement of the tt̄H signal strength. The first column corresponds to the pulls of each nuisance parameter and the second column corresponds to the $+1\sigma$ and -1σ impacts on the measurement of the signal strength

tt̄H signal rate. In the fit, the normalizations of the tt̄W(W) and tt̄Z processes are unconstrained; their correlations with the tt̄H signal strength are likewise evaluated.

Validation of the Statistical Model

The impact of the nuisance parameters in the measured tt̄H signal strength was evaluated; the result for the 30 most relevant nuisances can be found in Fig. 8.18, together with the associated pulls. The nuisance parameters with the highest impacts correspond to the theoretical uncertainties, the fake background estimation, the τ_h and lepton selection efficiency and the jet energy corrections. The fact that the leading uncertainties are very similar to the ones in the main analysis constitutes a validation of the statistical model. Further verification of the modelling of the data is provided by the goodness-of-fit, shown in Fig. 8.19. The observed χ^2, with a p-value of 0.32, lies at the core of the distribution of the 1000 generated pseudo-experiments, concluding that the statistical model describes well the observations.

Data and Simulated Event Yields

The expected and observed event yields for the signals and backgrounds in the signal regions of the analysis are listed in Table 8.6. For simplicity, no subcategorization is

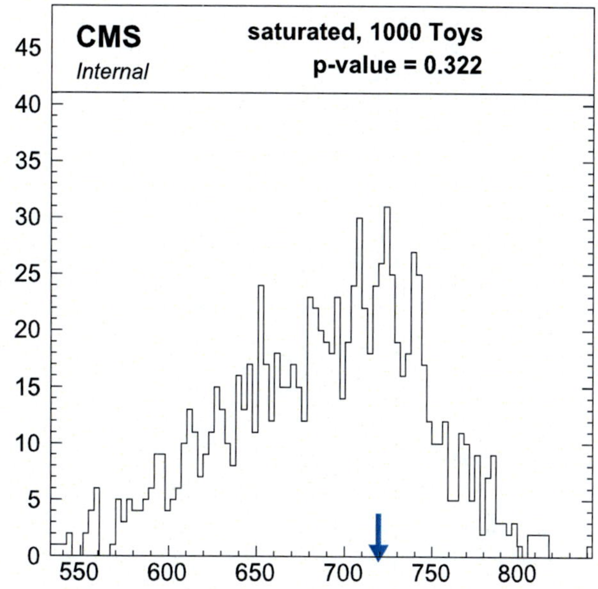

Fig. 8.19 Distribution of the χ^2 goodness of fit variable computed with the saturated model on 1000 pseudo-experiments (black line) compared to the observed value (blue arrow)

applied in the table. The values of these yields are post-fit, meaning they are computed for the best values of nuisance parameters and after applying the scale factors on the $t\bar{t}W$ and $t\bar{t}Z$ normalizations. The table shows an excellent agreement between the number of observed events and the sum of the expected signal and background events in all signal regions. The post-fit distributions of the discriminating observables in each channel can be found in Figs. 8.20, 8.21, 8.22, and 8.23; a good agreement in shape is found in all categories.

Measurement of the Signal Strength

The measured signal strength of the $t\bar{t}H$ process in the individual signal regions and for the combination of all of them can be found in Fig. 8.24. The measured rate of the $t\bar{t}H$ process amounts to

$$\mu_{t\bar{t}H} = 0.91 \pm 0.21\ \text{(stat)}\ ^{+0.21}_{-0.15}\ \text{(syst)}\,. \tag{8.34}$$

The associated uncertainty is compatible with the expected value obtained from an Asimov dataset of $\mu^{\text{exp.}}_{t\bar{t}H} = 1.00\ ^{+0.23}_{-0.22}$ (stat) $^{+0.23}_{-0.17}$ (syst). An excess of events over the background-only hypothesis is found, corresponding to an observed (expected) significance of 3.8σ (4.0σ), well above the evidence threshold.

To allow for a fair comparison, the signal strength in the main analysis is derived for the case where the tH signal strength is fixed to its SM value; it amounts to $\mu_{t\bar{t}H} = 0.99\ ^{+0.18}_{-0.18}$ (stat) $^{+0.18}_{-0.13}$ (syst). Hence, both results are in agreement within uncertainties. The control analysis has ~20% less sensitivity than the main analysis, as expected from the poorer separation power of the methods used and the limited categories included in the fit. Upon comparison by categories, the sensitivity of the $2\ell ss + 0\tau_h$ and $3\ell + 0\tau_h$ categories is worse in the control analysis, as it does not

Table 8.6 Number of observed events in the 4 categories of the control analysis compared to the event yields expected from the $t\bar{t}$H signal and from the main background processes, computed for the values of nuisance parameters and of the parameters of interest obtained from the maximum likelihood fit. The quoted uncertainties include the statistical and systematic components

Process	$3\ell + 0\tau_h$	$2\ell ss + 1\tau_h$	$4\ell + 0\tau_h$	$2\ell ss + 0\tau_h$
$t\bar{t}$H	58.5 ± 14.4	35.9 ± 8.2	1.9 ± 0.5	213.0 ± 51.0
tH	3.3 ± 0.2	2.4 ± 0.4	–	16.0 ± 0.8
$t\bar{t}Z + t\bar{t}\gamma^*$	121.5 ± 10.4	24.6 ± 4.3	5.4 ± 0.5	265.4 ± 24.4
$t\bar{t}W + t\bar{t}WW$	123.0 ± 13.8	41.3 ± 9.5	0.2 ± 0.0	814.6 ± 95.6
WZ	58.4 ± 7.8	11.0 ± 2.4	–	164.8 ± 18.8
ZZ	7.4 ± 2.1	1.9 ± 0.3	0.5 ± 0.1	15.2 ± 2.1
Fakes	130.5 ± 11.8	30.1 ± 3.9	–	1062.2 ± 80.6
Flips	–	0.7 ± 0.1	–	82.0 ± 15.5
Conversions	5.1 ± 1.1	1.3 ± 0.5	–	41.0 ± 11.0
Rares	43.0 ± 9.7	15.8 ± 3.6	0.8 ± 0.2	210.4 ± 46.1
Other Higgs boson processes	4.0 ± 0.6	0.6 ± 0.1	–	24.7 ± 2.8
Total expected background	496.2 ± 14.8	129.7 ± 10.8	6.9 ± 0.6	2696.2 ± 80.4
Data	600	169	12	2869

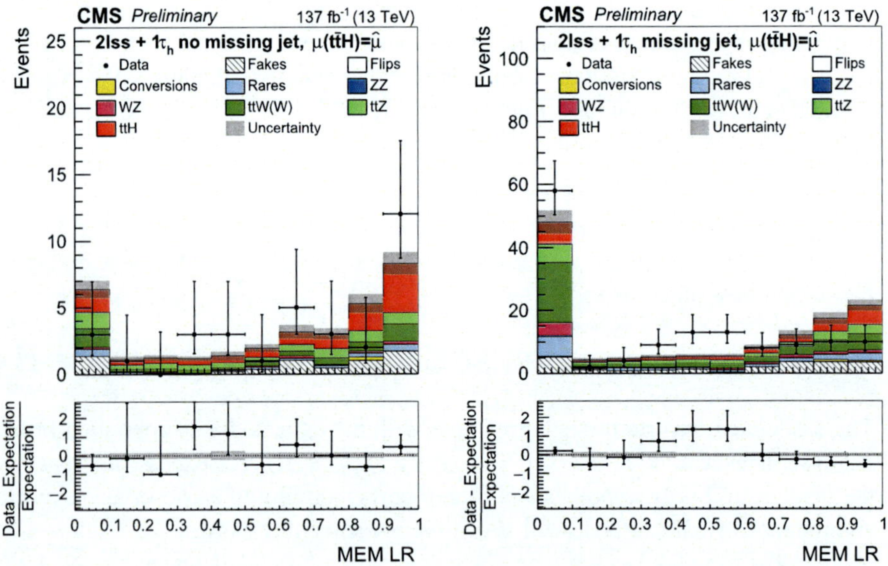

Fig. 8.20 Post-fit distributions of the MEM likelihood ratio used for signal extraction in the $2\ell ss + 1\tau_h$ category of the control analysis, for the no-missing-jet (left) and missing-jet (right) subcategories. The uncertainty bands include the statistical and systematic components

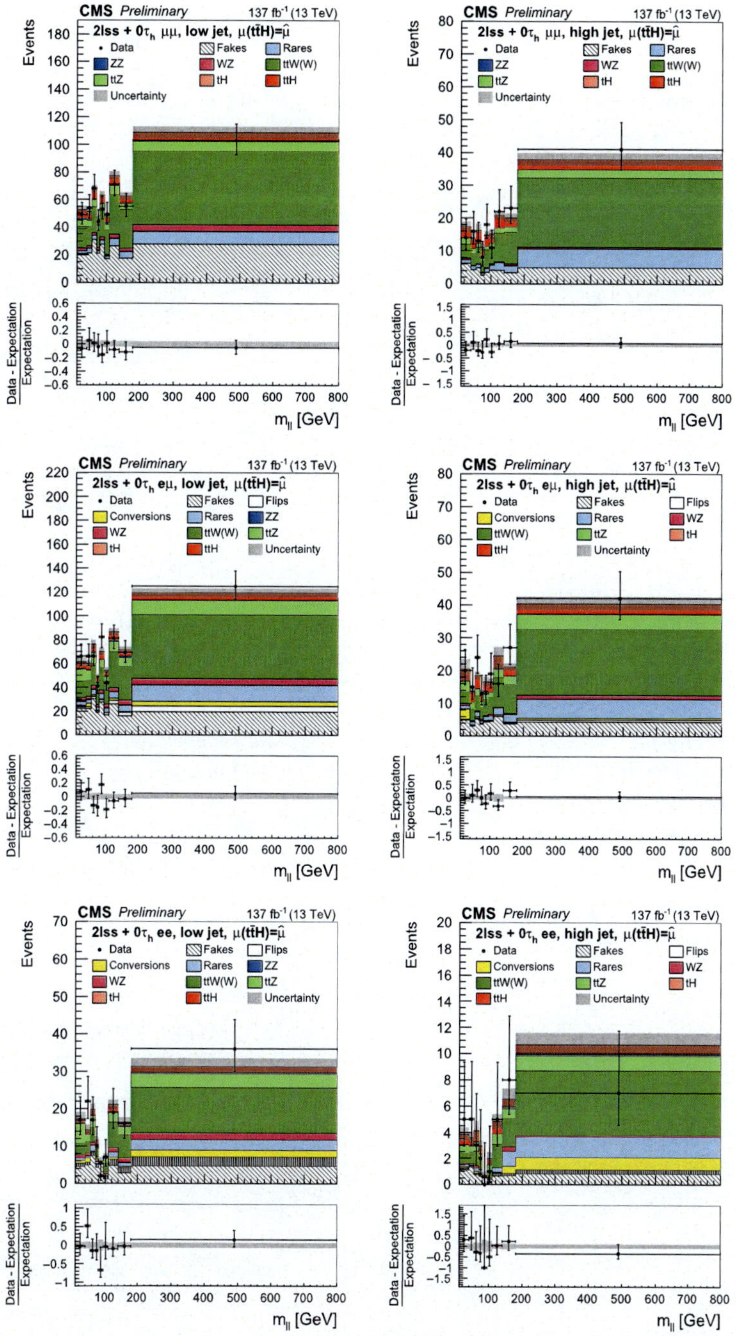

Fig. 8.21 Post-fit distributions of the invariant mass of the leptonic system used for the signal extraction in the $2\ell ss + 0\tau_h$ category of the control analysis. The left (right) column corresponds to the low-jet (high-jet) region. The first, second and third rows correspond to $\mu\mu$, $e\mu$ and ee subcategories, respectively. No subcategorization in terms of lepton charge is applied. The uncertainty bands include the statistical and systematic components

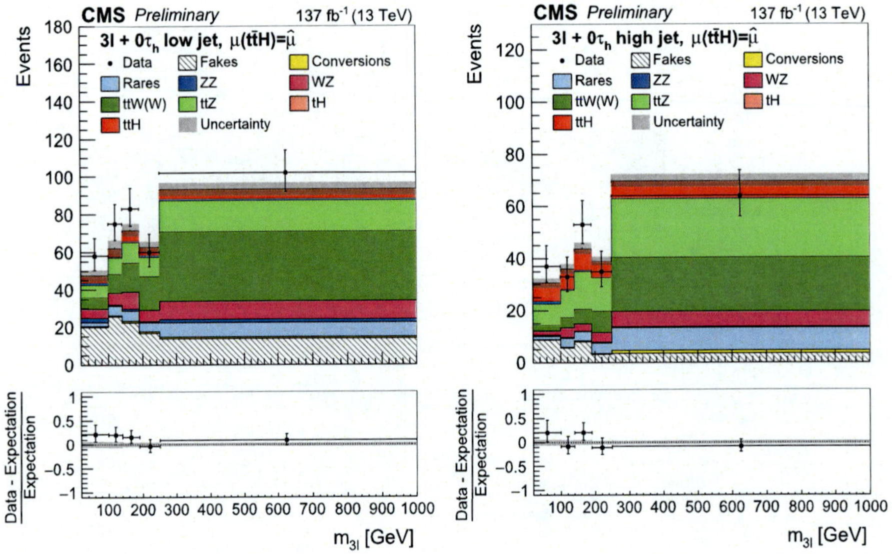

Fig. 8.22 Post-fit distributions of the invariant mass of the leptonic system used for the signal extraction in the $3\ell + 0\tau_h$ category of the control analysis. The left (right) column corresponds to the low-jet (high-jet) region. No subcategorization in terms of total charge is applied. The uncertainty bands include the statistical and systematic components

Fig. 8.23 Post-fit distribution of the invariant mass of the leptonic system used for the signal extraction in the $4\ell + 0\tau_h$ category of the control analysis. The uncertainty bands include the statistical and systematic components

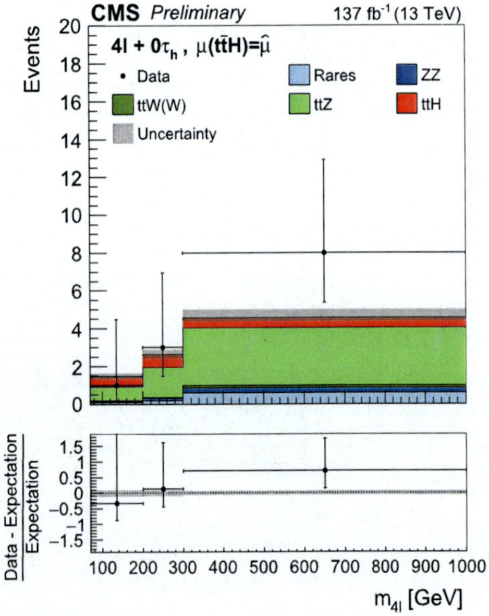

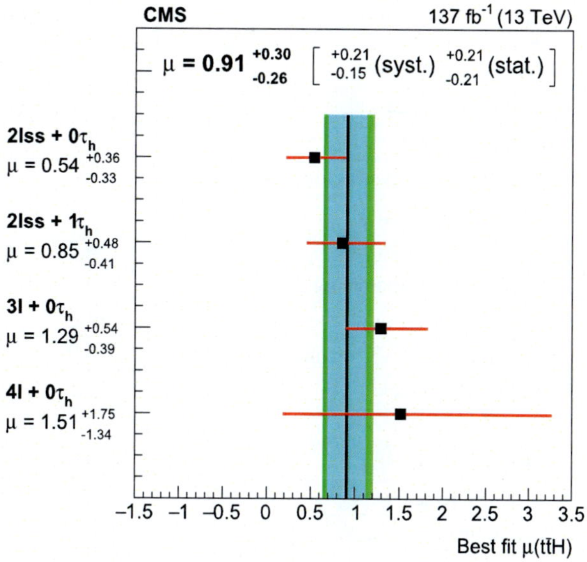

Fig. 8.24 Measured signal strength ($\mu = \sigma/\sigma_{\mathrm{SM}}$) of the $\mathrm{t\bar{t}H}$ process under the SM hypothesis, separately for each signal region considered in the control analysis and for the combination of all of them together with the control regions

profit of the high performance of the DNNs in the signal extraction. In the $2\ell ss + 1\tau_h$ category, where the MEM is used, the sensitivity is comparable to the main analysis. Nonetheless, the current implementation of the MEM is less convenient in the context of the global analysis, where the tH signal is also measured, since this would require the development of a dedicated weight for the tH hypothesis. In the $4\ell + 0\tau_h$ category, the sensitivities of the main and control analysis are comparable, with uncertainties surpassing 100% in both cases due to the low statistical power of this channel.

The impact of the systematic uncertainties on the measurement of the $\mathrm{t\bar{t}H}$ signal strength is summarized in Table 8.7, where the uncertainties have been grouped by source. Similarly to the main analysis, the largest impacts originate from the uncertainty on the reducible background estimate, the normalization of the processes estimated by the simulation, the theoretical uncertainties and the statistical uncertainty on the observed events. Being the leading uncertainties very similar to the ones in the main analysis, the control analysis demonstrates again the validity of the statistical model.

Signal and Background Correlations

As the normalizations of the $\mathrm{t\bar{t}W}$ and $\mathrm{t\bar{t}Z}$ backgrounds are left unconstrained, their production rates have been measured simultaneously with the $\mathrm{t\bar{t}H}$ signal strength in the maximum likelihood fit. The obtained best fit values are

$$\begin{aligned} \theta_{\mathrm{t\bar{t}W(W)}} &= 1.08 \,^{+0.21}_{-0.18}\,, \\ \theta_{\mathrm{t\bar{t}Z}} &= 0.89 \,^{+0.15}_{-0.13}\,. \end{aligned} \tag{8.35}$$

The values obtained in the main analysis after fixing the tH rate to its SM expectation amount to $\theta_{\mathrm{t\bar{t}W(W)}} = 1.45 \,^{+0.23}_{-0.21}$ and $\theta_{\mathrm{t\bar{t}Z}} = 1.02 \,^{+0.15}_{-0.14}$. Hence, the errors associated

Table 8.7 Main sources of systematic uncertainty and their impact on the measurement of the $t\bar{t}H$ signal strength. The quantity $\Delta\mu_{ttH}/\mu_{ttH}$ corresponds to the change in uncertainty when fixing the nuisances associated to it in the fit

Uncertainty source	$\Delta\mu_{ttH}/\mu_{ttH}[\%]$
Trigger efficiency	1.0
Electron and muon selection efficiency	6.6
τ_h selection efficiency	0.3
b-tagging efficiency	4.8
Jet energy scale and resolution	4.6
Reducible background estimate	11.3
Normalization of MC estimation processes	13.0
Theoretical sources	9.1
Prefiring	1.2
Integrated luminosity	1.5
MC and sideband statistical uncertainty	4.3
Data statistical uncertainty	24.6

to the estimation of these background rates are very comparable in the main analysis and in the control analysis. The values of $\theta_{t\bar{t}Z}$ are compatible within uncertainties between the main and the control analysis. However, the excess observed for $t\bar{t}W$ in the main analysis is softened out in the control analysis, where a ~25% lower value of $\theta_{t\bar{t}W(W)}$ is found. The effect is ascribed to the lower separation power of the $t\bar{t}H$ and $t\bar{t}W$ processes with the single variable approach compared to the DNN.

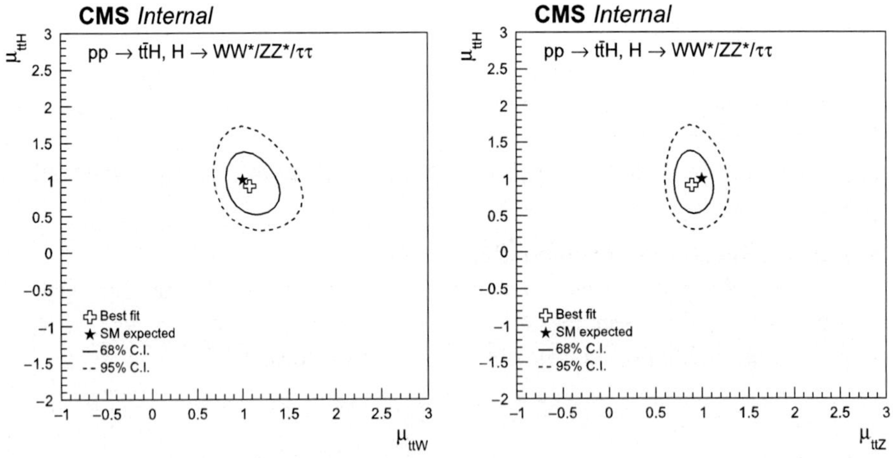

Fig. 8.25 Two-dimensional contours of $-2\Delta\ln\mathcal{L}$ as a function of the $t\bar{t}H$ and $t\bar{t}W$ production rates (left) and as a function of the $t\bar{t}H$ and $t\bar{t}Z$ production rates (right). For each case, the two production rate not shown on either the x or y axis is profiled

The correlations between the measured $t\bar{t}H$ signal rate and the $t\bar{t}W$ and $t\bar{t}Z$ production rates are found in Fig. 8.25, where the two-dimensional contours of the likelihood function corresponding to the 68% and 95% confidence levels are shown. The signal strengths of the signal and the backgrounds are conform the SM expectation; the correlations amongst them are moderate, demonstrating the performance of the MEM and the single variables in their separation.

8.5 Conclusions

An alternative measurement of the production of the Higgs boson in association with two top quarks in multilepton final states is conducted in the context of the control analysis. The results are derived following the same analysis strategy as in the main analysis, making use of an integrated luminosity of 137 fb^{-1} at $\sqrt{s} = 13$ TeV, corresponding to the full Run 2 dataset. Instead of the 10 categories considered in the main analysis, the control analysis is reduced to 4: the purely leptonic categories ($2\ell ss + 0\tau_h$, $3\ell + 0\tau_h$ and $4\ell + 0\tau_h$), where single variables are used for the signal extraction, and the $2\ell ss + 1\tau_h$ category, where the output of the Matrix Element Method is used. Being these categories the most sensitive, the control analysis constitutes a baseline analysis where the validity of the machine learning approaches, the modelling of the processes and the statistical interpretations can be probed. The results of the control analysis are derived in the context of the SM, focusing on the measurement of the $t\bar{t}H$ signal alone, contrary to the main analysis, where the tH signal is also measured and BSM interpretations are included.

As for the main analysis, the measured cross section of the $t\bar{t}H$ process in the control analysis is consistent with the SM, with a signal strength of $\mu_{t\bar{t}H} = 0.91^{+0.30}_{-0.26}$. The observed (expected) significance of the $t\bar{t}H$ process with respect to the background-only hypothesis is 3.8σ (4.0σ), well above the evidence threshold. The sensitivity of the control analysis is ~20% lower than in the main analysis, as expected from the poorer separation power of the single variable analysis and the restricted number of categories considered.

My personal contribution to the $t\bar{t}H$ multilepton analysis stands in the implementation of the MEM-based signal extraction in the $2\ell ss + 1\tau_h$ category. I have optimized the framework, tested the performance and derived the full set of results of the control analysis; besides, I contributed to the derivation of some of the SM results of the main analysis. The $2\ell ss + 1\tau_h$ final state profits from the highest sensitivity in the measurement of the $t\bar{t}H$ cross section in the $H \rightarrow \tau^+\tau^-$ decay final state, as it presents an optimal compromise between statistical power and signal purity. Establishing a direct link between the theory and the experiment, the MEM approach is physics-driven and constitutes a test of the theoretical model itself (the matrix element, the mass of the resonances, the parton distribution functions, etc.). The results derived in this thesis show how the MEM provides a comparable sensitivity to the DNN approach used in the $2\ell ss + 1\tau_h$ category of the main analysis. In addition, it does not require any training and it could be applied to other categories.

The results presented here serve as a benchmark of the potential of the method for future $t\bar{t}H$ multilepton analyses featuring larger datasets. For that purpose, the implementation of a tHq-specific signal hypothesis to measure the tH cross section is convenient; the MEM weight could be derived as a differential probability as a function of the varying κ_t modifier. The sensitivity of the analysis could potentially be improved by using the output of the MEM to define subcategories with different signal-over-background ratio, instead of fitting the MEM LR distributions directly. Another possible usage of the MEM weights could be as input to a machine learning algorithm, provided enough statistics are available, possibly resulting in a more enhanced signal-background separation with respect to the simpler LR. Alternatively, one could use the output of machine learning algorithms to define signal-depleted and signal-enriched regions where the MEM weights are computed separately, not only improving the separation of the signal from the backgrounds but also the separation of the backgrounds themselves.

A practical challenge associated with the use of the MEM is that the calculated likelihood is just an approximation given the finite resolution of the detector, the higher order corrections and the assumptions made in the kinematic reconstruction. Were the technological advances in terms of computing to continue, the MEM could profit from the inclusion of NLO corrections, information on the jet substructure or the hadronization processes, for instance. Ultimately, the resolution effects could be improved by deriving separate transfer functions for each subdetector component, instead of using a single transfer function per physics object.

References

1. CMS Collaboration (2017) Search for the associated production of a Higgs boson with a top quark pair in final states with a τ lepton at $\sqrt{s} = 13$ TeV. Technical report CMS-PAS-HIG-17-003, CERN, Geneva
2. Strebler T (2017) Probing the Higgs coupling to the top quark at the LHC in the CMS experiment. PhD thesis at Universitè Paris Saclay, Sep 2017
3. D0 Collaboration (2004) A precision measurement of the mass of the top quark. Nature 429:638–642
4. CMS Collaboration (2018) Measurement of the associated production of a Higgs boson with a top quark pair in final states with electrons, muons and hadronically decaying τ leptons in data recorded in 2017 at $\sqrt{s} = 13$ TeV. Technical Report CMS-PAS-HIG-18-019, CERN, Geneva
5. D0 Collaboration (2009) Observation of single top-quark production. Phys Rev Lett 103:092001
6. CDF Collaboration (2009) Observation of electroweak single top-quark production. Phys Rev Lett 103:092002
7. CDF Collaboration (2007) Precision measurement of the top-quark mass from dilepton events at CDF II. Phys Rev D 75:031105
8. CMS Collaboration (2015) Search for a standard model Higgs boson produced in association with a top-quark pair and decaying to bottom quarks using a matrix element method. Eur Phys J C 75(6):251
9. ATLAS Collaboration (2015) Search for the Standard Model Higgs boson produced in association with top quarks and decaying into bb in pp collisions at at $\sqrt{s} = 13$TeV with the ATLAS detector. Eur Phys J C 75(7):349

10. CMS Collaboration (2018) An event candidate for the production of a top quark and top anti-quark pair in conjunction with a Higgs Boson in the CMS detector, Jun 2018. https://cds.cern.ch/record/2621446
11. Neyman J, Pearson ES (1933) On the problem of the most efficient tests of statistical hypotheses. Philos Trans R Soc London Ser Contain Pap Math Phys Character 231:289–337
12. Alwall J et al (2014) The automated computation of tree-level and next-to-leading order differential cross sections, and their matching to parton shower simulations. J High Energy Phys 07:079
13. Lepage GP (1978) A new algorithm for adaptive multidimensional integration. J Comput Phys 27(2):192–203
14. Mastrolorenzo L (2016) Search for the Higgs boson decaying into τ lepton pairs with the Matrix Element Method and τ trigger optimization in the CMS experiment at the LHC. PhD thesis at Ecole Polytechnique, Sep 2016
15. CMS Collaboration (2017) Particle-flow reconstruction and global event description with the CMS detector. JINST 12(10):P10003–P10003
16. Grasseau G, Beaudette F, Martin Perez C, Zabi A, Chiron A, Strebler T, Hautreux G (2019) Deployment of a Matrix Element Method code for the ttH channel analysis on GPUs platform. EPJ Web Conf 214:06028, 01 2019

Chapter 9
Conclusion

On the 3rd December 2018, the LHC's second run came to an end after three years of outstanding operation. The increasingly high collision rates delivered in the core of CMS posed significant challenges to the detector, which performed with excellent reliability, demonstrating its versatility through the changing experimental conditions. Considerable efforts were put in place in the trigger system to deal with the harsher data-taking conditions, notably the upgrade of its Level-1 system at the beginning of the run, but also the increased level of sophistication and resilience of the reconstruction algorithms throughout data-taking. The response capability of the system had a beneficial impact on the sensitivity of a large variety of CMS analyses based on the full Run 2 data, including the one presented in this thesis. After the landmark Higgs boson discovery in Run 1, much more was learned about this unique scalar in Run 2, notably how it couples to the heaviest, third generation of quarks and leptons through the Yukawa interaction. The observation of the $t\bar{t}H$ process in 2018 was specially relevant, as it probed the interaction strength between the two most massive particles of the SM. This milestone was not expected to be within the reach of the CMS experiment until much more data have been gathered, testimony of the great progress of the experiment in refining the object reconstructions and analysis techniques.

The first part of this thesis work contributed to the success of the Run 2 operations via the optimisation of the Level-1 trigger algorithm for the reconstruction of hadronically decaying τ leptons during 2017 and 2018 data-taking. After the Level-1 upgrade in 2015, necessary changes to the algorithm were implemented towards the end of the run to maintain the selection efficiencies and trigger rates in a more complicated environment; the changes mainly concern the calibration and the evaluation of the isolation of the L1 τ_h candidates. The results show the stable triggering capabilities of the system throughout the whole detector and pileup levels delivered, granted by an improved energy resolution and QCD-induced jet background rejec-

C. Martin Perez, *Search for the Higgs Boson Produced in Association with Top Quarks with the CMS Detector at the LHC*, Springer Theses,
https://doi.org/10.1007/978-3-030-90206-3_9

tion. The trigger operations towards the subsequent Run 3 are aimed at enhancing the selectivity of exotic and rare processes, whose selection is limited by the current trigger configuration. New analysis-targeted τ_h triggers were studied within this thesis: by reproducing the topologies of the Higgs boson decaying to τ pairs in its single or double production at trigger level, looser thresholds can be set to the triggers, enhancing the extraction of the signals. As illustration, an acceptance increase of the $t\bar{t}H$ signal as high as 38% is achieved with the inclusion of a new di-τ_h+jet trigger, complementary to the classic di-τ_h trigger used during Run 2 data-taking. Additional optimisations of the algorithm have been identified and are being implemented as this thesis comes to an end, such as the improvement of the τ_h selection efficiency for decays with multiple pions or the development of custom L1 objects to target the pions resulting from W boson decays.

The ultimate L1 trigger upgrade is foreseen for the HL-LHC era; it will follow the current LHC operations in 2027 and will deliver luminosities 4 times the nominal LHC value. A record amount of data of 3000 fb^{-1} will be collected after its 10 years of operation, elucidating many open questions on the SM and on its possible BSM interpretations. To maximize the physics potential, a brand new detector (HGCAL) based on highly granular calorimetry will replace the radiation-damaged ECAL and HCAL detectors in the forward region, providing with fine transverse and longitudinal segmentation. The enhanced calorimetric shower discrimination of the upgraded detector, alongside the increased processing power of the L1 trigger, is a breeding ground for the new L1 τ_h trigger concept developed in this thesis. The algorithm is based on novel machine-learning techniques for calibration, pileup rejection and even decay mode identification for the first time at trigger level. Though under development, selection efficiencies comparable to Run 2 in a pileup environment of up to 200 are observed, revealing the potential of high-granularity triggering in the HL-LHC era. Additional improvements are planned for the algorithm, in particular the implementation of QCD-induced jet identification techniques to reject this overwhelming background so as to keep the rate under control.

Given the large integrated luminosity collected with the highly performing trigger of Run 2, the analysis of rare processes such as the tH and $t\bar{t}H$ productions is brought to unprecedented levels of precision. This thesis presented the most stringent measurement to date of the cross sections of the $t\bar{t}H$ and tH processes in final states containing multiple leptons (electrons or muons) and hadronically decaying τ leptons, sensitive to the Higgs decays into τ, W and Z pairs. For the first time, the $t\bar{t}H$ and tH processes are studied jointly to provide insight not only to the magnitude of the top Yukawa coupling, but also to its relative sign with respect to the coupling of the Higgs boson to electroweak bosons. The complexity of the several objects in the final states, along with the significant background contributions from many SM processes, calls for the use of multivariate methods to achieve an optimal separation of the signals and the backgrounds. Two separate approaches were used for this purpose, developed in parallel to achieve the maximal robustness of the results. The first approach used dedicated BDTs and DNNs in ten mutually exclusive categories; it is aimed at providing separation between the signals and the backgrounds, but also between the $t\bar{t}H$ and tH signals themselves. The second approach, to which I contributed the most, used four of the most sensitive categories to separate the $t\bar{t}H$ signal from the backgrounds using the Matrix Element Method and optimized single variables.

The results were found in agreement with the SM expectations: the ratio between the measured cross section and the SM prediction was found to be $0.92^{+0.26}_{-0.23}$ for the $t\bar{t}H$ process and $5.67^{+4.05}_{-3.98}$ for the tH process, as obtained from the first approach. In the case of the $t\bar{t}H$ search, an excess of events over the background-only hypothesis of 4.7σ (5.2σ) is observed (expected): the first observation of the $t\bar{t}H$ process in multileptonic final states is claimed. The compatibility of the results with the SM expectation was further demonstrated with the complementary second approach. In this case, a measured cross section for the $t\bar{t}H$ process of $0.91^{+0.30}_{-0.26}$ times the SM expectation was found; the 20% loss of sensitivity compared to the first approach is a result of the limited number of categories and the simpler methods used in the signal extraction. Beyond the SM interpretations were extracted from the $t\bar{t}H$ and tH combined search of the first approach. It was done in the context of the Inverted Top Coupling (ITC) scenario, which allows the top Yukawa coupling to acquire an opposite sign with respect to the Higgs coupling to the electroweak bosons. Assuming that the Higgs boson couples to the τ lepton with the values expected in the SM, the top Yukawa coupling was constrained to the intervals $-0.9 < y_t < -0.7$ or $0.7 < y_t < 1.1$ times the SM expectation for this coupling at 95% confidence level. Within this interpretation, both the SM and the ITC scenario are compatible with the data; however, the data favour the SM scenario.

In the era of high precision physics brought upon the HL-LHC, the $t\bar{t}H$ and tH productions will continue to play a key role in the determination of the top Yukawa coupling. Extrapolations of the current Run 2 results to the HL-LHC era estimate improvements on the precision of the measurements of the $t\bar{t}H$ and tH cross sections of from 55 to 70%, assuming an integrated luminosity of 3000 fb^{-1}. Upon combination with the $H \to b\bar{b}$ and $H \to \gamma\gamma$ final states, negative values of κ_t are foreseen to be excluded with a significance larger than 5σ, ruling out the ITC scenario. It remains to be verified whether the SM will continue to pass the solid experimental tests, or if clear indications of BSM physics will manifest. Come what may, the associated production of the Higgs boson with top quarks constitutes the main path to the exploration of the top Yukawa sector at the TeV scale, and could certainly open the way to unveiling the hidden mysteries of the field.

Curriculum Vitae

Research

- **Postdoctoral Research Associate in Experimental Particle Physics at the Mu3e experiment (PSI)** **Zürich (CH)**
  Institute for Particle Physics and Astrophysics, ETH Zürich Nov. 2020—present
- **Ph.D. in Experimental Particle Physics at the CMS experiment (LHC) Palaiseau (FR)**
  Laboratoire Leprince-Ringuet, CNRS, École Polytechnique—supervision by Dr. A. Zabi Oct. 2017–Oct. 2020
- **Master Thesis in Experimental Particle Physics at the CMS experiment (LHC)** **Zürich (CH)**
  Institute for Particle Physics, ETH Zürich—supervision by Dr. F. Nessi-Tedaldi Mar. 2017–Sept. 2017
  Title: "Radiation hard crystals for electromagnetic calorimetry in CMS HL-LHC".
- **Summer Student Programme in Accelerator Physics at FLASH and European-XFEL** **Hamburg (GE)**
  Deutsches Elektronen-Synchrotron (DESY) Jul. 2015–Sept. 2015
- **Summer and Technical Student Programmes in Science Communication at the FCC design study** **Geneva (CH)**
  European Organization for Nuclear Research (CERN) Jul. 2013–Apr. 2015

International Conferences

- **European Physical Society (High-Energy Physics)**, Ghent (BE), Jul. 2019. Parallel talk: *Measurements of ttH and tH production at CMS.*

C. Martin Perez, *Search for the Higgs Boson Produced in Association with Top Quarks with the CMS Detector at the LHC*, Springer Theses,
https://doi.org/10.1007/978-3-030-90206-3

- **Bachelor of Science in Physics (minor Fundamental Physics)** **Madrid (ES)**
  Universidad Complutense de Madrid *Sept. 2010–Jun. 2015*
- **Double Bachelor in Journalism and Media** **Madrid (ES)**
  Universidad Carlos III de Madrid *Sept. 2006–Sept. 2011*

Additional Skills

- **Teaching**: Teaching assistant in Laboratory of Particle Physics at Ecole Polytechnique and ETH Zürich (2017-present).
- **Languages**: Spanish (native), English (Certificate of Proficiency C2), French (Certificate DALF C1), German (A2).
- **Programming languages**: C++, Python, ROOT, MATLAB, Pandas, XGBoost, FLUKA.